AF594957

Entropy Based Fatigue, Fracture, Failure Prediction and Structural Health Monitoring

Entropy Based Fatigue, Fracture, Failure Prediction and Structural Health Monitoring

Editor

Cemal Basaran

MDPI • Basel • Beijing • Wuhan • Barcelona • Belgrade • Manchester • Tokyo • Cluj • Tianjin

Editor
Cemal Basaran
University at Buffalo
USA

Editorial Office
MDPI
St. Alban-Anlage 66
4052 Basel, Switzerland

This is a reprint of articles from the Special Issue published online in the open access journal *Entropy* (ISSN 1099-4300) (available at: https://www.mdpi.com/journal/entropy/special_issues/fatigue).

For citation purposes, cite each article independently as indicated on the article page online and as indicated below:

LastName, A.A.; LastName, B.B.; LastName, C.C. Article Title. *Journal Name* **Year**, *Volume Number*, Page Range.

ISBN 978-3-03943-807-5 (Hbk)
ISBN 978-3-03943-808-2 (PDF)

Contents

About the Editor

Cemal Basaran is Professor in the Dept. of Civil, Structural and Environmental Engineering at University at Buffalo, The State University of New York. He specializes in the computational and experimental mechanics of electronic materials. He has authored 145+ peer-reviewed journal publications, a textbook on Unified Mechanics Theory (©Springer-Nature, ISBN 978-3-030-57771-1), and several book chapters. His research includes the development of Unified Mechanics Theory, which is the unification of the laws of Newton and of thermodynamics at the ab initio level, and the nanomechanics of 2-Delectronic materials. Some of his awards include 1997 US Navy ONR Young Investigator Award, and 2011 ASME Electronic Packaging and Photonics Division, Excellence in Mechanics Award. He is a Fellow of the ASME. He has served and continues to serve on the editorial board of numerous peer-reviewed international journals, including *IEEE Components, Packaging and Manufacturing Tech, ASME Journal of Electronic* Packaging, and Entropy, among others. He has been the primary dissertation advisor to 24 PhD students. His research has been funded by NSF, ONR, DoD, State of New York, and many industrial sponsors including but not limited to Intel, Motorola, Northrop Grumman, Raytheon, Delphi, DuPont, Texas Instruments, Micron, Tyco Electronics, and Analog Devices. He serves as a research proposal reviewer for many national and international research funding agencies found around the globe, including but not limited to the UK, EU, France, China, Hong Kong, Saudi Arabia, Germany, Ireland, and Austria.

Editorial

Entropy Based Fatigue, Fracture, Failure Prediction and Structural Health Monitoring

Cemal Basaran

Department of Civil Structural and Environmental Engineering, University at Buffalo, SUNY, New York, NY 14260, USA; cjb@buffalo.edu

Received: 10 October 2020; Accepted: 14 October 2020; Published: 19 October 2020

This special issue is dedicated to entropy-based fatigue, fracture, failure prediction and structural health monitoring. The unification of laws of thermodynamics and Newtonian mechanics has been a pursuit of many scientists since the mid-19th century. Distinguished scientists from around the world who contributed to this special issue all show that unification of Newtonian mechanics with thermodynamics using entropy as a link eliminates the need for phenomenological continuum mechanics, where the second law of thermodynamics is usually imposed only as an external constraint, but is not satisfied at the material level, because derivative of displacement with respect to entropy is assumed to be zero. For example, the theory of elasticity assumes that there is no entropy generation at the material level. As a result, everything is reversible, which violates the second law of thermodynamics.

Group from Indian Institute of Technology Madras and University at Buffalo used unified mechanics theory for low cycle fatigue life prediction of Ti-6Al-4V alloys. Bin Jamal et al. [1] show that using unified mechanics theory fatigue life can be predicted using physics, rather than using the empirical curve fitting models. This is also the first peer-reviewed paper in literature to publish the laws of Newton and laws of thermodynamics in unified form at ab-initio level. The second law of unified mechanics theory is given by [1,2]

$$\mathbf{F} = m\frac{d[\mathbf{v}(1-\Phi)]}{dt} \tag{1}$$

where Φ is the Thermodynamic State Index (TSI), a linearly independent axis in addition to Newtonian space-time axes, that can have values between zero and one.

Scientists from Belarus State University contributed a noteworthy paper with their recent advances on mechanothermodynamics, which is essentially a theory almost identical to the unified mechanics theory. They both use entropy generation rate for degradation and unification of Newtonian mechanics and thermodynamics laws. Sosnovskiy and Sherbakov [3] formulate the main principles of the physical discipline of mechanothermodynamics that unites Newtonian mechanics and thermodynamics. Authors state that mechanothermodynamics combines two branches of physics, mechanics and thermodynamics, to take a fresh look at the evolution of complex systems. The analysis of more than 600 experimental results on polymers and metals are used for determining a unified mechanothermodynamics function of limiting states. They are also known as Fatigue Fracture Entropy (FFE) states.

A Purdue University group contributed their outstanding work on using maximum entropy models for fatigue damage in metals with application to low-cycle fatigue of aluminum 2024-T351. Young and Subbarayan [4] propose using the cumulative distribution functions derived from maximum entropy formalisms, utilizing thermodynamic entropy as a measure of damage to fit the low-cycle fatigue data of metals. The thermodynamic entropy is measured from hysteresis loops of cyclic tension–compression fatigue tests on aluminum 2024-T351. The plastic dissipation per cyclic reversal is estimated from Ramberg–Osgood constitutive model fits to the hysteresis loops and correlated to experimentally-measured average damage per reversal. The proposed model predicts fatigue life more accurately and consistently than several traditional models, including the Weibull distribution function and the Coffin–Manson relation. The formalism is founded on treating the failure process as a consequence of the increase in the entropy of the material due to plastic deformation. This argument leads to using inelastic dissipation as the independent variable (which provides the coordinate along TSI) for predicting low-cycle fatigue damage, rather than the more commonly used plastic strain. The entropy of the microstructural state of the material is modeled by statistical cumulative distribution functions, following examples in recent literature. They demonstrate the utility of a broader class of maximum entropy statistical distributions, including the truncated exponential and the truncated normal distribution. Authors show that not only are these functions demonstrated to have the necessary qualitative features to model damage, but they are also shown to capture the random nature of damage processes with greater fidelity.

University of Maryland, College Park scientists contributed an excellent study on measures of entropy to characterize fatigue damage in metallic materials. Yun and Modarres [5] show that Fatigue Fracture Entropy (FFE) is a material property independent of geometry or loading. This paper presents the entropic damage indicators for metallic material fatigue processes obtained from three associated energy dissipation sources. Authors state that, entropy, the measure of disorder and uncertainty, introduced from the second law of thermodynamics, has emerged as a fundamental and promising metric to characterize all mechanistic degradation phenomena and their interactions. Entropy has already been used as a fundamental and scale-independent metric to predict damage and failure. In this paper, three entropic-based metrics are examined and demonstrated for application to fatigue damage. Authors collected experimental data on energy dissipations associated with fatigue damage, in the forms of mechanical, thermal, and acoustic emission (AE) energies, and estimated and correlated the corresponding entropy generations with the observed fatigue damages in metallic materials. Three entropic theorems—thermodynamics, information, and statistical mechanics—support approaches used to estimate the entropic-based fatigue damage. Authors show that classical thermodynamic entropy provided a reasonably constant level of entropic endurance to fatigue failure. Finally, they indicate that an extension of the relationship between thermodynamic entropy and Jeffreys divergence from molecular-scale to macro-scale applications in fatigue failure resulted in an empirically-based pseudo-Boltzmann constant equivalent to the Boltzmann constant.

University of Texas at Austin researchers contributed an excellent paper on degradation-entropy generation methodology for system and process characterization and failure analysis. Osara and Bryant [6] formulated a new fatigue life predictor based on ab initio irreversible thermodynamics. The method combines the first and second laws of thermodynamics with the Helmholtz free energy, then applies the result to the degradation-entropy-generation relation to relate a desired fatigue measure—stress, strain, cycles or time to failure—to the loads, materials and environmental conditions (including temperature and heat) via the irreversible entropies generated by the dissipative processes that degrade the fatigued material. The formulations are then verified with fatigue data from the literature, for a steel shaft under bending and torsion.

Scientists from Northwestern Polytechnical University and Xi'an University of Architecture and Technology contributed an exceptional study titled an entropy-based failure prediction model for the creep and fatigue of metallic materials. Wang and Yao [7] state that it is well accepted that the second law of thermodynamics describes an irreversible process, which can be reflected by the entropy increase. Irreversible creep and fatigue damage can also be represented by a gradually increasing damage parameter. In the current study, an entropy-based failure prediction model for creep and fatigue is proposed based on the Boltzmann probabilistic entropy theory and continuum damage mechanics. A new method to determine the entropy increment rate for creep and fatigue processes is proposed. The relationship between entropy increase rate during creep process and normalized creep failure time is developed and compared with the experimental results. An entropy-based model is developed to predict the change of creep strain during the damage process. Experimental results of metals and alloys with different stresses and at different temperatures are utilized to verify their model. It shows that the theoretical predictions agree well with experimental data.

Universiti Kebangsaan Malaysia group, contributed a great study on prediction of fatigue crack growth rate based on entropy generation. Idris et al. [8] present the assessment of fatigue crack growth rate for dual-phase steel under spectrum loading based on entropy generation. According to the second law of thermodynamics, fatigue crack growth is related to entropy gain because of its irreversibility. In this work, the temperature evolution and crack length were simultaneously measured during fatigue crack growth tests until failure to ensure the validity of the assessment. Results indicate a significant correlation between fatigue crack growth rate and entropy. This relationship is the basis in developing a model that can determine the characteristics of fatigue crack growth rates, particularly under spectrum loading. Predictive results showed that the proposed model can accurately predict the fatigue crack growth rate under spectrum loading in all cases. The root mean square error in all cases is 10^{-7} m/cycle. In conclusion, they prove that entropy generation can accurately predict the fatigue crack growth rate of dual-phase steels under spectrum loading.

Researchers from Beihang University and Beijing Aeronautical Science & Technology Research Institute contributed a very interesting study on using copula entropy for quantifying dependence among multiple degradation processes. Sun et al. [9] studied multivariate degradation modeling to capture and measure the dependence among multiple features. In order to address this problem, this paper adopts copula entropy, which is a combination of the copula function and information entropy, to measure the dependence among different degradation processes. An engineering case study was utilized to illustrate the effectiveness of the proposed method. The results show that this method is valid for the dependence measurement of multiple degradation processes.

Scientists from Beihang University and North China University of Water Resources and Electric Power contributed an indirectly related paper on intelligent analysis algorithm for satellite health under time-varying and extremely high thermal loads. Li et al. [10] present a dynamic health intelligent evaluation model proposed to analyze the health deterioration of satellites under time-varying and extreme thermal loads. New definitions, such as health degree and failure factor and new topological system considering the reliability relationship, are proposed to characterize the dynamic performance of health deterioration. The dynamic health intelligent evaluation model used the thermal network method (TNM) and fuzzy reasoning to solve the problem of model missing and non-quantization between temperature and failure probability.

Nanjing University of Science and Technology and City University of Hong Kong teams participated with their paper titled effective surface nano-crystallization of $Ni_2FeCoMo_{0.5}V_{0.2}$ medium entropy alloy by rotationally accelerated shot peening. Liang et al. [11] reported the surface nano-crystallization of $Ni_2FeCoMo_{0.5}V_{0.2}$ medium-entropy alloy by rotationally accelerated shot peening (RASP). Transmission electron microscopy analysis revealed that deformation twinning and dislocation activities are responsible for the effective grain refinement of the high-entropy alloy. In order to reveal the effectiveness of surface nano-crystallization on the $Ni_2FeCoMo_{0.5}V_{0.2}$ medium-entropy alloy, a common model material, Ni, is used as a reference.

Conflicts of Interest: The author declares no conflict of interest.

References

1. Jamal M, N.B.; Kumar, A.; Lakshmana Rao, C.; Basaran, C. Low Cycle Fatigue Life Prediction Using Unified Mechanics Theory in Ti-6Al-4V Alloys. *Entropy* **2020**, *22*, 24. [CrossRef]
2. Basaran, C. *Introduction to Unified Mechanics Theory with Applications*; Springer-Nature: Cham, Switzerland, 2020.
3. Sosnovskiy, L.A.; Sherbakov, S.S. On the Development of Mechanothermodynamics as a New Branch of Physics. *Entropy* **2019**, *21*, 1188. [CrossRef]

4. Young, C.; Subbarayan, G. Maximum Entropy Models for Fatigue Damage in Metals with Application to Low-Cycle Fatigue of Aluminum 2024-T351. *Entropy* **2019**, *21*, 967. [CrossRef]
5. Yun, H.; Modarres, M. Measures of Entropy to Characterize Fatigue Damage in Metallic Materials. *Entropy* **2019**, *21*, 804. [CrossRef]
6. Osara, J.A.; Bryant, M.D. Thermodynamics of Fatigue: Degradation-Entropy Generation Methodology for System and Process Characterization and Failure Analysis. *Entropy* **2019**, *21*, 685. [CrossRef]
7. Wang, J.; Yao, Y. An Entropy-Based Failure Prediction Model for the Creep and Fatigue of Metallic Materials. *Entropy* **2019**, *21*, 1104. [CrossRef]
8. Idris, R.; Abdullah, S.; Thamburaja, P.; Omar, M.Z. Prediction of Fatigue Crack Growth Rate Based on Entropy Generation. *Entropy* **2020**, *22*, 9. [CrossRef]
9. Sun, F.; Zhang, W.; Wang, N.; Zhang, W. A Copula Entropy Approach to Dependence Measurement for Multiple Degradation Processes. *Entropy* **2019**, *21*, 724. [CrossRef]
10. Li, E.-H.; Li, Y.-Z.; Li, T.-T.; Li, J.-X.; Zhai, Z.-Z.; Li, T. Intelligent Analysis Algorithm for Satellite Health under Time-Varying and Extremely High Thermal Loads. *Entropy* **2019**, *21*, 983. [CrossRef]
11. Liang, N.; Wang, X.; Cao, Y.; Li, Y.; Zhu, Y.; Zhao, Y. Effective Surface Nano-Crystallization of Ni2FeCoMo0.5V0.2 Medium Entropy Alloy by Rotationally Accelerated Shot Peening (RASP). *Entropy* **2020**, *22*, 1074. [CrossRef]

Publisher's Note: MDPI stays neutral with regard to jurisdictional claims in published maps and institutional affiliations.

Article

Low Cycle Fatigue Life Prediction Using Unified Mechanics Theory in Ti-6Al-4V Alloys

Noushad Bin Jamal M [1], Aman Kumar [1], Chebolu Lakshmana Rao [1] and Cemal Basaran [2,*]

[1] Department of Applied Mechanics, Indian Institute of Technology, Madras 600036, India; noushadbj@gmail.com (N.B.J.M.); kumaraman2102@gmail.com (A.K.); lakshman@iitm.ac.in (C.L.R.)
[2] Civil, Structural and Environmental Engineering, University at Buffalo, State University of New York, New York, NY 10031, USA
* Correspondence: cjb@buffalo.edu

Received: 22 November 2019; Accepted: 22 December 2019; Published: 23 December 2019

Abstract: Fatigue in any material is a result of continuous irreversible degradation process. Traditionally, fatigue life is predicted by extrapolating experimentally curve fitted empirical models. In the current study, unified mechanics theory is used to predict life of Ti-6Al-4V under monotonic tensile, compressive and cyclic load conditions. The unified mechanics theory is used to derive a constitutive model for fatigue life prediction using a three-dimensional computational model. The proposed analytical and computational models have been used to predict the low cycle fatigue life of Ti-6Al-4V alloys. It is shown that the unified mechanics theory can be used to predict fatigue life of Ti-6Al-4V alloys by using simple predictive models that are based on fundamental equation of the material, which is based on thermodynamics associated with degradation of materials.

Keywords: entropy; fatigue; damage mechanics; unified mechanics; thermodynamics; Ti-6Al-4V; physics of failure

---

1. Introduction

Titanium alloys are popular for their superior mechanical properties, such as high yield strength, long fatigue life, toughness, low density, as well as corrosion resistance. About 80% of the global production of titanium alloys are used by aerospace industries [1]. One of the widely used titanium alloys is Ti-6Al-4V [2] which has a dual-phase crystal structure, namely, hexagonal close packed (HCP) and body centered cubic (BCC) structures. In the composition of Ti-6Al-4V alloy, titanium is the matrix material. Aluminium plays the role of stabilizing the HCP structure and vanadium preserve the BCC structure [3]. Many applications of Ti-6Al-4V alloys, such as aero engines, are subjected to cyclic loading [4]. Hence, it is essential to predict the fatigue life of such structural components, when they are subjected to varying amplitudes of cyclic loading during their service period. It is not always feasible to conduct fatigue experiments corresponding to all service conditions. Hence, predictive models based on fundamental physics of materials are helpful in predicting the fatigue life of structures.

A number of studies have been published to investigate the fatigue life of metals. Most of the damage prediction models are based on statistical test data analysis or on experimental curve fit [5–11]. Low cycle fatigue life prediction in Ti-6Al-4V alloys are generally done, based on stress [12], strain [5,6,13–16] or hysteresis loss [17]. Most of them are empirical curve-fit models [7,9,13,18–22] or mechanism based phenomenological models [23–25] such as fatigue crack initiation models [16]. A detailed review of such models, applied to metals, can be seen in the review article by Santecchia et al. [26]. A model, based on combined Newtonian mechanics and thermodynamics, instead of material-specific and loading-specific, can capture the mechanisms of fatigue damage without the need for curve fitting process.

If the system is less complicated and we want a quick solution we can opt for a one-dimensional model based on certain assumptions. However, validity of the model depends upon the accuracy of the assumptions made while formulation of one-dimensional analytical model. The interpretation of the results using one-dimensional model is also easy as it can be simple in its form and usage. A number of one-dimensional empirical curve-fit fatigue life prediction models can be seen in the literature [5–8,11,12,14–17]. Nevertheless, a physics-based one-dimensional model, which can be easily used to predict the fatigue life of Ti-6Al-4V, under appropriate assumptions, is still not found in the literature. If the system is very complicated to arrive at suitable one-dimensional fatigue life prediction model, we look for another appropriate and convenient method. It is known that, a three dimensional computational model can be incorporated with appropriate material nonlinearities (such as plastic flow), to account for the experimental observations [10,22] and to limit the assumptions in developing the model. However, a large number of cyclic loading simulation in a three dimensional numerical model is computationally very expensive [10]. Hence, it is very useful to have an appropriate physics-based procedure, in conjunction with three-dimensional numerical results, to account for all the nonlinearities associated with the computational model, even as we maintain the simplistic predictive capability of a one-dimensional model. Therefore, the present study is focused on both one-dimensional and three-dimensional, thermodynamics-based modeling of the deformation of standard test specimen to predict the fatigue life of Ti-6Al-4V.

Thermodynamics is a field of science that is developed to study change in the state of matter. The historical development of thermodynamics from its classical form to modern-age form has been reviewed by Haddad et al. [27,28]. Between 1872 and 1875, Boltzmann gave a mathematical expression to second law of thermodynamics for quantification of order/disorder in terms of a measure called *entropy*. In 1998, Basaran and Yan [29] introduced the unified mechanics theory, which unifies Newtonian mechanics with thermodynamics. In unified mechanics theory [29], in addition to nodal displacements, the entropy generation rate is also necessary to relate microstructural changes in the material with spatial and temporal coordinates. This concept [29] has been successfully implemented for a wide range of materials and has been experimentally and mathematically validated and reported in literature [18–20,25,30–65]. The entropy generation rate of any material under any external disturbances like mechanical, thermal, electrical, chemical, radiation, and corrosion can be calculated from principles of physics, using the fundamental equation, with no need for curve fitting phenomenological models or polynomials fit to experimental test data.

In the present study, unified mechanics theory is used to estimate the fatigue damage in Ti-6Al-4V, analytically with a one -dimensional (1-D) model as well as numerically with a three-dimensional (3-D) model, and this damage estimation procedure has been used to predict fatigue life under different loading conditions. Fundamental details of the unified mechanics theory-based fatigue life prediction are summarized in Section 2. The principles described in Section 2, are then applied to Ti-6Al-4V, by considering the plasticity as the dominant energy dissipation mechanism.

In order to establish the validity of the proposed model in cyclic loading, comparison of simulation with experimental results, under both the tensile and compressive loading are necessary. In Section 3, the details of implementation and validation of the proposed model, for both compressive and tensile monotonic loading is presented. After the validation of the proposed model, we introduce two different procedures, to estimate the low cycle fatigue life of Ti-6Al-4V alloys in Section 4. Finally, the observations from the presented work are discussed in Section 5, based on the observations made on the principles, procedure and results from the current study for the fatigue life prediction of Ti-6Al-4V alloys.

2. Unified Mechanics Theory-Based Life Prediction Model

2.1. Unified Mechanics Theory

Unified mechanics theory is just unification of Newton's universal laws of motion and laws of thermodynamics.

2.1.1. Second Law of Unified Mechanics Theory

Initial momentum of a mass, m, subjected to external force, $\mathbf{F}$ is defined by Newton's second universal law of motion. However, Newton's laws do not account for energy loss after the initial momentum. Energy loss takes place according to the first and second laws of thermodynamics. As a result, a marriage of laws of second law of Newton and laws of thermodynamic is given by:

$$\mathbf{F} = \frac{d\mathbf{P}}{dt} = \frac{d(m\mathbf{v})}{dt}(1 - \Phi) \tag{1}$$

where, $\mathbf{P}$ represents the momentum and $\mathbf{v}$ represents the velocity. Assuming a constant mass system,

$$\mathbf{F} = m\frac{d[\mathbf{v}(1 - \Phi)]}{dt} \tag{2}$$

where, Φ is the Thermodynamic State Index (TSI), which is normalized non-dimensional form of the second law of thermodynamics. TSI (Φ) starts at zero and reaches one when the system reaches maximum entropy and minimum entropy generation rate. The value of TSI (Φ) is calculated from the fundamental equation of the material, which accounts for all entropy generation mechanisms in the system under the given load towards a pre-defined failure. The fundamental equation must satisfy the conservation of energy, the first law of thermodynamics at every step. Therefore, TSI (Φ) just introduces laws of thermodynamics in to the laws of Newton.

2.1.2. Third Law of Unified Mechanics Theory

All forces between two objects exist in equal magnitude and opposite direction (Action–Reaction). However, resulting deformation, according to Hook's law, in two objects will change over time because of degradation. The resulting equation can be given by:

$$\mathbf{F}_{12} = \mathbf{F}_{21}[1 - \Phi] \tag{3}$$

where, the subscripts 12 and 21 represents the action and reaction, respectively. Based on Hooke's law, the reaction, $\mathbf{F}_{21}$ can be given by the following:

$$\mathbf{F}_{12} = \frac{dU_{21}}{d\mathbf{u}_{21}} = \frac{\left[d\left[\frac{1}{2}\,k_{21}[1 - \Phi]\,\mathbf{u}_{21}^2\right]\right]}{d\mathbf{u}_{21}} \tag{4}$$

where, U_{21} is the strain energy of the reactionary member, k_{21} is the stiffness of the reactionary member, $\mathbf{u}_{21}$ is the displacement in the reactionary member. If we assume that for the increment of displacement, $d\mathbf{u}_{21}$ derivative of TSI with respect to $d\mathbf{u}_{21}$ is smaller than derivative of displacement $\mathbf{u}_{12}$ by an order of magnitude as the differential in displacement $d\mathbf{u}_{21}$ goes to zero in the limiting case, we can write the following simple relation:

$$\mathbf{F}_{12} = k_{21}[1 - \Phi]\,\mathbf{u}_{21} \tag{5}$$

In unified mechanics theory, it has been shown that the degradation of the stiffness follows the laws of thermodynamics [8,18,20,22,27,29–33,35–54,56–59,66–69]. Combining laws of Newton and thermodynamics requires the modification of Newtonian space-time coordinate system. A new thermodynamic axis must be added to be able to define the thermodynamic state of a point. As a result, the motion of any particle can be defined only in a five-dimensional space that has five linearly independent axes. None of these axes can represent the information of other axes. Hence, entropy generation can be mapped onto a non-dimensional coordinate called Thermodynamics State Index (TSI) which is necessary to locate the thermodynamic state of the particle. Coordinates of a point can be defined by Newton's laws of motion in the space-time coordinate system. However, thermodynamic state coordinate cannot be defined by space-time coordinate system.

Figure 1 shows the coordinate system in unified mechanics theory. Let us assume there is a 5-year-old boy and 100-year-old man. Using the space-time Cartesian coordinate system, their location can be defined by x, y, z coordinates and age on the time axis. However, this does not give any information about their thermodynamic state. Let us assume that a 5-year-old boy has stage 4 cancer is expected to die in a few days and a 100-year-old is expected to die in few days. This information cannot be represented in x, y, z-time- space coordinate system as shown in Figure 1. However, on TSI axis, 5-year-old boy and 100-year-old will have the same thermodynamic state index coordinate at $\Phi = 0.999$.

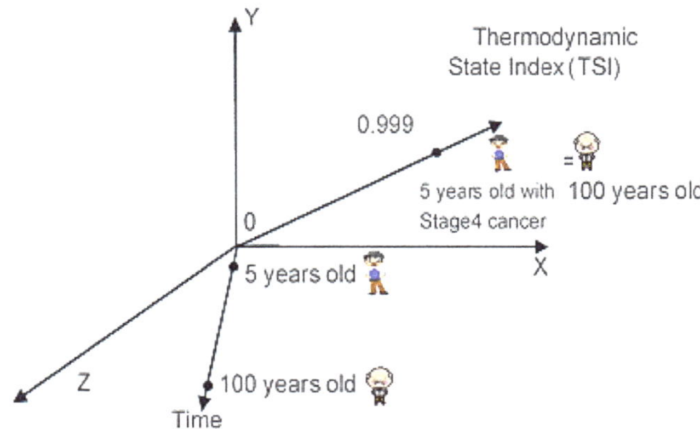

Figure 1. Coordinate system in unified mechanics theory.

Another example can be given for Newton's second law. If a soccer ball is given an initial acceleration with a force of **F**, it will move but eventually will come to a stop. Depending on the path it follows, it will come to a stop. Again, the initial acceleration of the ball is governed by the second law of Newton and slowing down process is governed by the laws of thermodynamics, which is represented by $(1 - \Phi)$ term. Detailed derivation of TSI can be seen in the literature [29]. We provide a simple summary in the following section.

2.1.3. Thermodynamic State Index (TSI) for Damage in Low Cycle Fatigue of Materials

Entropy and Helmholtz free energy are related by the thermodynamic principles [66] as follows:

$$\Psi = e - Ts \tag{6}$$

where Ψ represents the specific Helmholtz free energy, and e, T, s are the specific internal energy, temperature and specific entropy, respectively. Specific entropy is also related to the disorder parameter through Boltzmann's equation [29,30] as follows:

$$s = k_B \ln(\mathcal{W}) \tag{7}$$

Total entropy for a volume can be given by:

$$S = \frac{N_A k_B \, ln(\mathcal{W})}{m_s} \tag{8}$$

where, N_A, k_B, m_s are the Avogadro number, Boltzmann's constant and molar mass, respectively and $\mathcal{W}$ represent the disorder parameter [29,30,38,39,66]. Relation between the number of microstate, probability of microstates and disorder parameter is discussed extensively in the literature [70–72]. Using Equation (8), the TSI is given by:

$$\Phi = \Phi_c\left(1 - \exp\left(-\Delta s \frac{m_s}{R}\right)\right) \tag{9}$$

where, Φ_c, is a user defined parameter, representing the predefined failure criterion. R is gas constant. Δs is a measure of the total change in entropy at a point. Unified mechanics theory states that when a system undergoes thermodynamic change from state A to state B, the remaining useful life can be defined by a factor in each stage of its life, called thermodynamic state index (TSI), $\Phi \in [0,1]$. The ultimate failure is represented by a value of TSI equal to 1. Since, the value of Δs is to be evaluated on the basis of mechanisms of dissipation processes involved in a thermodynamic process, the value of Φ_c will be governed by a user-defined ultimate failure criterion.

2.2. Analytical Approach for the Prediction of Damage and Fatigue Life

From Equation (9), the TSI is governed by the change in entropy towards a predefined failure. All the dissipation processes that are related to failure lead to increase in entropy. Therefore, an appropriate measure of dissipation is needed to estimate the life of a process. In Ti-6Al-4V alloys, we consider only the mechanical process of dissipation, under monotonic as well as cyclic loading conditions. Hence the plastic dissipation is considered to be the dominant mechanism in the mechanical loading conditions. Entropy generation in plastic dissipation process can be calculated from a mechanical loading experiment in the following way:

$$\Delta s = \frac{1}{\rho T}\int_{t_1}^{t_2} \boldsymbol{\sigma} : d\boldsymbol{\varepsilon}^{\mathbf{p}} \tag{10}$$

where, ρ, is the mass density of the material, $\boldsymbol{\sigma}$ and $\boldsymbol{\varepsilon}^{\mathbf{p}}$ are the stress and plastic strain, respectively. T represents the temperature. Integral limits t_1 and t_2 represents the time bounds of the mechanical loading process, over which we quantify the change in entropy. For one dimensional case, the total plastic strain, $\varepsilon^p(t)$ is calculated as follows:

$$\varepsilon^p(t) = \varepsilon^{total}(t) - \frac{\sigma_{y_0}}{E} \tag{11}$$

where, $\varepsilon^{total}(t)$ is the total strain at the time of loading, t. σ_{y_0} and E are the yield stress and Young's modulus, respectively. In the case of monotonic loading, the plastic dissipation is calculated from the engineering stress-strain graphs. In order to accomplish this, the plot is divided into elastic and plastic regime of loading. The area under the plastic region is computed by trapezoidal integration rule, and the cumulative entropy is evaluated in each stage. This accumulated entropy is used to predict the TSI at each and every strain level. A schematic representation of computing the incremental plastic dissipation is given in Figure 2. Accumulated entropy at n-th strain increment is computed from the Equation (10) as follows:

$$\Delta s_n = \frac{1}{\rho T}\sum_{i=1}^{i=n} \sigma_i : \Delta\varepsilon_i^p \tag{12}$$

Using Equations (9), (11) and (12), one dimensional approximation of damage measure is calculated under the assumptions that the damage is uniform within the cross section of the dog-bone test sample, and there are no other geometric or boundary effects in the sample. It is also assumed that the heat generation entropy production is small when compared with the entropy generation due to plastic deformation. In case of low cycle fatigue loading, the plastic dissipation is calculated as the area under the stress-strain hysteresis loop. Each cyclic hysteresis loop of engineering stress-strain graph, which represents the incremental dissipation. Hence, the accumulated entropy can be calculated by summation of incremental entropy. For a strain-controlled experiment, the accumulated entropy is a function of stress. Since the stress level at a given stage of cyclic loading is governed by the thermodynamic state index (TSI), Φ of the material, the TSI can be used to calculate the incremental dissipation from any known stage of loading, as follows:

$$\Pi^p_{i+1} = (1 - \Phi_i)\Pi^p_i \tag{13}$$

where, Π^p_i and Π^p_{i+1} represents the hysteresis area at i-th and (i+1)-th cyclic loading, respectively and Φ represents the TSI. Hence, the entropy change at any loading stage can be calculated from the initial loading hysteresis area as follows:

$$\Delta s_n = \frac{1}{\rho T}\sum_{i=1}^{i=n} \Pi^p_i \tag{14}$$

$$\Phi_{i+1} = \Phi_c\left[1 - \exp\left(-\Delta s_i \frac{m_s}{R}\right)\right] \tag{15}$$

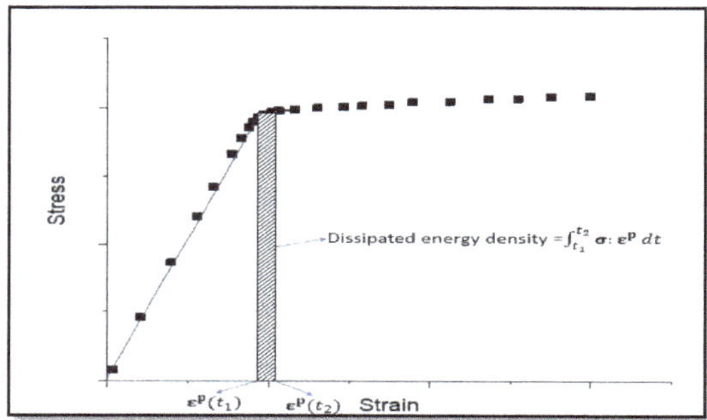

Figure 2. Schematic representation of computing plastic dissipation from the engineering stress-plastic strain graph.

It is important to point out that the entire thermodynamic response of the material point is mapped onto the TSI axes. Under no circumstances, the material point can exist outside the domain of [0,1]. The above approach has limitations that the one-dimensional approximation should be valid when the prediction is compared with experimental observations. To account for all the boundary and geometric effects related to stiffness, instabilities due to buckling, local cracking, stress concentrations, geometric nonlinearities, etc., we have developed a three-dimensional computational model. The detailed derivation is given in Section 2.3 below.

2.3. Computational 3-D Model for the Prediction of Damage

2.3.1. Derivation of the Computational Model

In this section, a three-dimensional model is derived, based on the unified mechanics theory. Entropy balance equation [4,20,29,30], can be written as follows:

$$\frac{dS}{dt} \geq -\frac{div\,\mathbf{J_q}}{T} + \frac{1}{T}\sigma : \mathbf{D} - \frac{\rho}{T}\frac{dW^e}{dt} + \frac{\rho r}{T} \tag{16}$$

The following equation, as written in indicial notation, is known as Clausius-Duhem inequality [67,73,74]:

$$\gamma = \frac{1}{T}\sigma_{ij}D_{ij} - \frac{\rho}{T}\frac{dW^e}{dt} - \frac{1}{T^2}J_{qi}T_{,i} + \frac{\rho r}{T} \geq 0 \tag{17}$$

where, i and j are the indices, representing the spatial coordinates. γ is the specific entropy generation rate. $\boldsymbol{\sigma}$ denotes the stress tensor and $T_{,i}$ represents the spatial derivative of temperature, namely, the gradient of temperature. $\mathbf{J_q}$ and r, represents the heat flux transfer and internal heat generation, respectively. For small strain problems, rate of deformation tensor $\mathbf{D}$ is equal to strain rate tensor $\dot{\varepsilon}$. According to Hooke's law, stress is related to the strain through a constitutive tensor as follows:

$$\sigma_{ij} = C_{ijkl}\varepsilon^{e}_{kl} \tag{18}$$

where, C_{ijkl} is the fourth order tangential constitutive tensor at the given stage of loading. ε^{e}_{kl} is the elastic part of strain tensor. Based on assumption of the additive decomposition of strain tensor [67], we can write the following equation for small strain problems:

$$\varepsilon^{total}_{ij} = \varepsilon^{e}_{ij} + \varepsilon^{p}_{ij} \tag{19}$$

where, ε^{total}_{ij} is the component of total strain tensor. For a given material point, based on unified mechanics theory one can write the following modified version of Equation (18), as follows:

$$\sigma_{ij} = (1-\Phi)C^{0}_{ijkl}\varepsilon^{e}_{kl} \tag{20}$$

where the tangential constitutive tensor C_{ijkl} is related to the virgin state of the same, C^{0}_{ijkl} (undamaged state) through TSI, Φ. For linear isotropic materials, undamaged constitutive tensor C^{0}_{ijkl} can be written as follows:

$$C^{0}_{ijkl} = \lambda\delta_{ij}\delta_{kl} + \mu(\delta_{ik}\delta_{jl} + \delta_{il}\delta_{jk}) \tag{21}$$

where, λ and μ are the Lame's parameters and δ_{ij} is the identity tensor. The following inverse relations can also be written for a linear elastic isotropic material:

$$\varepsilon^{e}_{ij} = \frac{1+\nu}{E}\sigma_{ij} - \frac{\nu}{E}\sigma_{kk}\delta_{ij} \tag{22}$$

where, E and ν are the elastic modulus and Poisson's ratio, respectively. The rate form of the Equation (19), can be written as follows,

$$\dot{\varepsilon}^{total}_{ij} = \dot{\varepsilon}^{e}_{ij} + \dot{\varepsilon}^{p}_{ij} \tag{23}$$

From incremental theory of plasticity, one can write the evolution equation for the fluxes, using the continuity of dissipation potential function, $\mathcal{F}^{p}$ (yield surface) [67] as follows:

$$\dot{\varepsilon}^{p}_{ij} = \dot{\Gamma}\frac{\partial\mathcal{F}^{p}}{\partial\sigma_{ij}} \tag{24}$$

Effective stress at a point can be defined as follows:

$$\sigma'_{ij} = \frac{1}{1-\Phi}\sigma_{ij} \tag{25}$$

where, σ'_{ij} is the component of effective stress tensor. Noting that Δs is the only function that depends on time, the time rate of change of TSI can be obtained by differentiating Equation (9), yielding:

$$\dot{\Phi} = \frac{m_s}{R}\Phi_c\dot{\Delta s}\left(\exp\left(-\Delta s\frac{m_s}{R}\right)\right) \tag{26}$$

Assuming that the process is isothermal for each small load increment and all the dissipation mechanisms other than plastic deformation are negligibly small for the strain-controlled monotonic, quasi-static loading and low cycle fatigue loading in Ti-6Al-4V, we can write the entropy evolution as given in Equation (10). Hence the rate form of the entropy evolution from Equation (10) can be written as follows:

$$\dot{\Delta s} = \frac{1}{\rho T}\sigma_{ij}\dot{\varepsilon}^{p}_{ij} \tag{27}$$

With the above assumption in the absence of kinematic hardening, we consider the following additive decomposition form of the Helmholtz free energy function as:

$$\Psi(\boldsymbol{\varepsilon^{e}}, h; \Phi) = \Psi^{E}(\boldsymbol{\varepsilon^{e}}; \Phi) + \Psi^{I}(h) \tag{28}$$

where, Ψ^{E} is the elastic strain energy and Ψ^{I} is the free energy from isotropic hardening process. In the Equation (28), the hardening flux parameter h evolves with plastic strain. From the Equations (26) and (27), the plastic strain is a function of TSI.

Using Equations (20), (21) and (28), we get the following form of free energy:

$$\Psi(\boldsymbol{\varepsilon^{e}}, h; \Phi) = \frac{1}{2}(1-\Phi)(\lambda\varepsilon^{e}_{kk}\varepsilon^{e}_{mm} + 2\mu\varepsilon^{e}_{ij}\varepsilon^{e}_{ij}) + (1-\Phi)\frac{1}{r}Kh^{r+1} \tag{29}$$

We have assumed a power law model for isotropic hardening. Here, K and r are the material parameters which are to be found from the succeeding parts of the formulation and experimental data. The conjugate force is derived from Equation (29) as follows [75]:

$$\boldsymbol{\sigma} = \rho\frac{\partial}{\partial\boldsymbol{\varepsilon^{e}}}\Psi \tag{30}$$

$$\sigma_{ij} = (1-\Phi)(\lambda\varepsilon^{e}_{kk}\delta_{ij} + 2\mu\varepsilon^{e}_{ij}) \tag{31}$$

The yield function for Ti-6Al-4V can be given by:

$$\mathcal{F}^{p}(\boldsymbol{\sigma}, \mathrm{H}; \Phi) = \sigma'_{eq} - \left(\sigma_{y_o} + H\right) \tag{32}$$

where, σ'_{eq} is the Von-Mises equivalent stress. σ_{y_o} represents the initial yield stress and H represents the hardening stress. Von-Mises equivalent stress is given by the following equation:

$$\sigma'_{eq} = \sqrt{\frac{3}{2}S'_{ij}S'_{ij}} \tag{33}$$

where, the effective deviatoric stress tensor S'_{ij}, is given by the following equation,

$$S'_{ij} = \sigma'_{ij} - \frac{\sigma'_{kk}}{3}\delta_{ij} \tag{34}$$

Hence, from Equations (24) and (32), we get the following relation for plastic strain rate tensor, $\dot{\varepsilon}^{p}_{ij}$:

$$\dot{\varepsilon}^{p}_{ij} = \dot{\Gamma}\frac{\partial\sigma'_{eq}}{\partial\sigma_{ij}} \tag{35}$$

Further simplification can be done on Equation (35) using the Equations (33) and (34). We get the following form for plastic strain rate tensor, based on normality rule of incremental theory of plasticity:

$$\dot{\varepsilon}^{p}_{ij} = \dot{\Gamma}\left[\frac{1}{(1-\Phi)}\frac{3}{2}\frac{S'_{ij}}{\sigma'_{eq}}\right] \tag{36}$$

where, $\dot{\Gamma}$ is the consistency parameter. By taking the norm of Equation (36) and by doing some algebra, we get the following equation to quantify the parameter, $\dot{\Gamma}$:

$$\dot{\varepsilon}^{p}_{eq} = \sqrt{\frac{2}{3}\dot{\varepsilon}^{p}_{ij}\dot{\varepsilon}^{p}_{ij}} = \dot{\Gamma}\frac{1}{(1-\Phi)} \tag{37}$$

Equation (37) is an important observation that the field variable, h, representing the isotropic hardening process, is related to the plastic deformation. Hence, we get the following form for $\dot{h}$ and $\dot{\varepsilon}^{p}_{ij}$:

$$\dot{h} = \dot{\varepsilon}^{p}_{eq}(1-\Phi) \tag{38}$$

$$\dot{\varepsilon}^{p}_{ij} = \dot{\varepsilon}^{p}_{eq}\left[\frac{3}{2}\frac{S'_{ij}}{\sigma'_{eq}}\right] \tag{39}$$

From Equation (39), it can be observed that the magnitude of plastic strain is given by the equivalent plastic strain, ε^{p}_{eq}, and the direction of plastic loading is given by the term, $\left[\frac{3}{2}\frac{S'_{ij}}{\sigma'_{eq}}\right]$.

2.3.2. Algorithm for the Computational Model

In this section, let us consider that all the variables having a superscript, 'n' represents values that are updated based on the previous loading and those variables with superscript, 'n+1' denotes the values corresponding to the current state of loading. All the variables having subscript, 'tr' represents the trial values. For simplicity in representation, indicial representation of the tensorial quantities are avoided.

Total strain at any increment is given by:

$$\boldsymbol{\varepsilon}^{\text{total}^{n+1}} = \boldsymbol{\varepsilon}^{\text{total}^{n}} + \Delta\boldsymbol{\varepsilon}^{\text{total}} \tag{40}$$

Using Equation (19):

$$\boldsymbol{\varepsilon}^{en+1}_{tr} = \boldsymbol{\varepsilon}^{\text{total}^{n+1}} - \boldsymbol{\varepsilon}^{p^{n}} \tag{41}$$

Using Equation (20):

$$\boldsymbol{\sigma}^{n+1}_{tr} = (1-\Phi^{n})\mathbf{C}^{0}\boldsymbol{\varepsilon}^{en+1}_{tr} \tag{42}$$

$$\boldsymbol{\sigma}^{n+1} = (1-\Phi^{n+1})\mathbf{C}^{0}\left(\boldsymbol{\varepsilon}^{\text{total}^{n+1}} - \boldsymbol{\varepsilon}^{p^{n+1}}\right) \tag{43}$$

Let:

$$w = (1-\Phi) \tag{44}$$

then:

$$\boldsymbol{\sigma}^{n+1} = w^{n+1}\mathbf{C}^{0}(\boldsymbol{\varepsilon}^{\text{total}^{n+1}} - \boldsymbol{\varepsilon}^{p^{n}} - \Delta\boldsymbol{\varepsilon}^{p}) \tag{45}$$

Using Equations (25), (36), (40), (41), and (44) in Equation (45), we get the following:

$$\boldsymbol{\sigma}'^{n+1} = \frac{1}{w^{n}}\boldsymbol{\sigma}^{n+1}_{tr} - \frac{1}{w^{n+1}}\mathbf{C}^{0}\Delta\Gamma\left[\frac{3}{2}\frac{\mathbf{S}'^{n+1}}{\sigma'_{eq}}\right] \tag{46}$$

Let:

$$p' = \frac{\sigma'_{kk}}{3} \tag{47}$$

Therefore, from Equations (34), (46) and (47), we can write the following expression:

$$\mathbf{S}'^{n+1} + \frac{1}{w^n} p_{tr}\,\mathbf{I} - \frac{1}{3}\mathbf{C}^0 \frac{\Delta\Gamma}{w^{n+1}}\left[\frac{3}{2}\frac{\mathbf{S}'^{n+1}}{\sigma'^{\,n+1}_{eq}}\right] = \frac{1}{w^n}\boldsymbol{\sigma}^{n+1}_{tr} - \frac{1}{w^{n+1}}\mathbf{C}^0\Delta\Gamma\left[\frac{3}{2}\frac{\mathbf{S}'^{n+1}}{\sigma'^{\,n+1}_{eq}}\right] \tag{48}$$

Using Equation (21) in (48), we get the simplified form for the iteration equation in indicial notation, as follows:

$$\sigma'^{\,n+1^2}_{eq}\left\{\delta_{ik}\delta_{jl} + \frac{1}{w^{n+1}}C^0_{ijkl}\Delta\Gamma\left[\frac{3}{2}\frac{1}{\sigma'^{\,n+1}_{eq}}\right]\right\}\left\{\delta_{ik}\delta_{jl} + \frac{1}{w^{n+1}}C^0_{ijkl}\Delta\Gamma\left[\frac{3}{2}\frac{1}{\sigma'^{\,n+1}_{eq}}\right]\right\} = \frac{\sigma^{tr\,n+1^2}_{eq}}{w^{n2}} \tag{49}$$

Algorithmically derived Equation (49) can be solved by an iteration procedure to find the value of $\Delta\Gamma$, simultaneously with the update of w. A Newton-Raphson iteration scheme is employed in the integration scheme of the present study to solve the yield function given in Equation (32). Successively, the entropy is updated using Equation (27) and the damage is calculated using Equation (15).

3. Validation of the Computational Model for Monotonic Loading

Prior to the simulation of fully reversed cyclic loading, it is important to check the validity of the model under tensile as well as compressive loading. The computation models described in Section 2.3, is implemented in commercial finite element package, ABAQUS. User material subroutine is written to update the stresses according to the strain increments that are supplied to the subroutine as input. In order to validate the model for tensile as well as compressive loading cases in Ti-6Al-4V, we have used the experimental data, reported by Biswas et al. [2] and Carrion et al. [76].

3.1. Validation of the Numerical Model for Monotonic Tensile Loading

The true stress-strain graph reported in the literature [76] for Ti-6Al-4V alloy, is used for the comparison between experimental data and the numerical predictions of monotonic tensile loading. Mill Annealed hot rolled bars were used [76] in the study. The material parameters are taken from the literature [76], so as to match with the material used for the comparison. Details of the model parameters are given in Table 1. Using the common assumption that the gauge section of a dog bone sample experiences uniform strain, we consider 5 mm length in the computational model. Hence, it can reduce the computational cost as well. Diameter of the specimen is kept the same, like that of the experimentally reported sample by Carrion et al. [76], which is 6.35 mm in diameter. In ABAQUS, linear brick elements, C3D8R are used to mesh the numerical model. One end of the sample is defined with zero displacement (fixed) boundary condition and the other end is subjected to controlled displacement loading in the axial direction. After a mesh convergence analysis, an optimum seed size of 0.9 mm is fixed for all the simulations. A schematic representation of the computational geometry is shown in Figure 3.

Table 1. Material parameters used in the numerical model for tensile loading in Ti-6Al-4V alloy.

Material Parameter	Value	Unit
Young's modulus, E	106	GPa
Poisson's ratio, ν	0.31	
Density, ρ	4540	kg/m^3
Critical TSI, Φ_c	1	
Hardening parameter, K	968.00	MPa
Hardening exponent, r	0.64	
Yield strength, σ_{y_0}	992.00	MPa
Molar mass, m_s	0.047867	kg/mol
Reference temperature, T	298	K

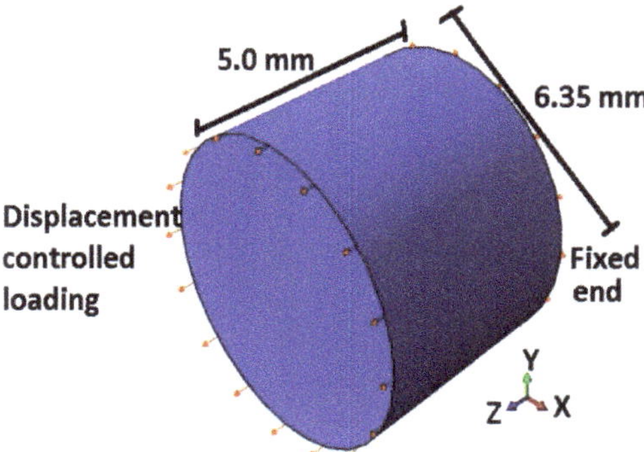

Figure 3. Schematics of numerical model for displacement controlled monotonic tensile loading in ABAQUS.

It can be observed from the Figure 4 that the true stress-strain graph, predicted for monotonic tensile loading in Ti-6Al-4V alloy, matches well with the experimental observations reported by Carrion et al. [76]. A smooth transition can be seen at point A, shown in Figure 4. This transition from elastic to plastic region can be due to the dislocation motion in the microstructure. Further, dislocation multiplication and interaction with each other and inclusions can be the possible reason behind strain hardening of the bulk material. Hence, the validation of the model under tensile loading can be considered as a basis for tensile loading in any kind of geometry or boundary conditions in the numerical investigation.

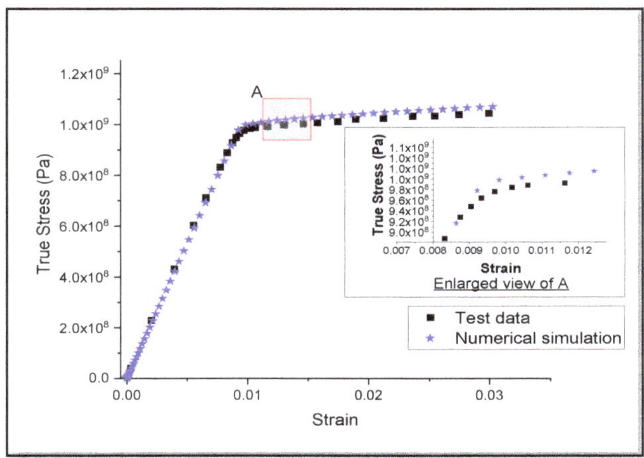

Figure 4. Comparison between monotonic tensile stress-strain graphs obtained from the test data [76] and numerical model.

A comparative plot between the numerical results for damage obtained from three-dimensional model and analytical results based on one dimensional approximation, as described in Section 2.2, is shown in Figure 5. It is observed that the level of matching between computational and experimental results for monotonic tensile loading is closer in the case of prediction of damage, based on the analytical approach and numerical analysis.

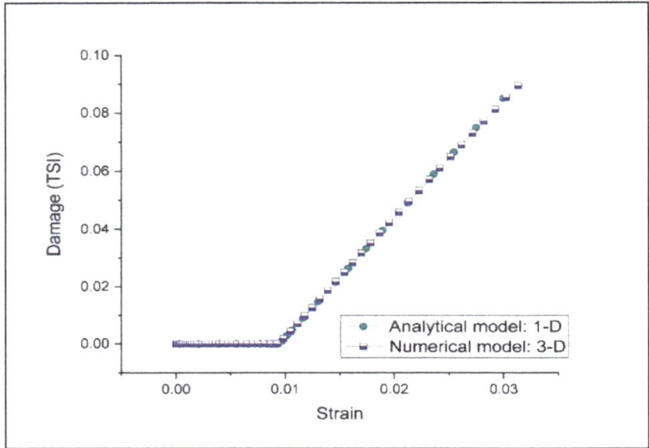

Figure 5. Comparison between the damage (TSI) prediction for monotonic tensile loading.

3.2. Validation of the 3-D Numerical Model for Monotonic Compressive Loading

Validation of the 3-D numerical model is done under compressive loading as well. Experimental result for a monotonic compression test, reported in the literature [2] is used to validate the proposed numerical model. The computational model parameters are taken from the literature [2], so as to match with the material used for the comparison. Even though the reported experimental results [2,76] are for Ti-6Al-4V alloys, it is noted that the materials are different in terms of their mechanical properties. Details of the model parameters used for the numerical simulation of monotonic compression test are listed in Table 2. We have considered the same dimensions in the numerical model, as that of the experimental samples [2]. Since, true stress-strain data is given in the literature [2], analytical procedure to compute TSI, requires an additional step. This method is adopted from well-known damage rule based on area reduction [77]. In the current study, damage parameter is represented by the TSI. Hence, the current area is related to the original area of undamaged section through the factor, TSI as follows:

$$A = (1 - \Phi)A_0 \tag{50}$$

where, A and A_0 represents the current area and initial area. The engineering stress and true stress are related by the principle of static equilibrium as follows,

$$\overline{\sigma}' A = \overline{\sigma} A_0 \tag{51}$$

where, $\overline{\sigma}'$ and $\overline{\sigma}$ represents the true stress and engineering stress respectively. Hence, in order to quantify the entropy, we have estimated the engineering yield stress data as follows:

$$\overline{\sigma}_y{}^{i+1} = \overline{\sigma}_{y'}{}^{i+1}(1 - \Phi^i) \tag{52}$$

where, $\overline{\sigma}_y{}^{i+1}$ and $\overline{\sigma}_{y'}{}^{i+1}$ represents the computed engineering stress and true stress at $(i+1)$-th strain, respectively. Φ^i is calculated based on the i-th strain data. Hence, in an incremental way, the TSI is computed using analytical procedure given in Section 2.2. Computation model in ABAQUS is discretized with linear brick finite elements C3D8R. One of the ends of the computational model is constrained from all the translations and the other end is subjected to displacement controlled compressive loading in the axial direction. A mesh convergence analysis is conducted and an optimum seed size of 0.9 mm is adopted in the simulations. A schematic representation of the computational geometry is shown in Figure 6.

Table 2. Material parameters used in the numerical model for compressive loading in Ti-6Al-4V alloy.

Material Parameter	Value	Unit
Young's modulus, E	118	GPa
Poisson's ratio, ν	0.31	
Density, ρ	4540	kg/m^3
Critical TSI, Φ_c	1	
Hardening parameter, K	550.00	MPa
Hardening exponent, r	0.65	
Yield strength, σ_{y_0}	1047.00	MPa
Molar mass, m_s	0.047867	kg/mol
Reference temperature, T	298	K

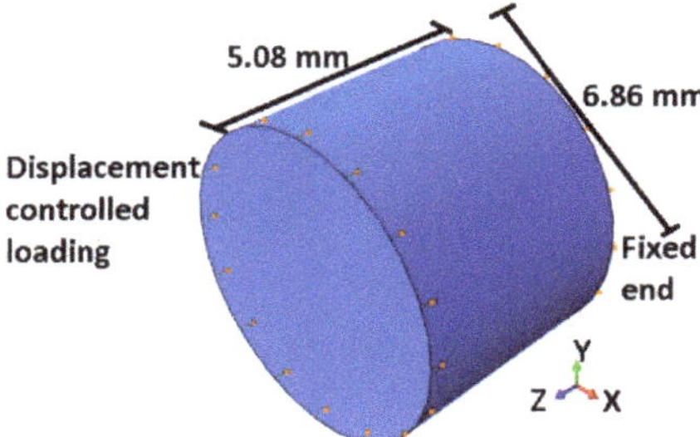

Figure 6. Schematics of numerical model for displacement controlled monotonic compressive. loading in ABAQUS.

Numerical results for monotonic compressive loading in Ti-6Al-4V alloy, shown in Figure 7, are found to be matching well with the reported experimental results [2]. Hence, the proposed model is taken as a basis to simulate compressive loading cases in the succeeding numerical investigations. Using the experimental [2] stress-strain graph, we have analytically calculated the TSI at every incremental plastic strain, based on the procedure stated in Section 2.2. As shown in Figure 8, both the analytical and numerical predictions for TSI matches very well.

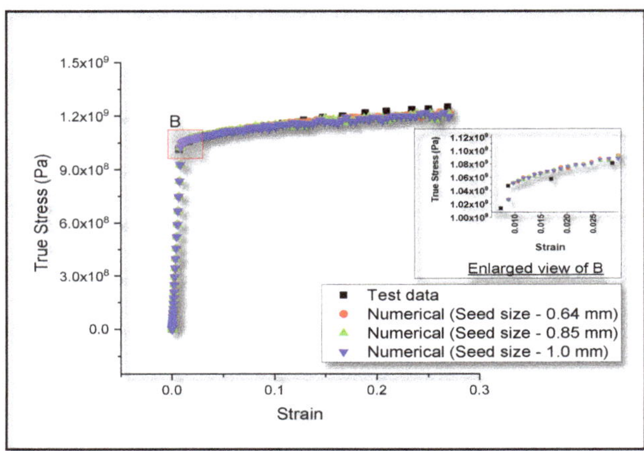

Figure 7. Comparison between monotonic compressive stress-strain graphs obtained from the test data [2] and numerical model.

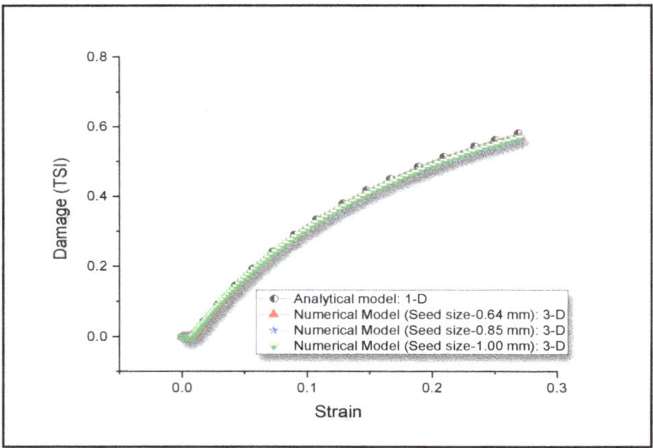

Figure 8. Comparison between the damage prediction for monotonic compressive loading.

4. Model Predictions for Low Cycle Fatigue Life

Carrion et al. [76] tested Ti-6Al-4V samples under tensile loading condition at a strain rate of the order of 10^{-3} s^{-1} at room temperature. Similar quasi-static loading condition is established in our numerical loading by controlling the step time of the numerical model in ABAQUS. The material model used in developing the 3-Dimensional numerical model is independent of the strain rate and the temperature and hence the strain rate hardening behavior and temperature effects, including the thermal dissipation are not considered in our study. Unified mechanics theory-based approach for damage calculation, described in Section 2, is used to predict the low cycle fatigue life of Ti-6Al-4V alloys. Details of the one-dimensional analytical model as well as the three-dimensional numerical model to predict fatigue life or Ti-6Al-4V are given in Section 4.1 below.

4.1. Analytical Approach for Fatigue Life Prediction

Experimental results [76] for the stabilized hysteresis loop is assumed to be closer to the first cycle hysteresis loop. Unified mechanics theory is used to evaluate the damage evolution under cyclic loading and the results are plotted in Figure 9.

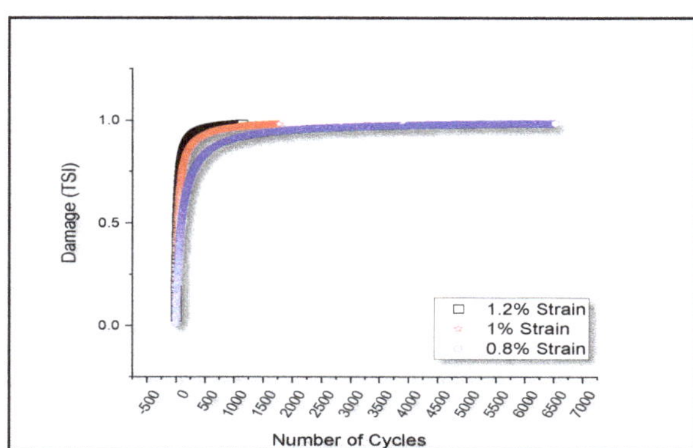

Figure 9. Analytical prediction of damage for different strain amplitudes of cyclic loading.

Low cycle fatigue life of the Ti-6Al-4V sample is predicted by fixing the TSI at failure as 0.98. This is necessary, as to prevent computational instabilities at the verge of failure that are not recorded by experimental results, are to be taken into account when we compare the mathematical model predictions with the experimental results. A MATLAB script is written to compute the fatigue life, from the stabilized hysteresis loop. The results are shown in comparison with the test data [76] and the corresponding numerical predictions at similar amplitudes, as shown in Figure 10.

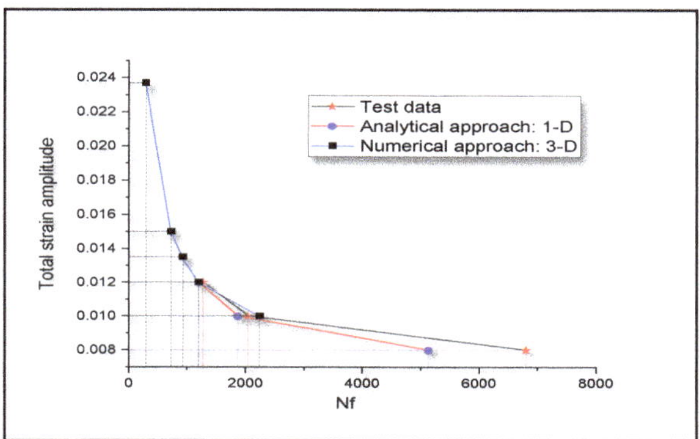

Figure 10. Low cycle fatigue life (Nf) prediction at different strain amplitudes of cyclic loading in comparison with the test data [76].

4.2. Computational Procedure for Fatigue Life Prediction

It is not feasible to conduct a large number of cyclic loading in the numerical model to predict fatigue life, especially when the amplitude of strain is very small because in ABAQUS this process would take weeks. In this section, we propose an alternate way of fatigue life prediction of Ti-6Al-4V alloys at different strain amplitudes, using a combined numerical-experimental procedure. If the hysteresis loop for a given strain amplitude is found out from the experiment, the same test is simulated by using the proposed model. Computational results after the first cycle of loading are used to find the scaling factor for incremental entropy in the computational model. The scaling factor is calculated as the ratio between the experimental hysteresis loop area for the stabilized loop and the numerically computed dissipation for the first cycle of loading. Then the computational model is used to evaluate the dissipation at different strain amplitudes of loading for a single cycle of loading. This hysteresis loop is used to predict the fatigue life at the given amplitude of strain, as per the procedure detailed in Section 2.3.

To compare the numerical predictions for fatigue life with experimental results [76], the same material data, as listed in Table 1, are used. It is assumed that the experimental results are free from any boundary effects or instabilities. Hence, the numerical analysis is done on the sample, with dimensions and boundary conditions as shown in Figure 3. Hysteresis loops at 1.2% strain amplitude for 50 cycles of loading are plotted in Figure 11a. A comparative hysteresis plot for first cycle and 50th cycle of loading is shown in Figure 11b. It can be observed from Figure 11a,b, that the hysteresis loop area decreases with cyclic loading. This reduction in hysteresis loop area is due to the reduction in strength of the material with the evolution of TSI. The fatigue life can be predicted by extrapolating the numerical results on TSI axis vs number of cycles. A comparison plot between test data and simulations for low cycle fatigue life prediction at different strain amplitudes is shown in Figure 10. In Figure 10, the average values of fatigue life test data [76] are plotted for stain amplitudes of 0.8%, 1.0% and 1.2% and compared with the analytical predictions. Fatigue life test data for other amplitudes of strain are not reported in the literature [76]. Results from the numerical approach, for the strain amplitudes 1.0% and 1.2% are also plotted and the model prediction is extended to a strain amplitude of 2.4%. Response at 0.8% strain amplitude was not computed with 3-D model because 0.8% strain amplitude is within the elastic region of loading.

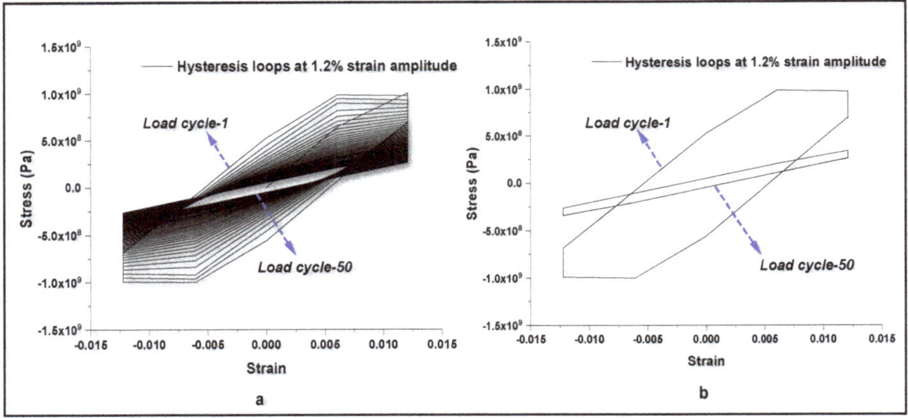

Figure 11. Numerical results on engineering stress-strain hysteresis loops for 1.2% strain amplitude of cyclic loading. (**a**) hysteresis loops at 1.2% strain amplitude for 50 cycles of loading; (**b**) comparative hysteresis plot for first cycle and 50th cycle of loading.

In Figure 10, it is clear that the one-dimensional analytical approach is underestimating the fatigue life by 1600 cycles at a total strain amplitude of 0.8%, while this discrepancy is less scattered in the test data [76]. This discrepancy in fatigue life prediction using one-dimensional model could be due to the unaccounted three-dimensional confinement effects in material response.

5. Conclusions

The work presented here is based on the unified mechanics theory, where the laws of Newtonian mechanics are combined with laws of thermodynamics, directly. The bridging factor in unified mechanics theory is the definition of thermodynamic state index, given in the Equation (9). The definition of damage proposed in the literature [29], is applied in the case of monotonic as well as low cycle fatigue loading conditions in Ti-6Al-4V alloys. Based on the principles of continuum mechanics, we have presented a numerical model, which account for the damage in case of plastic loading in Ti-6Al-4V. It is observed from the three-dimensional numerical and one-dimensional analytical results of the damage model prediction that, they match very well with the experimental observations in the case of monotonic tensile loading, as shown in Figures 4 and 5. In Figure 4, we have considered the stress-strain graph given in the literature [76] for validation and the corresponding damage prediction (value of TSI is around 0.1) is limited to a strain level of about 3%. Linear extrapolation of the damage curves plotted in Figure 5, can lead to wrong prediction of failure strain (to around 20% in the current study). Entropy at each time increment is dependent on the stress level. Hence, the accuracy of life prediction will be dependent on the constitutive model, used to predict the yielding of the material, in a three-dimensional numerical study.

The monotonic compressive stress-strain graph is matching well with the experimental results reported in the literature, as shown in Figure 7. The path traced by the damage prediction from one-dimensional analytical procedure and three-dimensional numerical procedure, as shown in Figure 8, also matches very well. In the case of compressive loading conditions, the results can be affected by the confining effects. The difference in nature of path traced by damage curves in compressive and tensile loading conditions could be due to the difference in confining effects in compressive loading, when compared with tensile loading. Similar observations for alloys can be seen in the literature [78,79]. Current study may be extended in future, for the detailed experimental and numerical investigations on such confining effects, under compressive loading. Since the current focus of the investigation is to introduce an efficient way of predicting the fatigue life of Ti-6Al-4V using computational tools in conjunction with the experiment, we have limited our study to fatigue life prediction.

Thermodynamics of life of any system, as postulated by the unified mechanics theory, is brought in to application level, for the case of low cycle fatigue life prediction in Ti-6Al-4V. From the comparative study on fatigue life prediction, as shown in Figure 10, the proposed procedures, described in Section 4, are found to be very efficient. Only one cycle experimental data is sufficient to predict the low cycle fatigue in Ti-6Al-4V alloys. Hence, the procedure stated in Section 4, will be useful for practical applications.

Author Contributions: Conceptualization, N.B.J.M. and A.K.; Methodology, N.B.J.M.; Software, N.B.J.M.; Validation, N.B.J.M. and A.K.; Formal Analysis, N.B.J.M.; Investigation, N.B.J.M., A.K., C.L.R. and C.B.; Resources, C.L.R. and C.B.; Data Curation, N.B.J.M.; Writing-Original Draft Preparation, N.B.J.M.; Writing-Review & Editing, N.B.J.M., A.K., C.L.R. and C.B.; Supervision, C.L.R. and C.B. All authors have read and agreed to the published version of the manuscript.

Funding: This research received no external funding.

Conflicts of Interest: The authors declare no conflict of interest.

References

1. Mouritz, A.P. *Introduction to Aerospace Materials*; Woodhead Publishing Limited: Cambridge, UK, 2012; ISBN 9781845695323.
2. Biswas, N.; Ding, J.L.; Balla, V.K.; Field, D.P.; Bandyopadhyay, A. Deformation and fracture behavior of laser processed dense and porous Ti6Al4V alloy under static and dynamic loading. *Mater. Sci. Eng. A* **2012**, *549*, 213–221. [CrossRef]
3. Banerjee, D.; Williams, J.C. Perspectives on titanium science and technology. *Acta Mater.* **2013**, *61*, 844–879. [CrossRef]
4. Altenberger, I.; Nalla, R.K.; Sano, Y.; Wagner, L.; Ritchie, R.O. On the effect of deep-rolling and laser-peening on the stress-controlled low- and high-cycle fatigue behavior of Ti-6Al-4V at elevated temperatures up to 550 °C. *Int. J. Fatigue* **2012**, *44*, 292–302. [CrossRef]
5. CHENG, A.S.; LAIRD, C. Fatigue Life Behavior of Copper Single Crystals. Part I: Observations of Crack Nucleation. *Fatigue Fract. Eng. Mater. Struct.* **1981**, *4*, 331–341. [CrossRef]
6. Ren, Y.M.; Lin, X.; Guo, P.F.; Yang, H.O.; Tan, H.; Chen, J.; Li, J.; Zhang, Y.Y.; Huang, W.D. Low cycle fatigue properties of Ti-6Al-4V alloy fabricated by high-power laser directed energy deposition: Experimental and prediction. *Int. J. Fatigue* **2019**, *127*, 58–73. [CrossRef]
7. Tanaka, K.; Mura, T. Dislocation Model for Fatigue Crack Initiation. *Am. Soc. Mech. Eng.* **1981**, *48*, 97–103. [CrossRef]
8. Guo, Q.; Zaïri, F.; Guo, X. An intrinsic dissipation model for high-cycle fatigue life prediction. *Int. J. Mech. Sci.* **2018**, *140*, 163–171. [CrossRef]
9. Sosnovskiy, L.A.; Senko, V.I. Tribo-fatigue. In Proceedings of the ASME International Mechanical Engineering Congress and Exposition, Tribology, Orlando, FL, USA, 5–11 November 2005; pp. 141–148.
10. Wang, T.; Samal, S.K.; Lim, S.K.; Shi, Y. Entropy Production-Based Full-Chip Fatigue Analysis: From Theory to Mobile Applications. *IEEE Trans. Comput. Des. Integr. Circuits Syst.* **2019**, *38*, 84–95. [CrossRef]
11. Hashin, Z. A RelnterpretatIon of the Palmgren- miner Rule for Fatigue Life. *J. Appl. Mech.* **2016**, *47*, 324–328. [CrossRef]
12. Smith, K.N.; Topper, T.H.; Watson, P. A stress–strain function for the fatigue of metals (stress-strain function for metal fatigue including mean stress effect). *J. Mater.* **1970**, *5*, 767–778.
13. Lemaitre, J.; Desmorat, R. *Engineering damage mechanics: ductile, creep, fatigue and brittle failures*; Springer: Berlin/Heidelberg, Germany; New York, NY, USA, 2005; ISBN 3540215034.
14. Bhattacharya, B.; Ellingwood, B. Continuum damage mechanics analysis of fatigue crack initiation. *Int. J. Fatigue* **1998**, *20*, 631–639. [CrossRef]
15. Ontiveros, V.; Amiri, M.; Kahirdeh, A.; Modarres, M. Thermodynamic entropy generation in the course of the fatigue crack initiation. *Fatigue Fract. Eng. Mater. Struct.* **2017**, *40*, 423–434. [CrossRef]

16. Harvey, S.E.; Marsh, P.G.; Gerberich, W.W. Atomic force microscopy and modeling of fatigue crack initiation in metals. *Acta Metall. Mater.* **1994**, *42*, 3493–3502. [CrossRef]
17. Kumar, J.; Sundara Raman, S.G.; Kumar, V. Analysis and Modeling of Thermal Signatures for Fatigue Damage Characterization in Ti–6Al–4V Titanium Alloy. *J. Nondestruct. Eval.* **2016**, *35*, 1–10. [CrossRef]
18. Sosnovskiy, L.A.; Sherbakov, S.S. Mechanothermodynamic entropy and analysis of damage state of complex systems. *Entropy* **2016**, *18*, 268. [CrossRef]
19. Kahirdeh, A.; Khonsari, M.M. Energy dissipation in the course of the fatigue degradation: Mathematical derivation and experimental quantification. *Int. J. Solids Struct.* **2015**, *77*, 74–85. [CrossRef]
20. Amiri, M.; Naderi, M.; Khonsari, M.M. An experimental approach to evaluate the critical damage. *Int. J. Damage Mech.* **2011**, *20*, 89–112. [CrossRef]
21. Sosnovskiy, L.A.; Sherbakov, S.S. Mechanothermodynamical system and its behavior. *Contin. Mech. Thermodyn.* **2012**, *24*, 239–256. [CrossRef]
22. Zhang, M.H.; Shen, X.H.; He, L.; Zhang, K.-S. Application of Differential Entropy in Characterizing the Deformation Inhomogeneity and Life Prediction of Low-Cycle Fatigue of Metals. *Materials* **2018**, *11*, 1917. [CrossRef]
23. Sosnovskiy, L.A.; Sherbakov, S.S. A Model of Mechanothermodynamic Entropy in Tribology. *Entropy* **2017**, *19*, 115. [CrossRef]
24. Sosnovskiy, L.; Sherbakov, S. *Mechanothermodynamics*; Springer: Berlin/Heidelberg, Germany, 2016; ISBN 978-3-319-24979-7.
25. Young, C.; Subbarayan, G. Maximum Entropy Models for Fatigue Damage in Metals with Application to Low-Cycle Fatigue of Aluminum 2024-T351. *Entropy* **2019**, *21*, 967. [CrossRef]
26. Santecchia, E.; Hamouda, A.M.S.; Musharavati, F.; Zalnezhad, E.; Cabibbo, M.; El Mehtedi, M.; Spigarelli, S. A Review on Fatigue Life Prediction Methods for Metals. *Adv. Mater. Sci. Eng.* **2016**, *2016*, 9573524. [CrossRef]
27. Haddad, W.M. Thermodynamics: The unique universal science. *Entropy* **2017**, *19*, 621. [CrossRef]
28. Haddad, W.M.; Chellabonia, V.; Nersesov, S.G. *Thermodynamics: A Dynamical Systems Approach*; Princeton University Press: Princeton, NJ, USA, 2005; ISBN 0-691-12327-6.
29. Basaran, C.; Yan, C.Y. A thermodynamic framework for damage mechanics of solder joints. *J. Electron. Packag. Trans. ASME* **1998**, *120*, 379–384. [CrossRef]
30. Basaran, C.; Nie, S. An irreversible thermodynamics theory for damage mechanics of solids. *Int. J. Damage Mech.* **2004**, *13*, 205–223. [CrossRef]
31. Basaran, C.; Tang, H. Implementation of a thermodynamic framework for damage mechanics of solder interconnect in microelectronic packaging. *ASME Int. Mech. Eng. Congr. Expo. Proc.* **2002**, *11*, 61–68.
32. Basaran, C.; Lin, M.; Ye, H. A thermodynamic model for electrical current induced damage. *Int. J. Solids Struct.* **2003**, *40*, 1738–1745. [CrossRef]
33. Gomez, J.; Basaran, C. A thermodynamics based damage mechanics constitutive model for low cycle fatigue analysis of microelectronics solder joints incorporating size effects. *Int. J. Solids Struct.* **2005**, *42*, 3744–3772. [CrossRef]
34. Gomez, J.; Basaran, C. Damage mechanics constitutive model for Pb/Sn solder joints incorporating nonlinear kinematic hardening and rate dependent effects using a return mapping integration algorithm. *Mech. Mater.* **2006**, *38*, 585–598. [CrossRef]
35. Tang, H.; Basaran, C. A damage mechanics-based fatigue life prediction model for solder joints. *J. Electron. Packag. Trans. ASME* **2003**, *125*, 120–125. [CrossRef]
36. Temfack, T.; Basaran, C. Experimental verification of thermodynamic fatigue life prediction model using entropy as damage metric. *Mater. Sci. Technol.* **2015**, *31*, 1627–1632. [CrossRef]
37. Wang, J.; Yao, Y. An entropy based low-cycle fatigue life prediction model for solder materials. *Entropy* **2017**, *19*, 503.
38. Basaran, C.; Zhao, Y.; Tang, H.; Gomez, J. A damage-mechanics-based constitutive model for solder joints. *J. Electron. Packag. Trans. ASME* **2005**, *127*, 208–214. [CrossRef]
39. Basaran, C.; Chandaroy, R. Thermomechanical analysis of solder joints under thermal and vibration loading. *J. Electron. Packag. Trans. ASME* **2002**, *124*, 60–66. [CrossRef]
40. Basaran, C.; Li, S.; Abdulhamid, M.F. Thermomigration induced degradation in solder alloys. *J. Appl. Phys.* **2008**, *103*, 123520-1–123520-9. [CrossRef]

41. Basaran, C.; Lin, M. Damage mechanics of electromigration induced failure. *Mech. Mater.* **2008**, *40*, 66–79. [CrossRef]
42. Basaran, C.; Lin, M. Damage mechanics of electromigration in microelectronics copper interconnects. *Int. J. Mater. Struct. Integr.* **2007**, *1*, 16–39. [CrossRef]
43. Basaran, C.; Nie, S. Time dependent behavior of a particle filled composite PMMA/ATH at elevated temperatures. *J. Compos. Mater.* **2008**, *42*, 2003–2025. [CrossRef]
44. Basaran, C.; Nie, S. A thermodynamics based damage mechanics model for particulate composites. *Int. J. Solids Struct.* **2007**, *44*, 1099–1114. [CrossRef]
45. Li, S.; Basaran, C. A computational damage mechanics model for thermomigration. *Mech. Mater.* **2009**, *41*, 271–278. [CrossRef]
46. Lin, M.; Basaran, C. Electromigration induced stress analysis using fully coupled mechanical-diffusion equations with nonlinear material properties. *Comput. Mater. Sci.* **2005**, *34*, 82–98. [CrossRef]
47. Shidong, L.; Abdulhamid, M.F.; Basaran, C. Simulating Damage Mechanics of Electromigration and Thermomigration. *Simulation* **2008**, *84*, 391–401. [CrossRef]
48. Yao, W.; Basaran, C. Electromigration damage mechanics of lead-free solder joints under pulsed DC: A computational model. *Comput. Mater. Sci.* **2013**, *71*, 76–88. [CrossRef]
49. Yao, W.; Basaran, C. Computational damage mechanics of electromigration and thermomigration. *J. Appl. Phys.* **2013**, *114*, 103708. [CrossRef]
50. Cuadras, A.; Crisóstomo, J.; Ovejas, V.J.; Quilez, M. Irreversible entropy model for damage diagnosis in resistors. *J. Appl. Phys.* **2015**, *118*, 165103-1–165103-8. [CrossRef]
51. Cuadras, A.; Romero, R.; Ovejas, V.J. Entropy characterisation of overstressed capacitors for lifetime prediction. *J. Power Sources* **2016**, *336*, 272–278. [CrossRef]
52. Cuadras, A.; Yao, J.; Quilez, M. Determination of LEDs degradation with entropy generation rate. *J. Appl. Phys.* **2017**, *122*, 145702-1–145702-7. [CrossRef]
53. Imanian, A.; Modarres, M. A thermodynamic entropy approach to reliability assessment with applications to corrosion fatigue. *Entropy* **2015**, *17*, 6995–7020. [CrossRef]
54. Imanian, A.; Modarres, M. A thermodynamic entropy-based damage assessment with applications to prognostics and health management. *Struct. Heal. Monit.* **2018**, *17*, 1–15. [CrossRef]
55. Jang, J.Y.; Khonsari, M.M. On the evaluation of fracture fatigue entropy. *Theor. Appl. Fract. Mech.* **2018**, *96*, 351–361. [CrossRef]
56. Liakat, M.; Khonsari, M.M. Entropic characterization of metal fatigue with stress concentration. *Int. J. Fatigue* **2015**, *70*, 223–234. [CrossRef]
57. Osara, J.A.; Bryant, M.D. A Thermodynamic Model for Lithium-Ion Battery Degradation: Application of the Degradation-Entropy Generation Theorem. *Inventions* **2019**, *4*, 23. [CrossRef]
58. Osara, J.A.; Bryant, M.D. Thermodynamics of Fatigue: Degradation-Entropy Generation Methodology for System and Process Characterization and Failure Analysis. *Entropy* **2019**, *21*, 685. [CrossRef]
59. Gomez, J.; Lin, M.; Basaran, C. Damage Mechanics Modeling of Concurrent Thermal and Vibration Loading on Electronics Packaging. *Multidiscip. Model. Mater. Struct.* **2006**, *2*, 309–326. [CrossRef]
60. Amiri, M.; Khonsari, M.M. On the role of entropy generation in processes involving fatigue. *Entropy* **2012**, *14*, 24–31. [CrossRef]
61. Wang, J.; Yao, Y. An entropy-based failure prediction model for the creep and fatigue of metallic materials. *Entropy* **2019**, *21*, 1104. [CrossRef]
62. Sun, F.; Zhang, W.; Wang, N.; Zhang, W. A copula entropy approach to dependence measurement for multiple degradation processes. *Entropy* **2019**, *21*, 724. [CrossRef]
63. Yun, H.; Modarres, M. Measures of Entropy to Characterize Fatigue Damage in Metallic Materials. *Entropy* **2019**, *21*, 804. [CrossRef]
64. Li, E.H.; Li, Y.Z.; Li, T.T.; Li, J.X.; Zhai, Z.Z.; Li, T. Intelligent analysis algorithm for satellite health under time-varying and extremely high thermal loads. *Entropy* **2019**, *21*, 983. [CrossRef]
65. Sosnovskiy, L.A.; Sherbakov, S.S. On the Development of Mechanothermodynamics as a New Branch of Physics. *Entropy* **2019**, *21*, 1188. [CrossRef]
66. Gunel, E.M.; Basaran, C. Damage characterization in non-isothermal stretching of acrylics. Part I: Theory. *Mech. Mater.* **2011**, *43*, 979–991. [CrossRef]

67. Naderi, M.; Amiri, M.; Khonsari, M.M. On the thermodynamic entropy of fatigue fracture. *Proc. R. Soc. A Math. Phys. Eng. Sci.* **2010**, *466*, 423–438. [CrossRef]
68. Naderi, M.; Khonsari, M.M. An experimental approach to low-cycle fatigue damage based on thermodynamic entropy. *Int. J. Solids Struct.* **2010**, *47*, 875–880. [CrossRef]
69. Abdulhamid, M.F.; Basaran, C. Influence of thermomigration on lead-free solder joint mechanical properties. *J. Electron. Packag. Trans. ASME* **2009**, *131*, 011002. [CrossRef]
70. Boltzmann, L. Ableitung des Stefan'schen Gesetzes, betreffend die Abhängigkeit der Wärmestrahlung von der Temperatur aus der electromagnetischen Lichttheorie. *Ann. Phys.* **1884**, *258*, 291–294. [CrossRef]
71. Sharp, K.; Matschinsky, F. Translation of Ludwig Boltzmann's paper "on the relationship between the second fundamental theorem of the mechanical theory of heat and probability calculations regarding the conditions for thermal equilibrium" Sitzungberichte der kaiserlichen akademie d. *Entropy* **2015**, *17*, 1971–2009. [CrossRef]
72. Planck, M. On the Law of Distribution of Energy in the Normal Spectrum. *Ann. Phys.* **1901**, *4*, 553. [CrossRef]
73. Lemaitre, J.; Chaboche, J.-L. *Mechanics of Solid Materials*; Cambridge University Press: Cambridge, UK, 1990; ISBN 0-521-32853-5.
74. Voyiadjis, G.Z.; Faghihi, D. Thermo-mechanical strain gradient plasticity with energetic and dissipative length scales. *Int. J. Plast.* **2012**, *30–31*, 218–247. [CrossRef]
75. Murakami, S. *Continuum Damage Mechanics*; Springer: Berlin/Heidelberg, Germany, 2012; ISBN 9789400726659.
76. Carrion, P.E.; Shamsaei, N.; Daniewicz, S.R.; Moser, R.D. Fatigue behavior of Ti-6Al-4V ELI including mean stress effects. *Int. J. Fatigue* **2017**, *99*, 87–100. [CrossRef]
77. Murakami, S. Mechanical modeling of material damage. *J. Appl. Mech. Trans. ASME* **1988**, *55*, 280–286. [CrossRef]
78. Spitzig, W.A.; Sober, R.J.; Richmond, O. Pressure dependence of yielding and associated volume expansion in tempered martensite. *Acta Metall.* **1975**, *23*, 885–893. [CrossRef]
79. Mahnken, R. Strength difference in compression and tension and pressure dependence of yielding in elasto-plasticity. *Comput. Methods Appl. Mech. Eng.* **2001**, *190*, 5057–5080. [CrossRef]

Article

Prediction of Fatigue Crack Growth Rate Based on Entropy Generation

Roslinda Idris, Shahrum Abdullah *, Prakash Thamburaja and Mohd Zaidi Omar

Department of Mechanical and Manufacturing Engineering, Faculty of Engineering and Built Environment, Universiti Kebangsaan Malaysia, UKM Bangi 43600, Selangor, Malaysia; ri.roslinda@gmail.com (R.I.); p.thamburaja@ukm.edu.my (P.T.); zaidiomar@ukm.edu.my (M.Z.O.)

* Correspondence: shahrum@ukm.edu.my; Tel.: +60-3-8911-8411

Received: 28 October 2019; Accepted: 2 December 2019; Published: 19 December 2019

Abstract: This paper presents the assessment of fatigue crack growth rate for dual-phase steel under spectrum loading based on entropy generation. According to the second law of thermodynamics, fatigue crack growth is related to entropy gain because of its irreversibility. In this work, the temperature evolution and crack length were simultaneously measured during fatigue crack growth tests until failure to ensure the validity of the assessment. Results indicated a significant correlation between fatigue crack growth rate and entropy. This relationship is the basis in developing a model that can determine the characteristics of fatigue crack growth rates, particularly under spectrum loading. Predictive results showed that the proposed model can accurately predict the fatigue crack growth rate under spectrum loading in all cases. The root mean square error in all cases is 10^{-7} m/cycle. In conclusion, entropy generation can accurately predict the fatigue crack growth rate of dual-phase steels under spectrum loading.

Keywords: degradation-entropy generation theorem; dual-phase steel; entropy generation; fatigue crack growth rate; spectrum loading

---

1. Introduction

Fatigue involves crack initiation, propagation and final fracture. The fatigue cracking problem of mechanical structures and components are exposed to variable amplitude loading (VAL) [1–4]. The number of cycles required for crack growth to reach a certain distance or failure can be predicted based on the principle of fracture mechanics. Paris and Erdogan [5] first formulated the empirical correlations between the fatigue crack growth (FCG) rate da/dN and the stress intensity factor range ΔK using a simple power law. Since then, fracture mechanics has provided exceptional contributions to the improvement of FCG prediction. The mathematical modelling of the FCG rate for steel is necessary to predict the residual strength or remaining life. The FCG in steel is a complex process with irreversible changes at micro, meso and macrolevels. Other classical approaches in determining the FCG rate are highly empirical in nature [6,7]. On the one hand, an empirical model simply describes a data but does not derive from physical principles. On the other hand, a model that fits a set of data cannot explain the reason for the material response. Hence, a model based on fundamental physical principles, including thermodynamics, is more necessary than a model based on analogy [8]. Any material behaviour can be expressed as a mathematical model when the second law of thermodynamics is fulfilled with suitable selection state variables, analytical expressions of the state potential and dissipative potentials [9]. Zhurkov et al. [10] stated that physical mechanisms and apparent characteristics of fatigue depend on the structure of materials and physical conditions, chemical composition, and the kind of utilized load. They modified the inhomogeneous field mechanism (IFM) model for the development of distinct fatigue degradation. In thermodynamic interpretation, the microscopic physical mechanisms are not well considered in this work. The existence of entropy generation is as such a prerequisite for the development of fatigue degradation for the explanation to understanding the microscopic physical mechanisms.

Given that the fatigue process is usually accompanied by energy transformation, developing a thermodynamic framework to study its characterisation is reasonable. Energy dissipation is an irreversible phenomenon, which in turn depicts the theory of entropy as an ideal tool in studying the fatigue process [11]. Meneghetti [12] studied the theoretical framework and the equivalent experimental techniques to calculate the dissipated heat energy in a structural volume surrounding the crack tip. The dissipated heat energy in a unit volume of the material per cycle (Q) can be adopted as an index of fatigue damage [13], whereas the average heat energy per cycle $\bar{Q}^*$ can serve as an elastic–plastic fracture mechanics parameter [14]. The thermodynamics framework has been recently applied in mechanical fatigue [15–18], which is not unexpected because fatigue degradation is an irreversible process that gradually ages the system until failure by fracture. Additionally, fatigue is a dissipation process, in which the accumulation of disorder is basically related to the generation of entropy based on the second law of thermodynamics [15,19–21] Thermal methods were used to study the dynamic crack behaviour of materials through the thermal signal time domain [22–25]. Furthermore, a cumulative entropy generation can provide estimation when crack initiation commences [26].

The entropy production that progressively accumulates is associated with the degradation or aging of the technique used and destroys the device until failure. The relationship between degradation and entropy generation because of irreversible processes was established within the system [27]. Particularly, material damage in different conditions, such as mechanical cyclic load, can be evaluated using the amount of entropy generation during degradation [28]. The driving motivation for degradation-entropy generation (DEG) theorem is that entropy monotonically increases, and free energy monotonically decreases for every ordinary process. Particularly, the entropy-generating irreversible process is present in all aging phenomena. Following the entropy increase, manufactured components return to their natural conditions through degradation, their integrity is consequently degraded and they eventually become non-functional [29]. However, within the framework of fracture mechanics, the entropy produced from crack growth has not been thoroughly investigated.

FCG for realistic structures involves VAL that leads to load interaction effects, and the result of FCG rate is difficult to predict. The entropy generated during the fatigue failure process can serve as a measure of degradation. Therefore, the FCG model called the entropy generation of fatigue crack (EGFC) for steel based on DEG theorem is proposed. An effective evaluation method for the FCG rate by using DEG theorem is essential, especially when employing the FCG test under spectrum loading. This study aims to understand how the DEG theorem can predict the FCG rate of a specimen, especially under spectrum loadings. The outcomes of the FCG test experiments and the theory revealed that fatigue degradation and entropy generation was closely related and that this relationship should be considered when evaluating the FCG rate based on the DEG theorem. Moreover, the results contributed fundamental improvements to the studies related to FCG rate without the need for conventional techniques based on empirical models, thereby easing the FCG rate prediction and lessening the necessary tests. Consequently, the generalisation of the predictions for any kind of spectrum loading will be relatively straightforward. For instance, the FCG rate was directly predicted from evaluation of the fatigue DEG. The advantage of using this method is the possibility of FCG prediction under spectrum loading. This step can be achieved through the measurement of the degradation coefficient of a given material under constant amplitude loading (CAL) by applying the concept that the total entropy generation is not dependent on load. The concept of DEG theorem is also reliable to evaluate the FCG rate.

2. Theoretical Model Based on Entropy Generation

2.1. Entropy Generation in an Irreversible Process

In an irreversible process, heat energy loss or dissipation can be observed because of intermolecular collisions and friction, which does not allow the recovery of energy when the process is reversed. Entropy can be defined as the irreversibility degree of a process. Sufficient amount of functional energy is lost due to dissipation or friction, which is disadvantageous to combustion reactions. Entropy is a non-existent form of energy that performs a beneficial work for the thermodynamic process. In conventional thermodynamics, reversible process entropy was described by Clausius as the ratio of heat energy transferred from the system to the absolute temperature. The equations of classical thermodynamics are not applicable for irreversible processes. The second law of thermodynamics could represent an inequality by introducing appropriate terms to account for the entropy's irreversible production. This inequality can be expressed as

$$\Delta\gamma = \sum_{m} \gamma_m + \sum_{n} \frac{Q_n}{T_n} + \Delta\gamma_{irr} \tag{1}$$

where $\Delta\gamma$ is the rise in the entropy of the system, $\sum_{m} \gamma_m$ is the net sum of the entropies transmitted into the system through transfer of matter, $\sum_{n} \frac{Q_n}{T_n}$ is net sum of the entropies transmitted into the system through heat transfer, Q is the transferred heat energy transferred, T is the absolute temperature and $\Delta\gamma_{irr}$ is the entropy generated by the irreversible processes happening within the system. The system endures an irreversible process and raises the entropy. The alteration in entropy caused by the irreversible processes that happens inside the system, which is always positive regardless if no matter or heat energy is transmitted into or out of the system. A simple example is the energy degradation utilised for mechanical works, wherein the dissipation of the internal energy occurs in the material body. If the dissipated energy dE_d have the same impact as that of the absorption dQ, which is in the form of heat, because of a constant equilibrium temperature T, then the entropy production rate can be presented as [30]

$$\left(\frac{d\gamma}{dt}\right)_{irr} = \frac{1}{T}\frac{dE_d}{dt} \tag{2}$$

where $\left(\frac{d\gamma}{dt}\right)_{irr}$ and $\frac{dE_d}{dt}$ are the irreversible entropy production and energy dissipation rates, respectively.

Naderi and Khonsari [31] stated that entropy is produced by the plastic work divided by temperature. The plastic work is mostly (around 90% for steels) dispersed into heat, and the remaining tiny portion in the material partakes in microstructural evolution [32]. The measurement of entropy production aims to assess the material damage. Part of the plastic work that dissipates through heat to the environment does not influence the degradation and damage and thus should be neglected when measuring the entropy generation during plastic deformation. In this study, the phrase 'dissipated energy' (E_d) is used instead of the plastic work W_p in computing the entropy production.

The rise in entropy is caused by the alteration in internal energy (i.e., the total energy in the thermodynamic structure during solids' deformation). Internal energy involves the strain energy absorbed during heat generation, the plastic work under strain hardening, the deformed solid and the sound released because of cracking [33]. The total energy can be obtained from the area under the load deformation curve of a solid. The total energy without the elastic strain energy renders the dispersed energy, which is entirely recoverable. This dissipated energy divided by the temperature is the entropy produced in the absence of heat transfer. There is a rise in stresses until reaching the strength limit on loading a material. When the stress reaches the tensile strength at a point, the material might crack. Entropy is generated by the solid that has a growing crack despite exhibiting an elastic mechanical response, signifying that irreversible crack propagation is an irreversible thermodynamic process [34].

2.2. Entropy Generated during FCG

FCG is an irreversible procedure that is related to the increase in entropy according to the second law of thermodynamics. The rate of entropy generation stated by Equation (2) with regards to the number of load cycles, which is relevant to fatigues, can be rewritten as

$$\left(\frac{d\gamma_{irr}}{dN}\right) = \frac{1}{T}\frac{dE_d}{dN} \tag{3}$$

where γ_{irr} it the irreversible entropy, E_d represents the dissipated energy, T is the absolute temperature (i.e., the surface temperature of specimen) and N is the number of loading cycles. Hence, the total entropy can be computed by integrating Equation (3) from 0 until the point of fracture N_f.

$$\gamma_f = \int_0^{N_f}\left(\frac{E_d}{T}\right)dN \tag{4}$$

where γ_f is the total entropy production of a fatigue failure.

The entropy pertaining of a thermodynamic system that undergoes cracking involves all the energy lost either because of the propagation or crack formation or through other dissipative mechanisms. For ductile materials (e.g., metals), energy is dissipated when a new crack surface is formed or when the plastic zone that lies ahead of the crack tip is lost. In this case, plasticity is the dominant dissipative mechanism.

3. Methodology

3.1. Entropy Generation of the EGFC Model Development Using DEG Theorem

Bryant et al. [27] introduced the DEG theorem, in which only one dissipative process p is responsible for the degradation of a system that generates entropy. According to the theorem, the crack length w indicates a measure of the degradation of the system that is dependent on the dissipative process, where $w = w\ (p)$. Similar to entropy generation, the degradation rate $D = dw/dt$ can be obtained by employing the chain rule.

$$D = \frac{dw}{dt} = \left(\frac{\partial w}{\partial p}\frac{\partial p}{\partial \zeta}\right)\frac{\partial \zeta}{\partial t} = YJ \tag{5}$$

where $Y = \frac{\partial w}{\partial p}\frac{\partial p}{\partial \zeta}$ is the degradation force. The degradation of a given system is dependent on the equivalent p as the representative of entropy generation. Given that thermodynamic flow J is probably the common parameter in Equation (5), the degradation coefficient can be obtained as:

$$B = \frac{Y}{X} = \frac{(\partial w/\partial p)(\partial p/\partial \zeta)}{(\partial_i S/\partial p)(\partial p/\partial \zeta)} = \frac{\partial w}{\partial_i S}\bigg| \tag{6}$$

where B describes the interaction of the degradation and entropy generation with p.

The crack length is considered as w, where $w = a$ in Equation (5). Therefore, degradation can be denoted as $a = a\{W_p(N)\}$, where W_p is the plastic energy generation at the tip of the crack and is probably the main dissipative process with the number of cycles N as the phenomenological variable. Numerous studies have attempted to the plastic energy generation W_p as a function of the FCG rate [35–37]. According to Equation (5), it can be written as $da/dt = YJ$, where $J = dN/dt$ and $Y = (da/dW_p)x(dW_p/dN)$. With the assumption that the plastic energy is dissipated as entropy generation, $dW_p = Td_iS$. Hence, the entropy generation can be defined as

$$\gamma = \frac{d_iS}{dt} = \frac{\partial_i S}{\partial p}\frac{\partial p}{\partial N}\frac{\partial N}{\partial t} = \frac{f}{T}\frac{dW_p}{dN} \tag{7}$$

where f is the frequency of the test and T is the surface temperature. By substituting $X = (1/T)\left(dW_p/dN\right)$ into Equation (5), and following expression was obtained.

$$D = \frac{da}{dt} = YJ = BXJ = B\frac{f}{T}\frac{dW_p}{dN} \tag{8}$$

As previously mentioned, B in Equation (8) measures how crack growth and entropy generation interrelate on the dissipative level of the plastic deformation process. The rate of crack growth given with respect to the number of load cycles can be expressed as

$$\frac{da}{dN} = \frac{da}{fdt} = \frac{B}{T}\frac{dW_p}{dN} \tag{9}$$

In terms of energy balance, the total energy needed to propagate a crack with a unit distance in a specific material is independent of the energy dissipation mechanism. Therefore, the energy absorbed per unit growth of the crack is equal to the dissipated plastic energy of the cyclic plastic zone per cycle [38]. This concept can be mathematically expressed as

$$W_c \delta a = \frac{dW_p}{dN} \tag{10}$$

where W_c is the plastic dissipation energy until fracture. By replacing the value of W_c, the crack growth rate da/dN is obtained as

$$\delta a = \frac{da}{dN} = \frac{1}{W_c}\frac{dW_p}{dN} \tag{11}$$

Combining Equations (9) and (11), we obtain

$$\frac{da}{dN} = \frac{1}{W_c}\frac{T}{B} \tag{12}$$

Several methods have been developed to assess W_c. In this study, the FCG rate was controlled using the crack tip opening displacement (CTOD). Dependency on ΔK^2 can be obtained, and the theories based on crack opening displacement will result to the Paris law exponent $m = 2$ [39]. Therefore, a correlation introduced by Skelton et al. [40] was employed.

$$\frac{da}{dN} = \frac{\Delta K^2(1-v)}{2\pi E W_c} \tag{13}$$

where ΔK is the stress intensity factor, v is the Poisson's ratio and E is Young's modulus. The EGFC model can be obtained by substituting Equation (12) into Equation (13).

$$\frac{da}{dN} = B\frac{1}{T}\frac{\Delta K^2(1-v)}{2\pi E} \tag{14}$$

Equation (14) represents the Paris–Erdagon law of crack growth, where the constant C is expressed as:

$$C = B\frac{1}{T}\frac{(1-v)}{2\pi E} \tag{15}$$

Equation (15) indicates the relationship between C and B, which is associated with entropy generation.

3.2. Prediction of Fatigue Life

In this study, the total fatigue life of each FCG test was estimated by integrating the Equation (14) using Simpson's rule [41]. Three neighbouring crack lengths, namely, a_j, a_{j+1} and a_{j+2} were used to calculate the number of cycles based on Simpson's rule. The number of cycles for crack length to propagate from distance a_j to a_{j+2}, can be obtained as:

$$\Delta N_{j+2} = \int_{a_j}^{a_{j+2}} [y]da = \frac{a_j\left(r^2-1\right)}{6r}\left[y_j r(2-r) + y_{j+1}(r+1)^2 + y_{j+2}(2r-1)\right] \tag{16}$$

where j is the number sequence, y_j is the difference between the numbers of cycles for crack length interval, r is the interval between the crack length and y represents the dN/da in Equation (14).

$$y = \frac{dN}{da} = T\frac{1}{B}\frac{2\pi E}{\Delta K^2(1-\nu)} \tag{17}$$

3.3. Materials and Specimens Preparation

In this study, a low carbon steel sample was used because it has the largest temperature range within the Eutectoid temperatures, minimum temperature of austenite A_1 and lower-bound temperature for austenite A_3 in the iron–carbon phase diagram amongst other types of carbon steel. A dual-phase steel was synthesised by subjecting low carbon steel samples to inter-critical annealing from a temperature above A_1 but below A_3 between the two phases' (i.e., ferrite + austenite) regions for a given period, followed by water quenching. The chemical composition of this steel is shown in Table 1. Two successive heat treatment processes were conducted to obtain the dual-phase material. In the first process, the as-received specimens were annealed at 760 °C for 90 min, followed by water quenching (inter-critically annealed) to achieve the martensite phase. The temperature range was selected on the basis of the highest fatigue strength [42]. The second process involves tempering at 400 °C for 2 h and then cooling at room temperature to eliminate the residual stresses and improve the toughness. This tempering temperature was chosen because low carbon steel is tempered after heat treatment between 200 °C and 600 °C [43–45]. All specimens (as-received and dual-phase specimens) were mechanically polished to remove all damaged layers.

Table 1. Chemical compositions of the steel (wt.%).

Elements	C	Mn	Si	P	S	Al
wt %	0.192	1.61	0.384	0.0162	0.0085	0.0314

3.4. Tensile and FCG Rate Tests

The tensile specimens were prepared according to ASTM E8 in sub-size dimensions with a gauge length of 25 mm (Figure 1). Tensile tests were also conducted according to ASTM E8 procedures at room temperature using a universal testing machine with a cross-head speed of 1.8 mm/min (equivalent to a strain rate of 0.001 s^{-1}) to investigate the mechanical properties of the dual-phase steel samples.

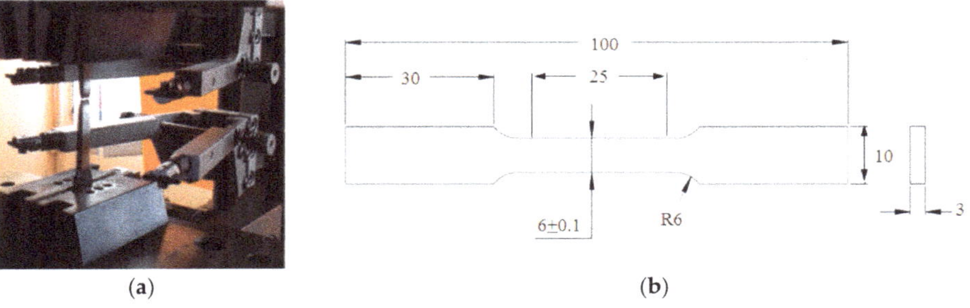

Figure 1. Tensile test; (**a**) experimental setup, (**b**) specimens' geometric dimensions (mm).

The geometric dimensions of the compact tension (CT) specimens were set according to ASTM E647, where thickness (B) = 12 mm and width (W) = 50 mm (Figure 2). Wire cut electric discharge machining (EDM) was used to cut the specimens. The residual compressive stresses from the milling process [46] were reduced by cutting a 4 mm sharp notch using EDM.

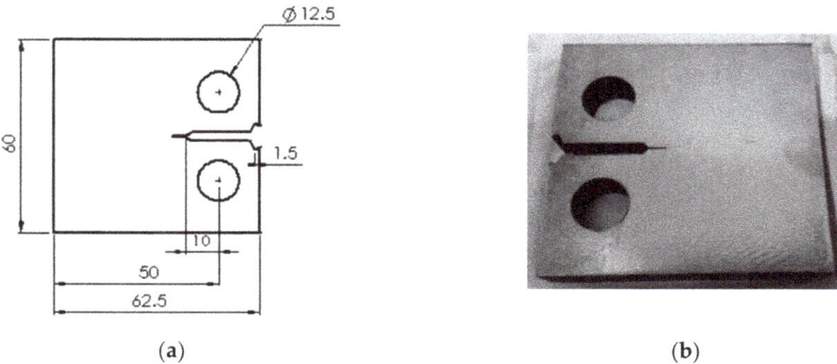

(a) (b)

Figure 2. Compact tension (CT) specimen for fatigue crack growth (FCG) test; (**a**) specimens' geometric dimensions (mm), (**b**) actual machined specimen.

The FCG tests were performed according to ASTM E647 procedures by utilising a servo-hydraulic universal test with a 100-kN capacity load cell. All experiments were conducted in an ambient setting with a load ratio R = 0.1 and loading frequency of 10 Hz sinusoidal waveform [47–49]. Three additional experimental tests were executed to represent the three types of loading. The FCG test classification was divided into three types: CAL, VAL and a two-step sequence loading involving high–low (H-L) and low–high (L-H) loadings. R was maintained constant throughout the experiment (Figure 3). In the two-step H–L loading experiment, the first loading step casted a considerable impact on the subsequent crack growth (i.e., the second loading step) when either R or the minimum load is equal in both steps [50].

The FCG tests commences with a fatigue pre-cracking process that allowed the formation of a crack via the sharp notch, which represents each CT specimen. The fatigue pre-cracking was conducted to produce the sharpened fatigue crack with sufficient straightness and size. The value of K_{max} was set as 32 MPa.m$^{1/2}$, along with the sinusoidal cyclic loading with R = 0.1 and a 10-Hz frequency to allow the desired crack to appear. The pre-crack value will surpass 0.10B almost all the time, adopting the value of the thickness h or 1.0 mm (0.040 in.), whichever is greater. The crack size on both sides (back and front) of the specimen was measured to ensure that the crack symmetry is maintained as indicated in the ASTM standard. The pre-crack length was set as 0.10B (1.2 mm) in this study. All specimens were then subjected to the FCG test under VAL and CAL conditions until fracture.

All the CT specimens were subjected to mode I opening loading. The compliance method with CTOD was applied to measure the fatigue crack length by employing a clip gauge at the notch mouth.

$$\alpha = a/W = 1.0010 - 4.6695u_x + 18.46{u_x}^2 - 236.82{u_x}^3 + 1214.9{u_x}^4 - 2143.6{u_x}^5 \quad (18)$$

$$u_x = \left\{\left[\frac{EVB}{P}\right]^{1/2} + 1\right\}^{-1} \quad (19)$$

where a is the crack length, W is the specimen width, B is the specimen thickness, E is Young's modulus and V is the CTOD. ΔK was calculated as

$$\Delta K = \frac{\Delta P}{B\sqrt{W}}\frac{(2+\alpha)}{(1-\alpha)^{\frac{3}{2}}}\left(0.886 + 4.64\alpha - 13.32\alpha^2 + 14.72\alpha^3 - 5.6\alpha^4\right) \quad (20)$$

where ΔP is the applied load range and α is the relative crack length (a/W).

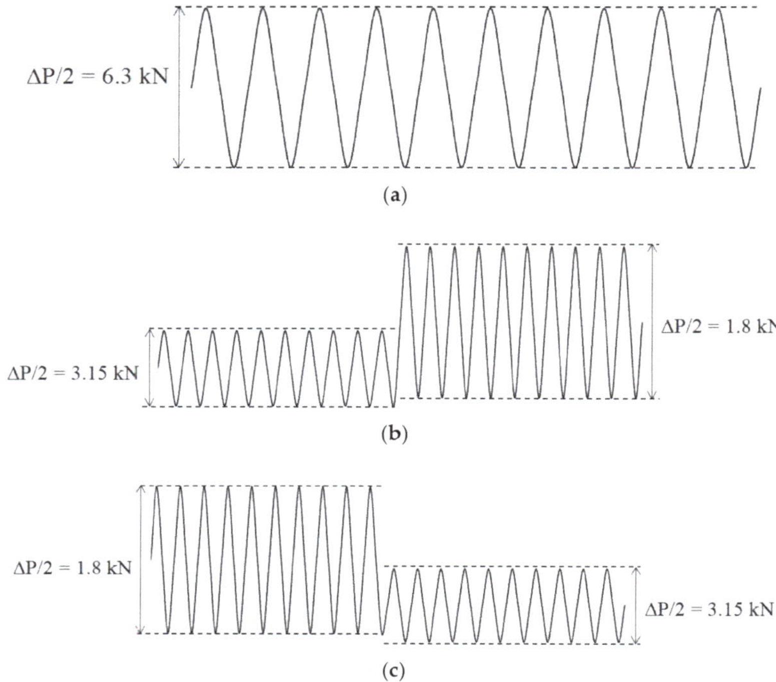

Figure 3. Schematic of the applied two-stage block loading sequences: (**a**) constant amplitude loading (CAL); (**b**) L-H; (**c**) H-L.

3.5. Temperature Measurement

Two thermocouples were used to determine the temperature evolution in the test samples. Naderi and Khonsari [51] utilised these thermocouples to oversee the specimen's thermal response. The delamination of the thermocouples was mitigated using magnetic thermocouples to ensure that good contact is maintained during the initiation at the specimen surface. In the FCG test, the surface temperature evolution of the specimen with respect to the crack growth was recorded using a magnetic thermocouple. The second thermocouple was positioned near the specimen to record the ambient temperature. This step ensures that no temperature change will occur in the surface of the specimen and that the ambient temperature will not exert great influence. A data acquisition tool was utilised to gather thermocouple data. Given that positioning the thermocouple at the spot where the crack occurs without disturbing the crack movement is difficult, a high-speed high-resolution infrared (IR) thermal imager was deployed to calibrate the thermocouples and minimise the error in the computation of entropy production, particularly at the start of the test, and to capture the crack growth's thermal image. The temperature was at the crack tip. Figures 4 and 5 illustrates the experimental setup of the FCG test for temperature measurement and the experimental process flow, respectively.

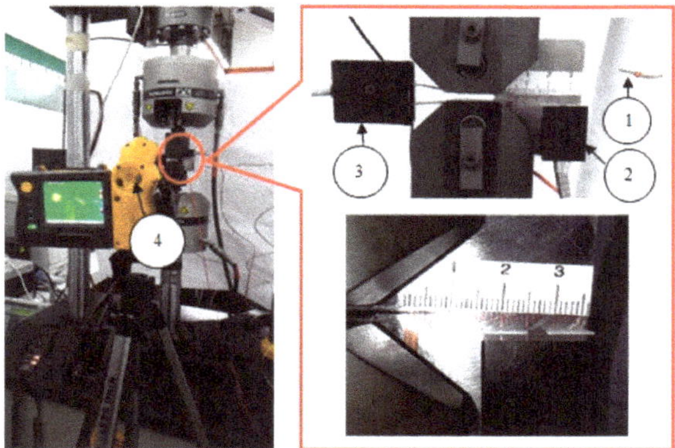

Figure 4. Schematic for FCG (fatigue crack growth) rate test: (1) Thermocouple-measured ambient temperature; (2) thermocouple-measured surface temperature; (3) clip gauge, (4) high-resolution IR thermal imager.

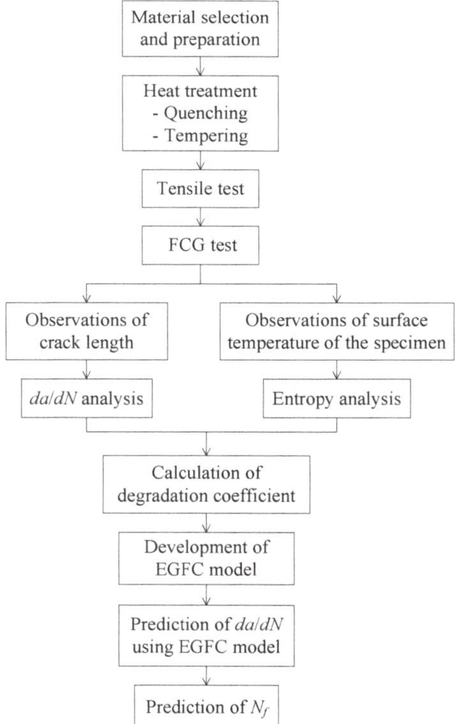

Figure 5. Flow of FCG rate prediction.

4. Results and Discussion

Experimental results are presented to determine the validity of the model in predicting the FCG rate. The proposed model can be useful in producing accurate predictive results of the measured FCG rate under spectrum loading.

4.1. Mechanical Properties

The heat treatment caused a more effective change in the mechanical property of the dual-phase steels than in the as-received steels. Table 2 presents the differences in the mechanical property of the as-received and dual-phase steel specimens. The ductility values of the latter were lower than those of the former. This result can be attributed to the existence of harder and less ductile ferrite matrix in the microstructure of the dual-phase steel specimen compared with the as-received steels. The as-received specimens had high ductility because of their softer ferrite–pearlite structures compared with the martensitic structures in the dual-phase steel [52].

Table 2. Mechanical properties of the steels.

Properties	Measured Value	
	As-Received	**Dual-Phase**
0.2% Yield strength (MPa)	388	495
Tensile strength (MPa)	536	597
Elongation at fracture (%)	30	13
Yield ratio (%)	72	83
Young's modulus (GPa)	204	185

The mechanical properties significantly changed after the heat treatment as proven by the increase in yield strength and ultimate tensile strength of the dual-phase steels. Both properties rapidly increased after heat treatment, reaching 495 and 597 MPa for the dual-phase and as-received steels, respectively. This finding specifies that the strength values of the dual-phase steels were higher than those of the as-received steel because of the presence of the harder second phase in the former [53]. The strength of the martensite produced in the soft ferrite matrix might differ from the structure formed when the steel was changed from austenite to 100% martensite [54,55].

4.2. FCG under Cyclic Loading

The FCG process is often classified into three stages: slow (stage I), stable (stage II) and rapidly growing regions (stage III). Figure 6 presents the fatigue life of the as-received and dual-phase steels based on crack length measurement during the FCG test. The figure shows that the crack initially grew at a slow rate and started to accelerate when the crack length increased after many cycles. The final point for each curve is the final fracture during the FCG testing. On the one hand, the analysis of FCG for VAL condition under L-H loading endured the longest life at 123,000 cycles, whereas that for CAL condition had the shortest life at 53,674 cycles. On the other hand, the analysis of FCG for VAL condition under H-L had a fatigue life of 112,683 cycles, which is within the range of the two analyses. The total cycles until failure depended on the type of the applied load. For high stresses, the crack growth rate represented by the slope of the curve was high at a given crack length and the FCG life (i.e., total number of applied cycles) was short. The highest stress level was observed at the shortest crack length during fracture. The magnitude of the applied stress and the fracture resistance of the material influenced the life until fracture of a given initial crack size.

The fatigue life under VAL conditions in both loadings was longer than that under CAL conditions because of the interaction of overload and underload cycle loadings. The sequence loading enhanced the crack growth retardation, which was simplified. This trend is similar to the findings of other research [56–58], wherein a slower crack growth was observed under low to high sequence loading compared with high-to-low sequence loading. Great crack growth retardation signifies long sample life. Other research suggested that the level and sequence of the load cycles can cause retardation or acceleration of crack growth, which can influence the fatigue life [59]. The retardation of crack growth is related to the changes in the size of plastic zone at the crack tip [60–62]. Other researchers proved that large plastic zones exert great effect on FCG retardation [63]. The crack growth in the large plastic zone caused the retardation of FCG rates. When the crack grew out of the overload plastic zone, a normal crack growth rate was observed upon achieving the original size of the plastic zone. This phenomenon is due to the increase in the material resistance passing through the plastic zone, whereas the overload is due to retardation of the FCG rate.

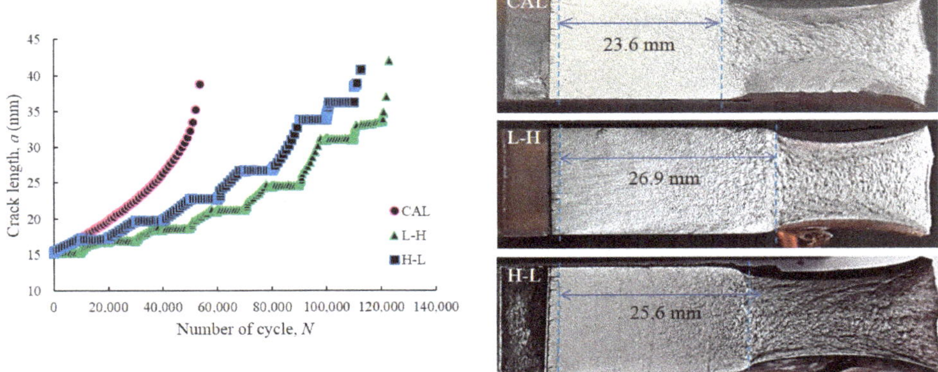

Figure 6. Fatigue life of dual-phase steels based on crack length observations.

4.3. Temperature Evolution for Cyclic Loadings

The representative entropy accumulation of the dual-phase specimen was plotted as a function of the number of cycles in the FCG test under CAL condition (Figure 7). The plot shows that the evolution of temperature went through three distinct stages [64]. In the first stage, the surface temperature rapidly rose at the start of the test because of the reaction of the material to any unexpected movement, defect and dislocation caused by surface intrusion and extrusion. The temperature stabilised in the second stage, and then abruptly increased before the occurrence of failure, which denotes the third stage.

In the first stage, limited temperature evolution implies a minimal number of cycles, which is $\approx$10% of the specimen's lifespan. [65] In this stage, surface temperature increased with energy density because of hysteresis impact and the accelerated heat generation versus heat loss caused by the specimen via radiation and convection. Moreover, a significant rise in temperature occurs because of the enhanced energy release because of the crack initiation events in the localised plastic zone, which increased the plastic zone's size at the notch tip. In the second stage, the response of the specimen towards cyclic stress, along with strain stability, resulted in the balance between energy dissipation and generation, which facilitated stable temperature readings. During this stage, a slight decrease in temperature was observed as new surfaces formed because of the generation of microcracks, which leads to the heat loss in the surroundings [66]. Finally, the third stage took up around 5–10% of the entire fatigue life. A sudden rise in temperature signifies a small number of cycles, which depicted rapid crack propagation prior to failure. The failure resulted in a large plastic deformation near the macrocrack tip, which was generated because of the stress concentrations near the cracked tips [67].

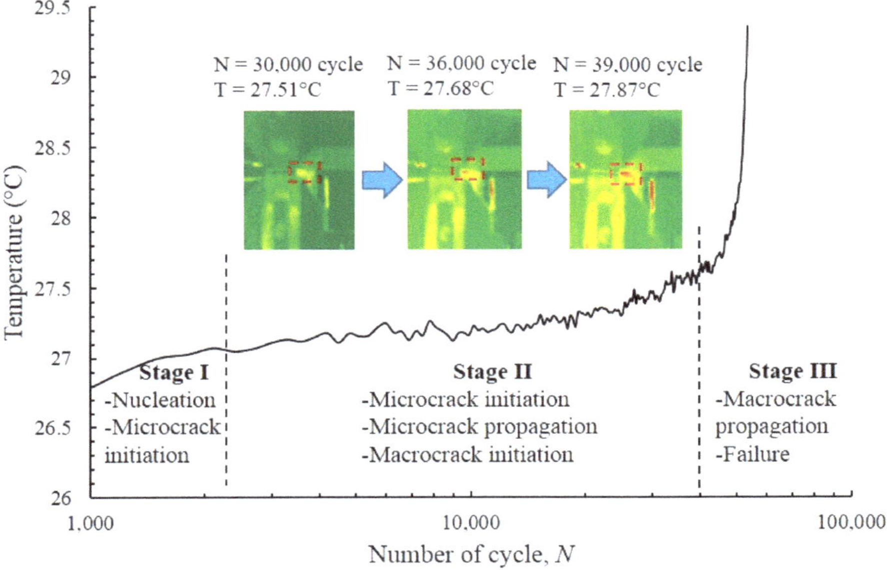

Figure 7. Typical temperature evolution during CAL.

Figure 8 demonstrates the disparity in temperature evolution under VAL and CAL conditions, where H-L and L-H loading were observed during the entire FCG test. The pattern of temperature evolution under the both conditions was similar, and the evolution process comprises three distinctive phases. With the application of the load, a fluctuation of the temperature evolution was observed under VAL, which included L-H and H-L loadings. However, irrespective of the fluctuation, the mean temperature still increased. The temperature oscillation was present during each fatigue cycle in the entire FCG test because of the thermoelastic effect. Figure 8 shows that the temperature fluctuations during the course of the FCG test until fracture occurs under VAL and CAL conditions were not significantly high and were approximately 3 °C. The temperature change percentage was below 1% (i.e., $\frac{\Delta Q}{Q_0} \approx 1\%$).

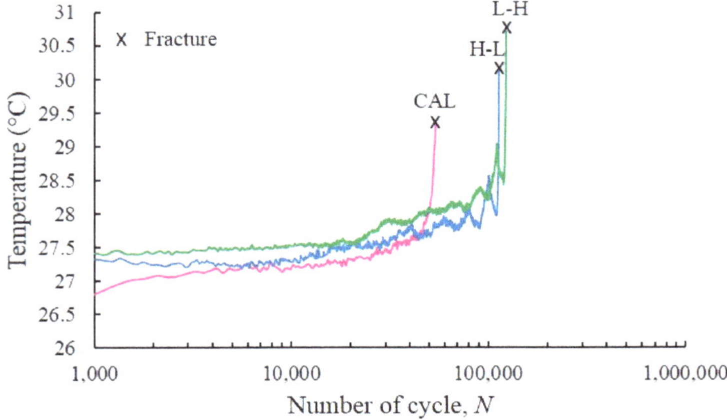

Figure 8. Temperature evolution during FCG test under spectrum loading conditions.

Figure 9 shows the temperature profiles at the initiation phase of the FCG test under VAL condition, which involve two steps of L-H and H-L. The mean temperature increased in proportion to the applied load applied under the condition when the load started to fluctuate in a sinusoidal wave. An increase in the temperature evolution was observed as the applied load increased when the load changed from L-H. A similar phenomenon was observed when the load changed from H-L, in which the temperature evolution showed a downward trend with the decrease in the applied load. The temperature evolution was affected by the changes in load application because the heat dissipation was dependent on the amount of load applied to the specimen. The temperature evolution increased as the applied load increased because the stress became concentrated at the crack tip [67].

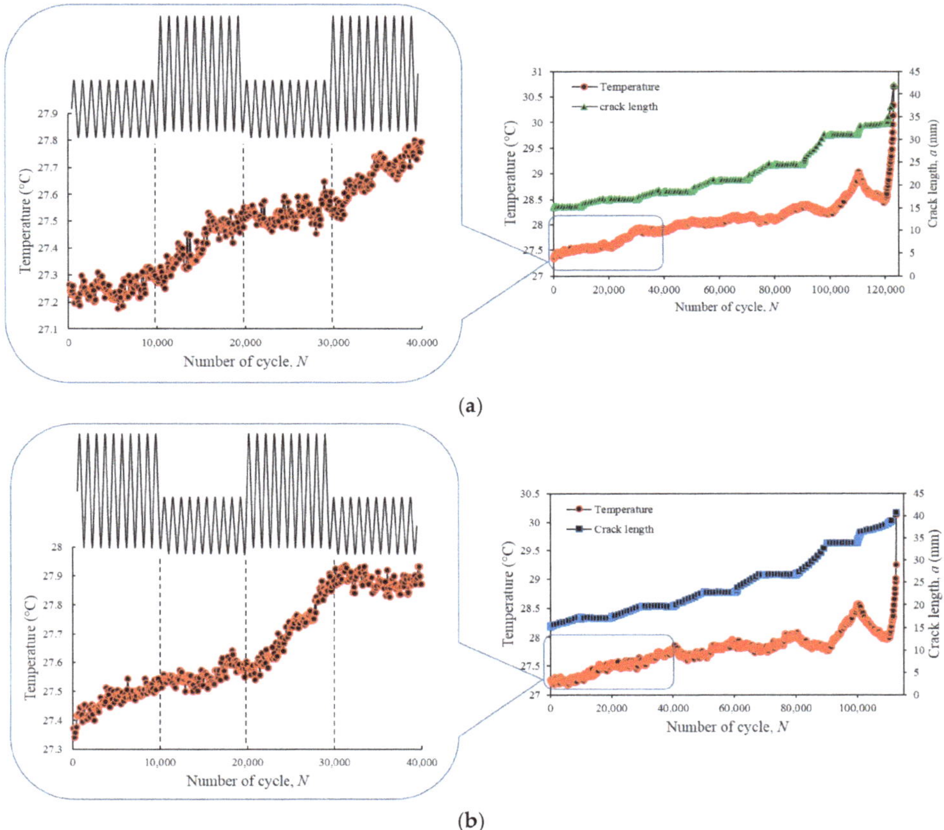

Figure 9. Temperature profile and crack length during the initial FCG test under sequence loading: (**a**) L-H; (**b**) H-L.

4.4. Entropy Generation

To determine the evolution of entropy generation, the complete FCG test was calculated. The equation for entropy generation was integrated at beginning of the test throughout a number of cycles until the fracture appeared. The outcomes of the overall entropy generation under VAL and CAL conditions are plotted in Figure 10. During the initiation of the test, the overall entropy generation was zero. However, upon reaching the fracture point, linear progression was observed. The overall entropy generation was 1.24, 2.84 and 2.61 MJ/m^3.K for dual-phase steel under CAL, L-H and H-L conditions, respectively. The variation in the number of cycles until failure with varying load application is the main reason behind the difference in the values of the total entropy generation under VAL and CAL conditions. The results indicate the entropy generation increases as the number of cycles until failure increases. The accretion of entropy generation was low when the load amplitudes were high (i.e., small number of cycles until failure) but increased when the load amplitude decreased.

According to the findings of other researchers, the total entropy generation of a particular material was not influenced by the load, frequency, or geometry; this value is constant during failure [28,31]. For materials with different characteristics, an entirely different total entropy generation at the fracture point can be expected. For example, stainless steel 304 or aluminium 6061 exhibits an overall entropy generation within the ranges of 60 and 4 MJ/m^3.K, respectively. Additionally, the increase in entropy at the point of failure can be considered as a characteristic of the material itself [31]. The results presented in Figure 10 differed from the previous studies on the entropy of fatigue life [68]. Other researchers reported a constant entropy generation at the end of the fracture. This difference lies in the energy dissipation calculation. In the present study, energy dissipation was determined by calculating the area under the load deformation curve minus the elastic strain energy. Naderi et al. [68] used the plastic strain energy density, which was calculated using the relationship presented by Morrow for the fully-reversed non-notched plastic strain that dominated the loading conditions.

This research presented the approximately linear slope of the total entropy generation that was plotted as a function of the number of cycles until failure of the material under CAL and VAL conditions. This observation revealed a monotonically rise in entropy generation until the failure point under both conditions. A methodology for the prediction of FCG rate for a given material under CAL and VAL conditions was developed through the linear relationship between entropy generation and fatigue failure. Furthermore, the prediction of FCG rate under VAL was difficult and complex. Therefore, a simple method was developed to predict the FCG rate of the specimen under both conditions using the CAL condition.

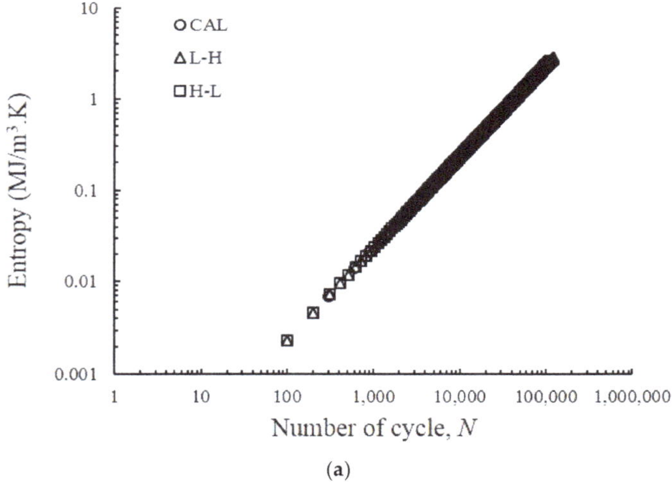

(a)

Figure 10. *Cont.*

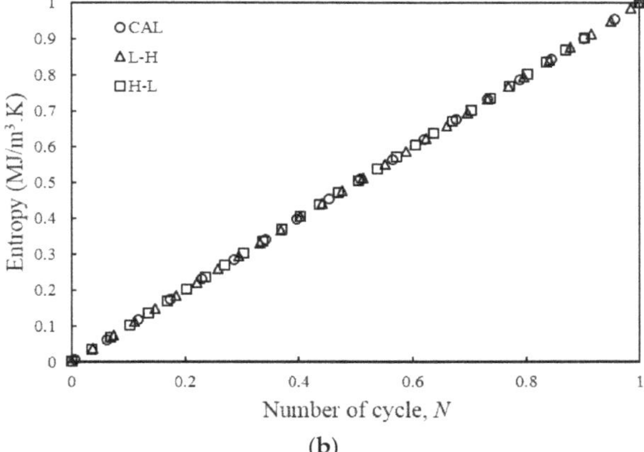

Figure 10. (**a**) Evolution of the total entropy generation in the FCG test under spectrum loading conditions; (**b**) normalised entropy generation.

4.5. Relationship between Degradation Coefficient and Entropy Generation

The relationship between the degradation rate and the entropy generation rate associated with B for dual-phase steels is shown in Figure 11. The degradation rate, which in this case is the crack length da/dN demonstrate a linear relationship with the components of entropy generation, that is, $\frac{da}{dN} = B\gamma$. The value of B for dual-phase steel under CAL condition is 1.09×10^{-7}. This value is used to predict the FCG rate for dual-phase steel under CAL and VAL conditions, determine the interaction between crack growth and entropy generation towards the degree of dissipative plastic deformation process and predict the FCG rate based on the DEG theorem.

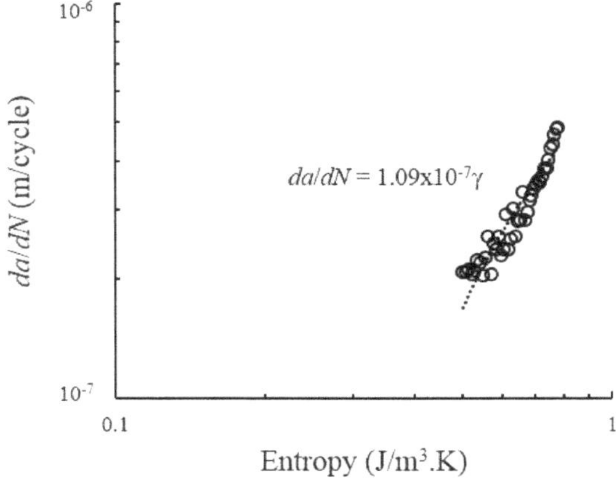

Figure 11. Relationship between degradation parameter rate and entropy generation under CAL condition.

4.6. FCG Prediction

The EGFC model was compared with the measured Paris regime crack growth data for dual-phase materials. The comparison between da/dN and the ΔK of the experimental data is represented by Equation (15) (Figure 12). The results showed a strong agreement between the EGFC model and the measured crack growth data. A visual illustration of the FCG rate is given in Figure 13 using the conventional scatter band. The FCG rate points were scattered around the 1:1 correlation line and was limited in the safety factor band (dashed line) of the 1:2 and 2:1 correlation lines. The scatter band showed the variations in the FCG rate prediction values for each type of loading. All data points for all types of loading fell within the range of the plotted accuracy band. Furthermore, all data points were considered accurate because of their proximity to the 1:1 correlation line. This finding shows that most of the predicted FCG rates are close to the experimental values. The results revealed an acceptable correlation between the entropy generation and the FCG rate under CAL and VAL conditions.

The accuracy of the EGFC model was compared with that of the experimental data determined by a statistical test of the variance of the given residuals. The RMSE values of the as-received and dual-phase steels are presented in Table 3. The RMSE values of the latter under CAL, L-H and H-L are 1.0291×10^{-7}, 1.9769×10^{-7} and 1.5409×10^{-7} m/cycle, respectively. This finding indicates that the EGFC model produced an accurate prediction of the FCG rate under CAL and VAL conditions and the experimental data for the measured crack growth had proximal scales with ΔK^2. The EGFC model was developed to predict the FCG for materials with an m value that is approximately or equal to two. Given that the FCG rate in this study was controlled by the CTOD, a dependency on ΔK^2 was observed [39], and a value of $m = 2$ was obtained from the theories based on crack opening displacement [40]. The predicted results were in good agreement when the measured data at the mid-range values of ΔK are used. Although the predicted slope had errors, the Paris regime data throughout the log-og plot was accurately represented by the proposed model. In conclusion, the EGFC model is applicable for CAL and VAL conditions. The minor errors can be attributed to the complete coupled thermomechanical solution of the problem, which was excluded from this study but will be investigated in the future.

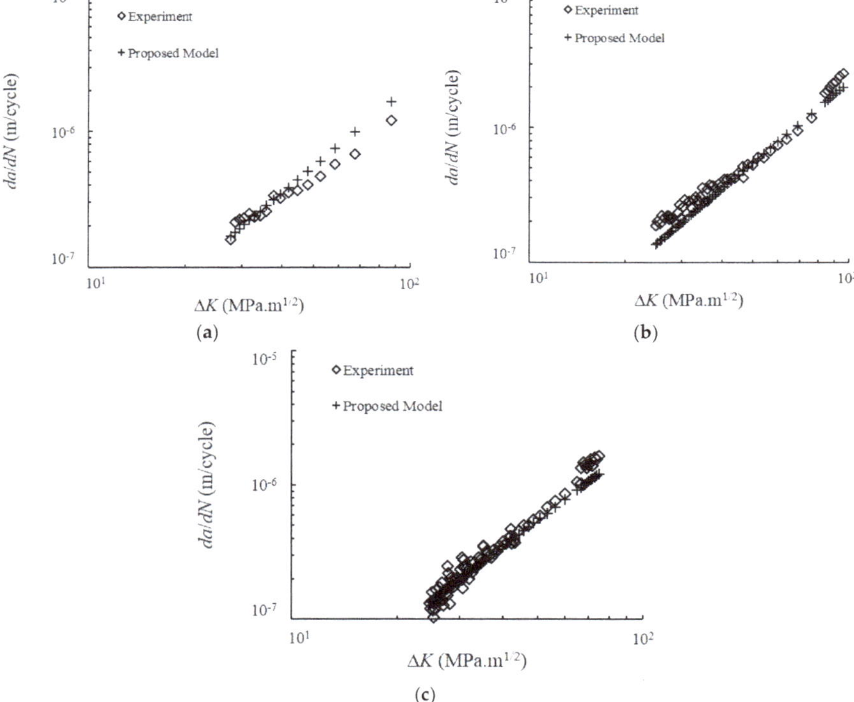

Figure 12. Comparison of the predicted crack growth rate using entropy generation of fatigue crack (EGFC) model with experimental data under spectrum loading conditions: (**a**) CAL; (**b**) L-H; (**c**) H-L.

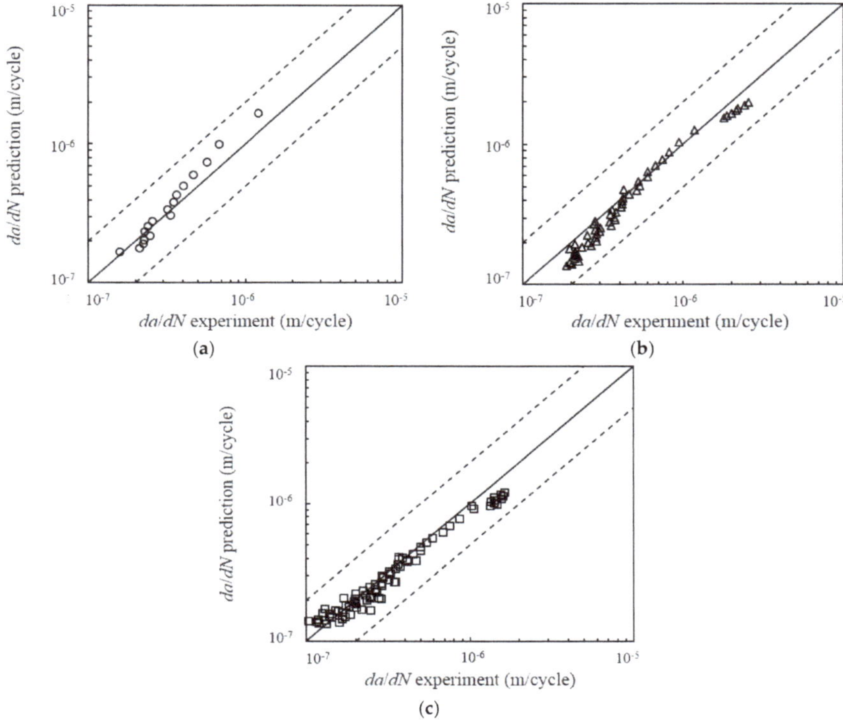

Figure 13. Comparison of the predicted FCG rate using EGFC model with experimental data using conventional scatter band for each type of load: (**a**) CAL; (**b**) L-H; (**c**) H-L.

Table 3. Accuracy of the EGFC (entropy generation of fatigue crack) model with respect to the experimental data.

Loading	RMSE (m/Cycle)
CAL	1.0291×10^{-7}
L-H	1.9769×10^{-7}
H-L	1.5409×10^{-7}

The coefficient of variations (CVs) of the measured FCG rate data were calculated and compared using traditional techniques based on the experiments and on the DEG theorem to further explore the effects of FCG rate prediction using entropy generation. The CV is a relative dispersion measure that describes the standard deviation as a percentage of the arithmetic mean of a set of observations [69].

$$CV = \frac{\sigma}{\mu} \times 100\% \tag{21}$$

where σ and μ are the FCG rate's standard deviation and arithmetic mean, respectively. When CV = 0, all values are the same regardless of the variability or uncertainty. High CV value indicates great data transmission. Table 4 presents the values of the CVs for FCG rates under different types of loading. Under all loading conditions, the CV of the FCG rate based on the DEG theorem was higher than those obtained using traditional techniques. In conclusion, entropy generation is sensitive to variations when predicting the FCG rate.

Table 4. CV values for FCG rate prediction.

Coefficient of Variance (CV)	FCG Rate (m/Cycle)	
	Proposed Model	Experiment
CAL	66.02%	61.87%
L-H	74.25%	51.65%
H-L	74.33%	65.61%

4.7. Fatigue Life Prediction

Figure 14 shows the correlation between fatigue life estimation and experimental outcomes. The fatigue life points scattered around the 1:1 correlation line are limited in the safety factor band of the 1:2 and 2:1 correlation lines. Results with a scattering factor of 2 on lifetimes are usually accepted in fatigue life analysis [70]. The finding shows that most of the fatigue life obtained from the experimental work are near to the predicted value. The plot indicates that the predicted fatigue life is in agreement with the actual fatigue lifespan of the specimens.

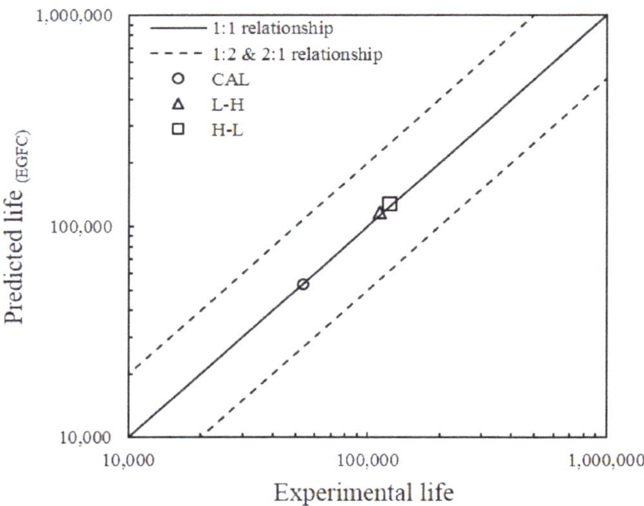

Figure 14. Correlation between the predicted and experimental results under different types of loads.

The accuracy of the predicted fatigue life can be obtained as

$$\delta\% = \frac{N_{pre.} - N_{exp.}}{N_{exp.}} \times 100\% \tag{22}$$

where $N_{pre.}$ is the predicted life and $N_{exp.}$ is the experimental life, which represents the specimen's number of cycles until failure during the FCG test. Table 5 presents the error values of the predicted and experimental data under CAL and VAL conditions. The results verifies that the predicted fatigue life is in good agreement with the experimental data under CAL and VAL conditions; the error is less than 5%. Moreover, the observed correlation between the FCG rate and the entropy generation is acceptable for the evaluation of the FCG rate based on DEG theorem when the FCG tests are conducted under CAL and VAL conditions. Therefore, even if the total entropy generation under CAL and VAL conditions are initially different because of the different number of cycles until failure for each type of loading, the severity of degradation and the lifespan of the specimen can be predicted and evaluated by assuming that the total entropy generation is independent of the load and when the degradation coefficient of a material under CAL is given [31,71].

Table 5. Accuracy of the predicted fatigue life based on the experimental fatigue life.

Type of Load	Predicted Life $_{(EGFC)}$	Experimental Life	Error (%)
CAL	53,535	53,674	0.3
L-H	115,592	112,683	2.6
H-L	128,407	123,056	4.3

5. Conclusions

This study investigated the entropy generation in dual-phase steels during FCG and proposed the EGFC model. Additionally, an expression for entropy generation in terms of temperature evolution was developed using the concept of the DEG theorem. On the basis of the DEG theorem, the EGFC model was developed to predict the FCG rate. This theorem showed that entropy generation and crack growth are closely related because of the degradation coefficient B, allowing the easy determination of the empirical Paris–Erdogan law of crack growth from the DEG theorem's considerations. Results showed that the EGFC model is in good agreement with the experimental results for dual-phase steels under spectrum loading conditions. The value of the RMSE in all cases is 10^{-7} m/cycle. Based on the results, the EGFC model is applicable in a variety of loadings, particularly those exhibiting a ΔK^2 dependence on the FCG rate or for materials with $m = 2$. Furthermore, the predicted fatigue life under CAL and VAL conditions can predict the actual fatigue life obtained from the experimental work with an error less than 5%. Finally, the entropy generation calculated from the surface temperature of a specimen under FCG test can be utilised to predict the FCG rate of dual-phase steel via intensity degradation coefficient.

Author Contributions: Formal analysis, R.I.; investigation, R.I.; methodology, R.I.; supervision, S.A., P.T. and M.Z.O.; validation, R.I.; writing—original draft, R.I.; writing—review and editing, R.I. and S.A. All authors have read and agreed to the published version of the manuscript.

Funding: This research received no external funding.

Acknowledgments: The authors wish to acknowledge Ministry of Higher Education Malaysia via Universiti Kebangsaan Malaysia (UKM) for the research funding.

Conflicts of Interest: The authors declare no conflict of interest.

References

1. Hu, X.; Quoc, T.; Wang, J.; Yao, W.; Ton, L.H.T.; Singh, I.V.; Tanaka, S. A new cohesive crack tip symplectic analytical singular element involving plastic zone length for fatigue crack growth prediction under variable amplitude cyclic loading. *Eur. J. Mech. A Solids* **2017**, *65*, 79–90. [CrossRef]
2. Zhu, L.; Jia, M. Estimation study of structure crack propagation under random load based on multiple factors correction. *J. Braz. Soc. Mech. Sci. Eng.* **2016**, *39*, 681–693. [CrossRef]
3. Li, S.; Zhang, Y.; Qi, L.; Kang, Y. Effect of single tensile overload on fatigue crack growth behavior in DP780 dual phase steel. *Int. J. Fatigue* **2018**, *106*, 49–55. [CrossRef]
4. Beden, S.M.; Abdullah, S.; Ariffin, A.K.; Al-Asady, N.A. Fatigue crack growth simulation of aluminium alloy under spectrum loadings. *Mater. Des.* **2010**, *31*, 3449–3456. [CrossRef]
5. Paris, P.C. Fracture mechanics and fatigue: A historical perspective. *Fatigue Fract. Eng. Mater. Struct.* **1998**, *21*, 535–540. [CrossRef]
6. Jiang, Y.; Chen, M. Researches on the fatigue crack propagation of pipeline. *Energy Procedia* **2012**, *14*, 524–528. [CrossRef]
7. Li, Y.; Wang, H.; Gong, D. The interrelation of the parameters in the Paris equation of fatigue crack growth. *Eng. Fract. Mech.* **2012**, *96*, 500–509. [CrossRef]
8. Haslach, H.W.J. *Maximum Dissipation Non-Equilibrium Thermodynamics and Its Geometric Structure*; Springer: New York, NY, USA, 2011.

9. Lemaitre, J.; Benallal, A.; Billardon, R.; Marquis, D. Thermodynamics and phenomenology. In *Continuum Thermomechanics*; Springer: Dordrecht, The Netherlands, 2000; pp. 209–223.
10. Zhukov, S.; Glaum, J.; Kungl, H.; Sapper, E.; Dittmer, R.; Genenko, Y.A.; Von Seggern, H. Fatigue effect on polarization switching dynamics in polycrystalline bulk ferroelectrics. *J. Appl. Phys.* **2016**, *120*, 1–14. [CrossRef]
11. Basaran, C.; Yan, C.Y. A Thermodynamic framework for damage mechanics of solder joints. *J. Electron. Packag.* **1998**, *120*, 379. [CrossRef]
12. Meneghetti, G.; Ricotta, M. Evaluating the heat energy dissipated in a small volume surrounding the tip of a fatigue crack. *Int. J. Fatigue* **2016**, *92*, 605–615. [CrossRef]
13. Meneghetti, G.; Ricotta, M. The heat energy dissipated in a control volume to correlate the crack propagation rate in stainless steel specimens. *Fract. Struct. Integr.* **2017**, *11*, 299–306. [CrossRef]
14. Meneghetti, G.; Ricotta, M. The heat energy dissipated in the material structural volume to correlate the fatigue crack growth rate in stainless steel specimens. *Int. J. Fatigue* **2018**, *115*, 107–119. [CrossRef]
15. Basaran, C.; Nie, S. An Irreversible Thermodynamics theory for damage mechanics of solids. *Int. J. Damage Mech.* **2004**, *13*, 205–223. [CrossRef]
16. Imanian, A.; Modarres, M. A thermodynamic entropy approach to reliability assessment with applications to corrosion fatigue. *Entropy* **2015**, *17*, 6995–7020. [CrossRef]
17. Imanian, A.; Modarres, M. A thermodynamic entropy-based damage assessment with applications to prognostics and health management. *Struct. Health Monit.* **2018**, *17*, 1–15. [CrossRef]
18. Wang, J.; Yao, Y. An entropy based low-cycle fatigue life prediction model for solder materials. *Entropy* **2017**, *19*, 503. [CrossRef]
19. Amiri, M.; Khonsari, M.M. Life prediction of metals undergoing fatigue load based on temperature evolution. *Mater. Sci. Eng. A* **2010**, *527*, 1555–1559. [CrossRef]
20. Amiri, M.; Khonsari, M.M. On the role of entropy generation in processes involving fatigue. *Entropy* **2012**, *14*, 24–31. [CrossRef]
21. Liakat, M.; Khonsari, M.M. Entropic characterization of metal fatigue with stress concentration. *Int. J. Fatigue* **2014**, *70*, 223–234. [CrossRef]
22. Thamburaja, P.; Jamshidian, M. A multiscale Taylor model-based constitutive theory describing grain growth in polycrystalline cubic metals. *J. Mech. Phys. Solids* **2014**, *63*, 1–28. [CrossRef]
23. Ancona, F.; De Finis, R.; Demelio, G.P.; Galietti, U.; Palumbo, D. Study of the plastic behavior around the crack tip by means of thermal methods. *Procedia Struct. Integr.* **2016**, *2*, 2113–2122. [CrossRef]
24. De Finis, R.; Palumbo, D.; Galietti, U. Mechanical behaviour of stainless steels under dynamic loading: An investigation with thermal methods. *J. Imaging* **2016**, *2*, 32. [CrossRef]
25. De Finis, R.; Palumbo, D.; Ancona, F.; Galietti, U. Energetic approach based on IRT to assess plastic behaviour in CT specimens. In Proceedings of the SPIE—The International Society for Optical Engineering, Anaheim, CA, USA, 9–13 April 2017; Volume 10214, pp. 1–12.
26. Ontiveros, V.; Amiri, M.; Kahirdeh, A.; Modarres, M. Thermodynamic entropy generation in the course of the fatigue crack initiation. *Fatigue Fract. Eng. Mater. Struct.* **2017**, *40*, 423–434. [CrossRef]
27. Bryant, M.D.; Khonsari, M.M.; Ling, F.F. On the thermodynamics of degradation. *Proc. R. Soc. A* **2008**, *464*, 2001–2014. [CrossRef]
28. Amiri, M.; Naderi, M.; Khonsari, M.M. An experimental approach to evaluate the critical damage. *Int. J. Damage Mech.* **2011**, *20*, 89–112. [CrossRef]
29. Amiri, M.; Modarres, M. An entropy-based damage characterization. *Entropy* **2014**, *16*, 6434–6463. [CrossRef]
30. Fathima, K.M.P.; Kishen, J.M.C. Prediction of Fatigue Life in Plain Concrete Using Entropy Production. *J. Eng. Mech.* **2015**, *141*, 1–10.
31. Naderi, M.; Khonsari, M.M. A thermodynamic approach to fatigue damage accumulation under variable loading. *Mater. Sci. Eng. A* **2010**, *527*, 6133–6139. [CrossRef]
32. Wang, X.G.; Ran, H.R.; Jiang, C.; Fang, Q.H. An energy dissipation-based fatigue crack growth model. *Int. J. Fatigue* **2018**, *114*, 167–176. [CrossRef]
33. Baker, G. Thermodynamics in solid mechanics: A commentary. *Philos. Trans. R. Soc. A* **2005**, *363*, 2465–2477. [CrossRef]
34. Eftis, J.; Liebowitz, H. On Surface Energy and The Continuum Thermodynamics of Brittle Fracture. *Eng. Fract. Mech.* **1976**, *8*, 459–485. [CrossRef]

35. Klingbeil, N.W. A total dissipated energy theory of fatigue crack growth in ductile solids. *Int. J. Fatigue* **2003**, *25*, 117–128. [CrossRef]
36. Ranganathan, N.; Chalon, F.; Meo, S. Some aspects of the energy based approach to fatigue crack propagation. *Int. J. Fatigue* **2008**, *30*, 1921–1929. [CrossRef]
37. Nittur, P.G.; Karlsson, A.M.; Carlsson, L.A. Numerical evaluation of Paris-regime crack growth rate based on plastically dissipated energy. *Eng. Fract. Mech.* **2014**, *124*, 155–166. [CrossRef]
38. Fu, D.L.; Zhang, L.; Cheng, J. An energy-based approach for fatigue crack growth. *Key Eng. Mater.* **2006**, *324–325*, 379–382. [CrossRef]
39. Bodner, S.R.; Davidson, D.L.; Lankford, J.A. A description of fatigue crack growth in terms of plastic work. *Eng. Fract. Mech.* **1983**, *17*, 189–191. [CrossRef]
40. Skelton, R.P.; Vilhelmsen, T.; Webster, G.A. Energy criteria and cumulative damage during fatigue crack growth. *Int. J. Fatigue* **1998**, *20*, 641–649. [CrossRef]
41. Yu, J.; Ziehl, P.; Zárate, B.; Caicedo, J. Prediction of fatigue crack growth in steel bridge components using acoustic emission. *J. Constr. Steel Res.* **2011**, *67*, 1254–1260. [CrossRef]
42. Tayanç, M.; Aytaç, A.; Bayram, A. The effect of carbon content on fatigue strength of dual-phase steels. *Mater. Des.* **2007**, *28*, 1827–1835. [CrossRef]
43. Zhang, Z.; Delagnes, D.; Bernhart, G. Ageing effect on cyclic plasticity of a tempered martensitic steel. *Int. J. Fatigue* **2007**, *29*, 336–346. [CrossRef]
44. Tu, M.; Hsu, C.; Wang, W.; Hsu, Y. Comparison of microstructure and mechanical behavior of lower bainite and tempered martensite in JIS SK5 steel. *Mater. Chem. Phys.* **2008**, *107*, 418–425. [CrossRef]
45. Li, H.; Gao, S.; Tian, Y.; Terada, D.; Shibata, A.; Tsuji, N. Influence of tempering on mechanical properties of ferrite and martensite dual phase steel. *Mater. Today Proc.* **2015**, *2*, S667–S671. [CrossRef]
46. Romeiro, F.; De Freitas, M.; Fonte, M. Fatigue crack growth with overloads/underloads: Interaction effects and surface roughness. *Int. J. Fatigue* **2009**, *31*, 1889–1894. [CrossRef]
47. Li, S.; Kang, Y.; Kuang, S. Effects of microstructure on fatigue crack growth behavior in cold-rolled dual phase steels. *Mater. Sci. Eng. A* **2014**, *612*, 153–161. [CrossRef]
48. Tang, Y.; Zhu, G.; Kang, Y.; Yue, L.; Jiao, X. Effect of microstructure on the fatigue crack growth behavior of Cu-Be-Co-Ni alloy. *J. Alloy. Compd.* **2016**, *663*, 784–795. [CrossRef]
49. Mansor, N.I.I.; Abdullah, S.; Ariffin, A.K. Effect of loading sequences on fatigue crack growth and crack closure in API X65 steel. *Mar. Struct.* **2019**, *65*, 181–196. [CrossRef]
50. Kalnaus, S.; Fan, F.; Jiang, Y.; Vasudevan, A.K. An experimental investigation of fatigue crack growth of stainless steel 304L. *Int. J. Fatigue* **2009**, *31*, 840–849. [CrossRef]
51. Naderi, M.; Khonsari, M.M. Real-time fatigue life monitoring based on thermodynamic entropy. *Struct. Health Monit.* **2011**, *10*, 189–197. [CrossRef]
52. Idris, R.; Abdullah, S.; Thamburaja, P.; Omar, M.Z. An Experimental investigation of tensile properties and fatigue crack growth behaviour for dual- phase steel. *J. Mech. Eng.* **2018**, *15*, 155–167.
53. Guan, M.; Yu, H. Fatigue crack growth behaviors in hot-rolled low carbon steels: A comparison between ferrite-pearlite and ferrite-bainite microstructures. *Mater. Sci. Eng. A* **2013**, *559*, 875–881. [CrossRef]
54. Maleque, M.A.; Poon, Y.M.; Masjuki, H.H. The effect of intercritical heat treatment on the mechanical properties of AISI 3115 steel. *J. Mater. Process. Technol.* **2004**, *153–154*, 482–487. [CrossRef]
55. Salih, A.A.; Omar, M.Z.; Syarif, J.; Sajuri, Z. An investigation on the microstructure and mechanical properties of quenched and tempered SS440C martensitic stainless steel. *Int. J. Mech. Mater. Eng.* **2012**, *7*, 119–123.
56. Schijve, J.; Skorupa, M.; Skorupa, A.; Machniewicz, T.; Gruszczynski, P. Fatigue crack growth in the aluminium alloy D16 under constant and variable amplitude loading. *Int. J. Fatigue* **2004**, *26*, 1–15. [CrossRef]
57. Borrego, L.P.; Costa, J.M.; Ferreira, J.M. Fatigue crack growth in thin aluminium alloy sheets under loading sequences with periodic overloads. *Thin-Walled Struct.* **2005**, *43*, 772–788. [CrossRef]
58. Mansor, N.I.I.; Abdullah, S.; Ariffin, A.K. Discrepancies of fatigue crack growth behaviour of API X65 steel. *J. Mech. Sci. Technol.* **2017**, *31*, 4719–4726. [CrossRef]
59. Maljaars, J.; Pijpers, R.; Slot, H. Load sequence effects in fatigue crack growth of thick-walled welded C-Mn steel members. *Int. J. Fatigue* **2015**, *79*, 10–24. [CrossRef]
60. Ray, A.; Patankar, R. Fatigue crack growth under variable-amplitude loading: Part I—Model formulation in state-space setting. *Appl. Math. Model.* **2001**, *25*, 979–994. [CrossRef]

61. Huang, X.P.; Zhang, J.B.; Cui, W.C.; Leng, J.X. Fatigue crack growth with overload under spectrum loading. *Appl. Fract. Mech.* **2005**, *44*, 105–115. [CrossRef]
62. Pereira, M.V.S.; Darwish, F.A.I.; Camarao, A.F.; Motta, S.H. On the Prediction of fatigue crack retardation using Wheeler and Willenborg models. *Mater. Res.* **2007**, *10*, 101–107. [CrossRef]
63. Pavlou, D.G. Prediction of fatigue crack growth under real stress histories. *Eng. Struct.* **2000**, *22*, 1707–1713. [CrossRef]
64. Maletta, C.; Bruno, L.; Corigliano, P.; Crupi, V.; Guglielmino, E. Crack-tip thermal and mechanical hysteresis in Shape Memory Alloys under fatigue loading. *Mater. Sci. Eng. A* **2014**, *616*, 281–287. [CrossRef]
65. Fargione, G.; Geraci, A.; La Rosa, G.; Risitano, A. Rapid determination of the fatigue curve by the thermographic method. *Int. J. Fatigue* **2002**, *24*, 11–19. [CrossRef]
66. Lee, H.T.; Chen, J.C.; Wang, J.M. Thermomechanical behaviour of metals in cyclic loading. *J. Mater. Sci.* **1993**, *28*, 5500–5507. [CrossRef]
67. Yang, B.; Liaw, P.K.; Wang, H.; Jiang, L.; Huang, J.Y.; Kuo, R.C.; Huang, J.G. Thermographic investigation of the fatigue behavior of reactor pressure vessel steels. *Mater. Sci. Eng. A* **2001**, *314*, 131–139. [CrossRef]
68. Naderi, M.; Amiri, M.; Khonsari, M.M. On the thermodynamic entropy of fatigue fracture. *Proc. R. Soc. A (Math. Phys. Eng. Sci.)* **2009**, *466*, 423–438. [CrossRef]
69. Herman, H.; Bucksch, H. *Dictionary Geotechnical Engineering*, 2nd ed.; Springer: New York, NY, USA, 2003.
70. Velay, V.; Bernhart, G.; Delagnes, D.; Penazzi, L. A continuum damage model applied to high temperature fatigue lifetime prediction of a martensitic tool steel. *Fatigue Fract. Eng. Mater. Struct.* **2005**, *28*, 1009–1023. [CrossRef]
71. Idris, R.; Abdullah, S.; Thamburaja, P.; Omar, M.Z. The need to generate entropy characteristics for fatigue life prediction in low-carbon steel. *J. Braz. Soc. Mech. Sci. Eng.* **2018**, *40*, 409. [CrossRef]

Article

On the Development of Mechanothermodynamics as a New Branch of Physics

Leonid A. Sosnovskiy [1] and Sergei S. Sherbakov [2,3,*]

1 Scientific and Production Group TRIBOFATIGUE Ltd., 246050 Gomel, Belarus; tribo-fatigue@mail.ru
2 State Committee on Science and Technology of the Republic of Belarus, 220072 Minsk, Belarus
3 Department of Theoretical and Applied Mechanics, Belarusian State University, 220030 Minsk, Belarus
* Correspondence: sherbakovss@mail.ru

Received: 15 October 2019; Accepted: 27 November 2019; Published: 2 December 2019

Abstract: This paper aims to substantiate and formulate the main principles of the physical discipline-mechanothermodynamics that unites Newtonian mechanics and thermodynamics. Its principles are based on using entropy as a bridge between mechanics and thermodynamics. Mechanothermodynamics combines two branches of physics, mechanics and thermodynamics, to take a fresh look at the evolution of complex systems. The analysis of more than 600 experimental results allowed for determining a unified mechanothermodynamical function of limiting states (critical according to damageability) of polymers and metals. They are also known as fatigue fracture entropy states.

Keywords: mechanothermodynamics; tribo-fatigue entropy; wear-fatigue damage; stress-strain state; limiting state; damage state; dangerous volume; interaction; irreversible damage

1. Introduction

Any scientific discipline is based on the understanding and mathematical description of the behavior of certain phenomena revealing specific properties of some existing or imaginary objects [1,2].

Hierarchical structure of objects can be found from the study of specific objects that give rise to relevant branches of mechanics. Figure 1 shows a simplified hierarchical structure of objects (in case gas and fluid continua are absent) and mechanothermodynamics as a new branch of knowledge [2].

The concept of a material object given as a dimensionless and structureless point capable of moving in time and space gave impetus to the development of Newtonian theoretical mechanics aimed at understanding and describing a great variety of motions of such a physically unreal object. This concept made theoretical mechanics a useful science. As a result, the motion of points like electrons or planets, i.e., extremely small microcosm objects and huge universe objects can be correctly analyzed. If "big points" have mass, then the interaction patterns of moving celestial bodies, etc., in mechanics of space flight, machines, and mechanisms, all that moves, are the subject of the analysis with the implication of theoretical Newtonian mechanics methods.

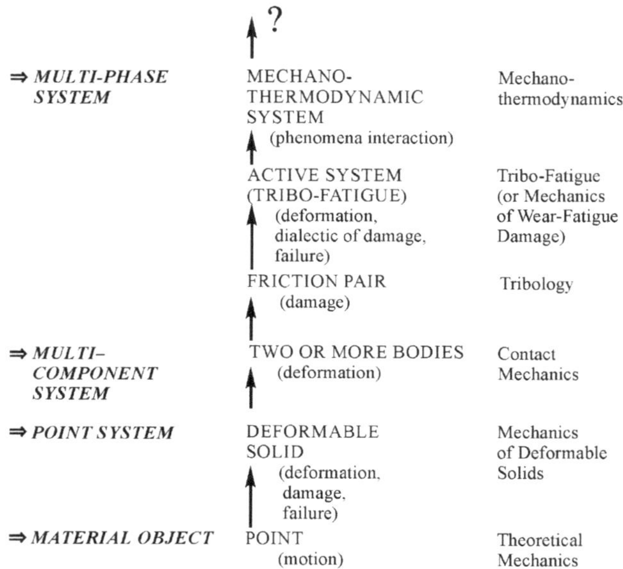

Figure 1. Objects of study in mechanics (from simple to complex) in a simplified hierarchical form.

An interconnected set of points may represent a continuum—a solid, for example. When solid points are capable of moving or shifting relative to each other at different loads, it becomes possible to develop the concept of a new object, let us say a deformable solid. Naturally, mechanics of deformable solids must be developed in order to examine its stress-strain state at any point and, finally, to understand and mathematically describe changes in size, motion, and distortion of a solid as a whole. A deformable solid may be considered as a specimen, material, or a structural element in relation to the study objectives. Mechanics of materials, composites, structures, soil, etc., damage and failure mechanics (under static, cyclic, impact, loads, etc.), mesomechanics, and micromechanics, etc., examined specific properties of these objects. Mechanical behavior and properties of reversible and irreversible points motions in deformable solids were found as well by theories of elasticity and plasticity, respectively. Deformable solids also had a diversity of specific properties: viscoelasticity, elasto-viscoplasticity, etc. It was discovered that mechanics of deformable solids is one of the most powerful research means to model behavior of objects at various conditions.

One of the components of numerous mechanical systems is a deformable solid. The compression of two solids together started the development of a new branch of deformable solid mechanics–contact mechanics. Then, it is a study of a friction pair, for which a relative motion of two bodies at contact load is considered. Later, tribology as a special scientific discipline emerged. Its main objective is to examine friction behavior between solid bodies and interface damage of materials of various friction pairs at rolling, sliding, impact, slippage, etc. A friction pair may be treated as a multicomponent system since the third body forms in the region of moving contact due to the appearance of tribo-destruction products and/or the presence of lubricant.

A "peculiar object" (active system) is more complicated than a friction pair [3]. In the twentieth century, the concept of an active system was introduced. An active system is defined by any mechanical system at cycling loading. Here, the friction process proceeds simultaneously at rolling, sliding, impact, etc. So, the active system may be considered as a friction pair, at least one element of which undergoes volumetric deformation. Such systems have complex wear-fatigue damage due to kinetic interactions of friction, fatigue, wear, corrosion, erosion, etc. Naturally, the appearance of a new object of study gave impetus to a scientific discipline shortly named tribo-fatigue ("tribo" is friction in Greek and "fatigue" is fatigue in French [3]) or mechanics of wear-fatigue damage [2] (mechanics of tribo-fatigue systems [4]).

Figure 1 displays the increase of complexity of objects that are studied by successive arrows. The last object is represented by a multi-phase system. It is a mechanothermodynamical (MTD) system uniting the laws of Newtonian mechanics and thermodynamics. The union of Newtonian mechanics and thermodynamics was formulated and experimentally proved for metals, alloys, and composites [5–14].

The above approaches and models for the energy and stress-strain states of complex systems at thermodynamic and mechanical loads are considered in the well-known works [15–18]. Damage and entropy concepts are important for building a model of an MTD system.

The main ideas of materials' behavior at fracture process conditions are discussed in Reference [19]. Study [20] considers features of mechanics of damage as a part of fracture mechanics and its applications. The basics of heterogeneous continuum physical mesomechanics, which develops on the border of physics of plasticity, continuum mechanics, and strength of materials, are given in Reference [21]. This discipline is concerned with stressed and damaged materials at macro-, meso- and micro-levels.

References [22,23] examine the constitutive relations for strain-induced damage at thermodynamic loads. They also discuss the use of failure mechanics of civil and mechanical engineering components in the brittle, fatigue, creep, and ductile conditions at thermomechanical loads. References [24,25] discuss the related tasks of formation plasticity and vibration theories for steady-state vibrations in elastoplastic bodies.

References [26,27] present a concise review of the main damage models for mechanics of continua and micromechanics, including evolution kinetics, and discuss further research areas. Reference [28] proposes a general development of continuum damage models. This model is defined by yield and empirical damage potential surfaces in space. It also considers damage mechanisms (cracking, isotropic damage, etc.) reducing material strength.

The stress-based limiting criterion for the conditions of linear and spatial strain states is described in Reference [29] using the results of experimental and theoretical studies. A thermodynamic model of friction and non-associated flow for geotechnical materials is given in Reference [30]. Models of large strain elastic-plastic behavior of ductile metals under anisotropic damage are investigated extensively in References [31,32]. References [33–35] deal with elastic, plastic, and damage behavior of materials in a thermodynamic statement using hardening internal state variables for both plasticity and damage. Some authors proposed damage theory of polycrystalline material [36,37], taking into account kinematic, thermodynamic, and kinetic coupling.

Reference [38] considers the model of microscopic damage of ellipsoidal voids that are capable of changing their shape for the materials at mixed hardening. The results of model materials X-ray tomography were used to study voids behavior in References [39–41]. Void growth and the shape change at large plastic deformation studied by means of scanning electron microscopy (SEM) is discussed in Reference [42]. Anisotropic damage progression for porous ductile metals with second phases is presented through mechanisms of void nucleation, growth, and coalescence in Reference [40]. Reference [43] presents the analytical and computational mesoscopic models for nucleation and interaction of microcracks near a macrocrack tip based on elasticity and dislocation theories. The framework allowing a combination of plasticity and damage models of inelastic behavior is proposed in Reference [44].

Generation of entropy in flow with silver and copper nanoparticles was studied in Reference [45]. Radiative mixed convective flow of viscous fluid to rotating disk was considered subject to viscous dissipation and Joule heating. It was shown that entropy generation rate increases for higher radiation parameter, Brinkman number, nanoparticle volume fraction, and Reynolds number. Entropy generation in magnetohydrodynamic radiative flow to the rotating disk of variable thickness was studied in Reference [46] and showed that entropy generation rate increases for higher radiation parameter but decreases for higher Eckert number. Another interesting study is devoted to entropy generation in nonlinear radiative flow of viscous nanomaterial towards a stretched surface [47]. An increasing trend was observed for both entropy generation and Bejan number due to the increase of thermophoresis variable and temperature difference parameter. Study of entropy generation in mixed convective flow of nanofluid between two stretchable rotating discs was made using Buongiorno nanofluid model [48] and showed that the entropy generation rate has inverse behavior in relation to the Hartman number. A study of magnetohydrodynamic radiative nanomaterial flow of Casson fluid towards a stretched surface [49] showed that entropy generation rates boost through the magnetic variable while the Bejan number decays.

References [1,50–52] contain the fundamentals of mechanothermodynamics and formulate two of its principles. The first principle states that damageability of all things has no conceivable boundaries. The second principle states that effective energy fluxes (entropies) at loads of different nature under irreversible changes in a MTD system are not additive, they interact dialectically. Corresponding entropy analysis [1] is made on the basic principles of tribo-fatigue [2–4] and thermodynamics [5]. The present study is dedicated to the analysis based on the energy presentations of mechanics, thermodynamics, and tribo-fatigue. It allowed us to reveal and study novel behavior and evolution patterns of a MTD system.

Current and perspective models and methods address the following specific features of mechanothermodynamics that differ it from thermodynamics:

1. An object (a system of interacting continuums, but not a continuum),
2. The state of the object (observed and limiting, but not just the observed),
3. Energy model (the allocation of the effective part in the irreversible component of the energy—the part spent on the production of damage, but not just the separation of energy into reversible and irreversible parts),
4. Non-additivity (the interaction of energy or entropy components caused by loads of different nature, but not their simple addition).

2. Thermomechanical State

We consider the thermomechanical task [15–18]. It will be used for the creation of energy and entropy models of MTD systems.

Continuum state of an elementary volume dV is described in the following way [16,17]:

$$\sigma_{ij,j} + \rho f_i = \rho \dot{v}_i, i = 1,\ 2,\ 3, \tag{1}$$

where, the σ_{ij} are the stresses, ρ is the density, the f_i are the volumetric forces, and the v_i are the velocities.

With the repeated index summation rule used, mechanical energy conservation of a continuum of volume V is obtained by multiplying scalar Equation (1) by a velocity vector v_i:

$$\int\limits_V v_i \sigma_{ij,j} dV + \int\limits_V \rho v_i f_i dV = \int\limits_V \rho v_i \dot{v}_i dV. \tag{2}$$

The right side of Equation (2) is kinetic energy K change in the continuum of volume V:

$$\int\limits_V \rho v_i \dot{v}_i dV = \frac{d}{dt}\int\limits_V \rho \frac{v_i v_i}{2} dV = \frac{d}{dt}\int\limits_V \rho \frac{v^2}{2} dV = \frac{dK}{dt}. \tag{3}$$

Using the known transformations with the consideration of Gauss–Ostrogradsky's theorem, we obtain the equation for continuum mechanical energy [16]:

$$\frac{dK}{dt} + \int\limits_V \sigma_{ij} \dot{\varepsilon}_{ij} dV = \int\limits_\Pi \sigma_{ij} l_j d\Pi + \int\limits_V \rho v_i f_i dV, \tag{4}$$

or,

$$\frac{dK}{dt} + \frac{\delta U}{dt} = \frac{\delta A}{dt},$$

where, ε_{ij} denotes the strain rate, Π the continuum surface, l the director cosines at the continuum surface, $\delta U/dt$ the internal force power, and $\delta A/dt$ the power of internal surfaces and volumetric forces.

In Expression (4), the symbol δ shows that in the general case, the increment (variation) cannot be an accurate differential.

In the thermomechanical statement, the rate of change in the internal energy U [16] is usually given by the integral:

$$\frac{dU}{dt} = \frac{d}{dt}\int_V \rho u dV = \int_V \rho \dot{u} dV, \tag{5}$$

where $u = \lim_{\Delta m \to 0} \frac{u(\Delta m)}{\Delta m}$ is the specific internal energy (internal energy density) of an elementary volume of mass Δm.

The rate of heat transfer to the continuum is expressed in the following form:

$$\frac{\delta Q}{dt} = -\int_\Pi c_i l_i d\Pi + \int_V \rho z dV, \tag{6}$$

where, c_i characterizes the heat flux per unit area of the continuum surface per unit time due to heat conduction and z– the constant of heat radiation per unit mass per unit time.

The pattern of change in thermomechanical continuum energy is then of the form:

$$\frac{dK}{dt} + \frac{dU}{dt} = \frac{\delta A}{dt} + \frac{\delta Q}{dt} \tag{7}$$

In Expression (7), transforming surface integrals into volume integrals yields the local form of the energy equation:

$$\frac{d}{dt}\left(\frac{v^2}{2} + u\right) = \frac{1}{\rho}\left(\sigma_{ij} v_i\right)_{,j} + f_i v_i - \frac{1}{\rho} c_{i,i} + z. \tag{8}$$

If we subtract the scalar product of Equation (1) and the velocity vector v_i from Equation (8), then the local energy equation will be obtained as follows:

$$\frac{du}{dt} = \frac{1}{\rho}\sigma_{ij}\dot{\varepsilon}_{ij} - \frac{1}{\rho} c_{i,i} + z = \frac{1}{\rho}\sigma_{ij}\dot{\varepsilon}_{ij} + \frac{dq}{dt} \tag{9}$$

where dq is the heat flux per unit mass.

According to Equation (9), the internal energy changes are equal to the sum of the stress power and the heat flux to the continuum.

In relation to the thermodynamic system, we define two characteristic functions of its state: absolute temperature T and entropy S that can be interpreted as the characteristic of the ordered (or chaotic) state of the thermodynamic system. Usually, the entropy is assumed to have an additivity property, i.e.,

$$S = \sum_i S_i. \tag{10}$$

Continuum mechanics [16,17] considers the specific entropy S per unit mass as:

$$S = \int_V \rho s dV. \tag{11}$$

References [16,17] show that the specific entropy increment ds can occur because of the interaction with the environment (the increment $ds^{(e)}$) or inside the system itself (the increment $ds^{(i)}$):

$$ds = ds^{(e)} + ds^{(i)}. \quad (12)$$

The quantity $ds^{(i)}$ is equal to zero in reversible processes and is above zero in irreversible processes.

If we express the heat flux per unit mass through dq, then in the case of reversible processes, the increment will be as follows:

$$Tds = dq. \quad (13)$$

By the second law of thermodynamics, we see that the rate of change in the total entropy S of the continuum of volume V cannot be smaller than the sum of the heat flux through the volume boundary and the entropy produced by external sources inside the volume (Clausius–Duhem's inequality) [16,17]:

$$\frac{d}{dt}\int_V \rho s dV \geq \int_V \rho e dV - \int_\Pi \frac{c_i l_i}{T} d\Pi \quad (14)$$

where, e is the local external entropy source power per unit mass. Formula (14) shows that the equality is valid for reversible processes and the inequality is valid for irreversible processes.

In Formula (14), transforming the surface integral into the volume integral arrives at a relation for a rate of internal entropy production per unit mass:

$$\gamma \equiv \frac{ds}{dt} - e - \frac{1}{\rho}\left(\frac{c_i}{T}\right)_j \geq 0 \quad (15)$$

Continuum mechanics assumes that we can decompose the stress tensor into two parts: the conservative part $\sigma_{ij}^{(C)}$ for reversible processes (elastic deformation, liquid pressure) and the dissipative part $\sigma_{ij}^{(D)}$ for irreversible processes (plastic deformation, liquid viscous stresses):

$$\sigma_{ij} = \sigma_{ij}^{(C)} + \sigma_{ij}^{(D)} \quad (16)$$

We can then present an expression for energy change rate (9) in the following form:

$$\frac{du}{dt} = \frac{1}{\rho}\sigma_{ij}\dot{\varepsilon}_{ij} + \frac{dq}{dt} = \frac{1}{\rho}\sigma_{ij}^{(C)}\dot{\varepsilon}_{ij} + \frac{1}{\rho}\sigma_{ij}^{(D)}\dot{\varepsilon}_{ij} + \frac{dq}{dt}. \quad (17)$$

If Equation (13) is assumed to be valid for irreversible processes, then the total entropy production rate is:

$$\frac{ds}{dt} = \frac{1}{\rho T}\sigma_{ij}^{(C)}\dot{\varepsilon}_{ij} + \frac{1}{\rho T}\sigma_{ij}^{(D)}\dot{\varepsilon}_{ij} + \frac{1}{T}\frac{dq}{dt}, \quad (18)$$

or

$$\frac{ds}{dt} = \frac{1}{\rho T}\left(\frac{du_M}{dt} + \frac{du_T}{dt}\right) = \frac{1}{\rho T}\left(\frac{du_M^{(C)}}{dt} + \frac{du_M^{(D)}}{dt} + \frac{du_T}{dt}\right)$$

Expression (18) for the total local entropy change rate in the continuum elementary volume can find wide use in practice.

In view of entropy additivity assumption (10), the sum in Expression (18) can be supplemented by other terms that allow the internal entropy production in the liquid (gas) volume due to different mechanisms to be taken into consideration. Similarly, for the continuum volume dV we can consider the internal chemical processes [15]:

$$dU = dQ + dA + dU_{sub} = TdS - pdV + \sum_1^n \mu_k dN_k, \quad (19)$$

$$dS = \frac{dU + pdV}{T} - \frac{1}{T}\sum_1^n \mu_k dN_k. \quad (20)$$

If dV is considered not as a finite, but elementary volume of continuum, then based on Equations (17), (19) and (20), we can write the changes in its specific energy and entropy in the following differential form:

$$du = \frac{1}{\rho}\sigma_{ij}d\varepsilon_{ij} + dq + \sum_k \mu_k dn_k; \tag{21}$$

$$ds = \frac{1}{\rho T}\sigma_{ij}d\varepsilon_{ij} + \frac{1}{T}dq + \frac{1}{T}\sum_k \mu_k dn_k, \tag{22}$$

where, n_k is the number of mols per unit mass.

For the continuum of volume V, on the basis of Equations (5) and (11), Expressions (21) and (22) will assume the form:

$$dU = \int_V \rho du dV = \int_V \sigma_{ij}d\varepsilon_{ij}dV + \int_V \rho dq dV + \int_V \rho \sum_k \mu_k dn_k \, dV; \tag{23}$$

$$dS = \int_V \rho ds dV = \frac{1}{T}\int_V \sigma_{ij}d\varepsilon_{ij}dV + \frac{1}{T}\int_V \rho dq dV + \frac{1}{T}\int_V \rho \sum_k \mu_k dn_k \, dV. \tag{24}$$

Having introduced the chemical entropy component (the last terms in Expressions (21)–(23)), we can have not only a more complete behavior of the continuum state, but we can also describe self-organization processes initiating stable structures with increasing the heat flux to the continuum.

Being quite common, the specified models of continuum energy and entropy states (Equations (17)–(24)) nevertheless do not permit one to satisfactorily describe some processes to occur in such a continuum as a deformable solid. However, a convenient idea of the additivity of energy and entropy components (Equation (11)) applicable to model elastic deformation is not suitable to describe non-linear processes. The available models do not also take into account an entropy growth due to the solid damageability as a specific characteristic of change in the structure organization. Following the tribo-fatigue ideas [2–4,51], the damageability is understood as any irreversible change in structure, continuum, shape, etc., of a deformable solid that leads to its limiting state. Although at plasticity modeling, the elasticity limit is not implicitly allowed for, the damageability at mechanical or contact fatigue proceeds in the course of linear elastic deformation. To describe it, we need a particular approach and must examine limiting fatigue characteristics of material. The below approach overcomes the above drawbacks.

3. Main Principles

References [2,4,50] show that in the general case, an MTD system is given as a thermodynamic continuum where solids are distributed (scattered), interacting with each other and with the continuum. Figure 2 illustrates the continuum fragment of limited size $\Omega(X, Y, Z)$. The continuum with a temperature θ and a chemical composition Ch () has two solid elements (A and B) interacting in the contact zone $S(x, y, z)$ that can move relatively to each other. Arbitrary mechanical loads perceived by one of them (by element A) in the x, y, z coordinate system are transformed into internal transverse forces Q_x, Q_y, Q_z, longitudinal forces N_x, N_y, N_z, as well as into bending moments N_x, N_y, N_z. Element B is pressed to element A by the loads that are reduced to the distributed normal pressure $p(x, y)$ and the tangential pressure $q(x, y)$. The origin of the coordinates is shifted to the point of original contact O of the two elements (before deformation). We can easily notice that the elements A and B form the tribo-fatigue system [4], presented in Reference [2] as a friction pair (it consists of the element A without internal forces ($N_i = 0$, $Q_i = 0$, $M_i = 0$, $i = x, y, z$) and of the element B). So, the tribo-fatigue system is the friction pair, in which, at least one of these elements perceives non-contact loads, and hence, it undergoes volumetric deformation. The advantage of such an MTD system is that the corresponding solutions reported in mechanics of deformable solids, contact mechanics, mechanics of tribo-fatigue systems (tribo-fatigue), and in tribology can be adopted to analyze the state of solids and the system components.

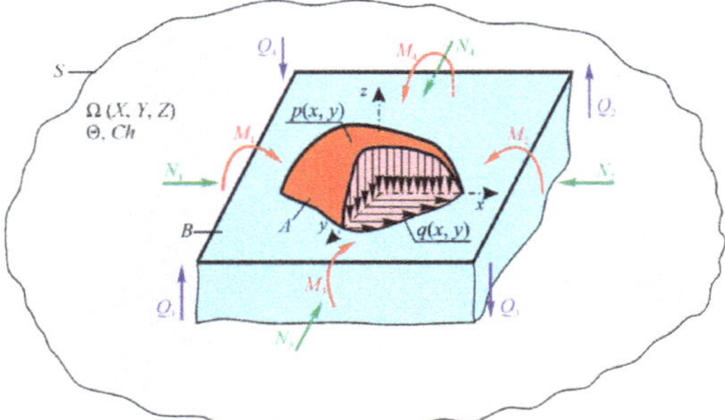

Figure 2. Mechanothermodynamical (MTD) system (**A**) denotes surface (contact) tractions; (**B**) denotes loaded body).

Now, it is the task to describe the MTD system energy state at mechanical and thermodynamic loads with the consideration of the environment influence.

The energy state of any system is of interest. However, in relation to the MTD system, of importance is to examine its damageability, and as a consequence, conditions of reaching the limiting state. Of special interest is the analysis of the so-called translimiting or supercritical conditions [2].

According to References [2–4], we can formulate the main ideas, being the basis of the developed theory.

I. Bearing in mind that the MTD system elements perceive different loads: mechanical, thermal, and electrochemical, the traditional analysis of their damageability and limiting state at mechanical stresses or strains [53–61] can be a basis of studies, but it appears insufficient, and hence, ineffective. This means that the MTD system states must be analyzed using more general energy concepts.

II. Since mechanical, thermodynamic [62–66], and electrochemical loads specify the damageability of MTD system solids, we should use a generalized idea of its complex damage due to these loads when they act simultaneously. Such damage will be called any irreversible change in shape, size, volume, mass, composition, structure, continuity, and hence, physical-mechanical properties of system elements. It is a corresponding change in the functions of the system as a whole.

III. Four particular phenomena: mechanical fatigue, friction, and wear, as well as thermodynamic and electrochemical processes, specify the complex damage onset and development. These phenomena are called particular ones in the sense that each of them can be implemented as independent and individual. This leads to the corresponding energy state and damageability by particular (individual) criteria.

IV. In the general case, all particular phenomena and MTD system processes develop simultaneously and in one zone. The MTD system states are attributed to not one of any of the above phenomena, but to their joint (collective) development, and consequently, to their interaction.

V. If all the energy U_Σ supplied to the MTD system is responsible for its physical state, then the condition of its damageability is specified by the effective (dangerous) part $U_\Sigma^{eff} << U_\Sigma$ spent for generation, motion, and interaction of irreversible damages.

VI. The effective energy U_Σ^{eff} at volumetric deformation of solids can be given in the form of the function of three energy components: thermal U_T^{eff}, force U_n^{eff}, and frictional U_τ^{eff}:

$$U_\Sigma^{eff} = F_\Lambda\left(U_T^{eff}, U_n^{eff}, U_\tau^{eff}\right), \tag{25}$$

where, F_Λ considers the irreversible kinetic interaction of particular damage phenomena. The components $U_T^{eff}, U_n^{eff}, U_\tau^{eff}$ of the effective energy U_Σ^{eff} do not possess the additivity property.

VII. We allow for the processes of electrochemical (in particular, corrosion) damage of solids by introducing the parameter $0 \le D_{ch} \le 1$ and study them as electrochemical damageability when acted upon by temperature ($D_{T(ch)}$), stress ($D_{\sigma(ch)}$), as well as by corrosion and friction ($D_{\tau(ch)}$). So, Function (25) assumes the form:

$$U_{\Sigma}^{eff} = F_{\Lambda}\left(U_{T(ch)}^{eff}, U_{n(ch)}^{eff}, U_{\tau(ch)}^{eff}\right). \tag{26}$$

VIII. The condition for the effective energy U_{Σ}^{eff} to attain its limiting value—the critical quantity U_0 in some area of limited size—in the dangerous volume of the MTD system, serves as the generalized criterion of the limiting (critical) state.

IX. It is considered that the energy U_0 is a fundamental constant for a given material and must not depend on testing conditions, input energy types, or damage mechanisms.

X. The three-dimensional (3D) area $V_{ij} \subset V_0$ of a deformable solid (V_0 is the working volume) with the critical state of the material of its components at all its points is called the dangerous volume.

XI. In the general case, the limiting (critical) state of the MDT system is attained not because effective energy components grow, and hence, because irreversible damages at individual different-nature loads are accumulated, but because they interact dialectically. Their direction is characterized by the development of spontaneous hardening-softening of materials at the considered operating conditions. Thus, when Function (26) is taken into account, the hypothesis of the limiting (critical) state of the MTD system can be presented in the following general form:

$$\Phi(u_{\sigma(ch)}^{eff}, u_{\tau(ch)}^{eff}, u_{T(ch)}^{eff}, \Lambda_{n\backslash k\backslash l}, m_k, u_0) = 0, \tag{27}$$

where, the m_k k = 1, 2, ... , are some characteristic properties (hardening-softening) of contacting materials, and the $\Lambda_{k\backslash l\backslash n} \gtreqless 1$ are the functions (parameters) of dialectic interactions of effective energies (irreversible damages) at different-nature loads. It means that at $\Lambda_k > 1$, the damageability increase is realized, at $\Lambda_l < 1$—its decrease, and at $\Lambda_n = 1$—its stable development.

XII. When item III is taken into consideration, hypothesis (27) must be multi-criterion from the physical viewpoint, i.e., it must describe not only the states of the system as a whole, but its individual elements through different criteria of performance loss (wear, fatigue damage, pitting, corrosion damage, thermal damage, etc.). In particular cases, we can attain the corresponding limiting (critical) states through one or two, three, or several criteria at a time.

XIII. Attaining the limiting state:

$$u_{\Sigma}^{eff} = u_0 \tag{28}$$

means that the MTD system completely loses its integrity, i.e., all of its functions. At the same time, its elements reach the critical damageability:

$$0 < \psi_u^{eff} = u_{\Sigma}^{eff}/u_0 \tag{29}$$

$$\psi_u^{eff}\left(\psi_{\sigma(ch)}, \psi_{\tau(ch)}, \psi_{T(ch)}, \Lambda_{k\backslash l\backslash n}, m_k\right) = 1 \tag{30}$$

XIV. If $t = t_0$ is the time of the system onset and $T_{\oplus}$ is the time when the system reaches the limiting (critical) state, then the failure time of its functions is consistent with the relative lifetime (longevity) $t/T_{\oplus} = 1$. But the system lifetime T_* as a material object is longer than its lifetime as a whole ($T_* >> T_{\oplus}$), since at $t > T_{\oplus}$, the long process of its degradation–disintegration is realized when a great number of remains, pieces, fragments, etc., are formed. This process develops when acted upon by not only possible mechanical loads, but mainly by the environment, up to the moment when the system as a material object dies at $t = T_*$. The system death is its complete disintegration into an infinitely large number of ultimately small particles (atoms). The below conditions describe the translimiting existence of the system as a gradually disintegrating material object:

$$\psi_u^{eff} \to \infty, \tag{31}$$

$$d_{\psi} \to 0, \tag{32}$$

where d_{ψ} is the average size of disintegration particles. The organic relationship $\psi_u^{eff}(d_{\psi})$ must exist between ψ_{Σ} and d_{ψ}. Then, the condition for the system death is:

$$t/T_* = 1 \tag{33}$$

XV. The disintegration particles of the "old system" are not destroyed and are spent to form and increase a number of "new systems". This is the essence of the MTD system evolution hysteresis.

4. Damageability Energy Theory and Limiting States

First, we concretize Function (25).

For the effective energy to be determined, we will consider the work of internal forces in the elementary volume dV of tribo-fatigue systems (A, B in Figure 2). In the general case, we can write the differential of work of internal forces and the temperature dT_{Σ} with the consideration of the disclosure rule of the biscalar product of the stress and strain tensors σ and ε:

$$\begin{aligned} du = \sigma_{ij}\cdot\cdot d\varepsilon_{ij} + kdT_{\Sigma} = \begin{pmatrix} \sigma_{xx} & \sigma_{xy} & \sigma_{xz} \\ \sigma_{yx} & \sigma_{yy} & \sigma_{yz} \\ \sigma_{zx} & \sigma_{zy} & \sigma_{zz} \end{pmatrix}\cdot\cdot\begin{pmatrix} d\varepsilon_{xx} & d\varepsilon_{xy} & d\varepsilon_{xz} \\ d\varepsilon_{yx} & d\varepsilon_{yy} & d\varepsilon_{yz} \\ d\varepsilon_{zx} & d\gamma_{zy} & d\varepsilon_{zz} \end{pmatrix} + \\ +kdT_{\Sigma} = \sigma_{xx}d\varepsilon_{xx} + \sigma_{yy}d\varepsilon_{yy} + \sigma_{zz}d\varepsilon_{zz} + \sigma_{xy}d\varepsilon_{xy} + \sigma_{xz}d\varepsilon_{xz} + \sigma_{yz}d\varepsilon_{yz} + kdT_{\Sigma}\,; \end{aligned} \tag{34}$$

where k is the Boltzmann constant.

We will proceed from the fact that in the general case, according to References [2,4], normal and shear stresses, which cause the processes of shear (due to friction) and tear (due to tension-compression), play a decisive role in forming wear-fatigue damage.

In this case, it makes sense to divide the tensor σ into two parts: σ_{τ} is the friction-shear stress tensor, or, briefly, the shear tensor, σ_n is the normal stress tensor (tension-compression), or, briefly, the tear tensor. In Equation (28), we will distinguish the tear part σ_n and the shear part σ_{τ} of the tensor σ as:

$$\begin{aligned} du = \sigma_{ij}^{(V,\,W)}\cdot\cdot d\varepsilon_{ij}^{(V,\,W)} + kdT_{\Sigma} = \left(\sigma_n{}^{(V,\,W)} + \sigma_{\tau}{}^{(V,\,W)}\right)\cdot\cdot d\varepsilon_{ij}^{(V,\,W)} + kdT_{\Sigma} = \\ = \sigma_n{}^{(V,\,W)}\cdot\cdot d\varepsilon_{ij}^{(V,\,W)} + \sigma_{\tau}{}^{(V,\,W)}\cdot\cdot d\varepsilon_{ij}^{(V,\,W)} + kdT_{\Sigma}. = du_n + du_{\tau} + du_T. \end{aligned} \tag{35}$$

In accordance with items III and IV, we must present the tensors σ_{ij} and ε_{ij} as follows:

$$\begin{aligned} \sigma_{ij} = \sigma_{ij}^{(V,\,W)} = \sigma_{ij}\left(\sigma_{ij}^{(V)}, \sigma_{ij}^{(W)}\right), \\ \varepsilon_{ij} = \varepsilon_{ij}^{(V,\,W)} = \varepsilon_{ij}\left(\varepsilon_{ij}^{(V)}, \varepsilon_{ij}^{(W)}\right). \end{aligned} \tag{36}$$

where, the volume loads (the general cases of 3D bending, torsion, tension-compression) give rise to the stress and strain tensors with the superscript V and the contact interaction of system elements to those with the superscript W.

We can present Expression (35) with regard to (36) as follows:

$$\begin{aligned} du = \sigma_{ij}^{(V,\,W)}\cdot\cdot d\varepsilon_{ij}^{(V,\,W)} + kdT_{\Sigma} = \left(\sigma_n{}^{(V,\,W)} + \sigma_{\tau}{}^{(V,\,W)}\right)\cdot\cdot d\varepsilon_{ij}^{(V,\,W)} + kdT_{\Sigma} = \\ = \sigma_n{}^{(V,\,W)}\cdot\cdot d\varepsilon_{ij}^{(V,\,W)} + \sigma_{\tau}{}^{(V,\,W)}\cdot\cdot d\varepsilon_{ij}^{(V,\,W)} + kdT_{\Sigma}\cdot = du_n + du_{\tau} + du_T. \end{aligned} \tag{37}$$

When there is a linear relationship between stresses and strains, Expression (36) will assume the form:

$$\sigma_{ij} = \sigma_{ij}^{(V,\,W)} = \sigma_{ij}^{(V)} + \sigma_{ij}^{(W)} = \begin{pmatrix} \sigma_{xx}^{(V)} + \sigma_{xx}^{(W)} & \sigma_{xy}^{(V)} + \sigma_{xy}^{(W)} & \sigma_{xz}^{(V)} + \sigma_{xz}^{(W)} \\ \sigma_{yx}^{(V)} + \sigma_{yx}^{(W)} & \sigma_{yy}^{(V)} + \sigma_{yy}^{(W)} & \sigma_{yz}^{(V)} + \sigma_{yz}^{(W)} \\ \sigma_{zx}^{(V)} + \sigma_{zx}^{(W)} & \sigma_{zy}^{(V)} + \sigma_{zy}^{(W)} & \sigma_{zz}^{(V)} + \sigma_{zz}^{(W)} \end{pmatrix}, \tag{38}$$

$$\varepsilon_{ij} = \varepsilon_{ij}^{(V,W)} = \varepsilon_{ij}^{(V)} + \varepsilon_{ij}^{(W)} = \begin{pmatrix} \varepsilon_{xx}^{(V)} + \varepsilon_{xx}^{(W)} & \varepsilon_{xy}^{(V)} + \varepsilon_{xy}^{(W)} & \varepsilon_{xz}^{(V)} + \varepsilon_{xz}^{(W)} \\ \varepsilon_{yx}^{(V)} + \varepsilon_{yx}^{(W)} & \varepsilon_{yy}^{(V)} + \varepsilon_{yy}^{(W)} & \varepsilon_{yz}^{(V)} + \varepsilon_{yz}^{(W)} \\ \varepsilon_{zx}^{(V)} + \varepsilon_{zx}^{(W)} & \varepsilon_{zy}^{(V)} + \varepsilon_{zy}^{(W)} & \varepsilon_{zz}^{(V)} + \varepsilon_{zz}^{(W)} \end{pmatrix}, \quad (39)$$

And Expression (37) will be as follows:

$$\begin{gathered} du = u = \tfrac{1}{2}\sigma_{ij} \cdot\cdot \varepsilon_{ij} + kT_\Sigma = \tfrac{1}{2}\left(\sigma_{ij}^{(V)} + \sigma_{ij}^{(W)}\right) \cdot\cdot \left(\varepsilon_{ij}^{(V)} + \varepsilon_{ij}^{(W)}\right) + \\ + kT_\Sigma = \tfrac{1}{2}\left[\left(\sigma_n^{(V)} + \sigma_n^{(W)}\right) + \left(\sigma_\tau^{(V)} + \sigma_\tau^{(W)}\right)\right] \cdot\cdot \left(\varepsilon_{ij}^{(V)} + \varepsilon_{ij}^{(W)}\right) + kT_\Sigma = \\ = \frac{1}{2}\left[\begin{pmatrix} \sigma_{xx}^{(V)} + \sigma_{xx}^{(W)} & 0 & 0 \\ 0 & \sigma_{yy}^{(V)} + \sigma_{yy}^{(W)} & 0 \\ 0 & 0 & \sigma_{zz}^{(V)} + \sigma_{zz}^{(W)} \end{pmatrix} + \begin{pmatrix} 0 & \sigma_{xy}^{(V)} + \sigma_{xy}^{(W)} & \sigma_{xz}^{(V)} + \sigma_{xz}^{(W)} \\ \sigma_{yx}^{(V)} + \sigma_{yx}^{(W)} & 0 & \sigma_{yz}^{(V)} + \sigma_{yz}^{(W)} \\ \sigma_{zx}^{(V)} + \sigma_{zx}^{(W)} & \sigma_{zy}^{(V)} + \sigma_{zy}^{(W)} & 0 \end{pmatrix}\right] \cdot\cdot \\ \cdot\cdot \begin{pmatrix} \varepsilon_{xx}^{(V)} + \varepsilon_{xx}^{(W)} & \varepsilon_{xy}^{(V)} + \varepsilon_{xy}^{(W)} & \varepsilon_{xz}^{(V)} + \varepsilon_{xz}^{(W)} \\ \varepsilon_{yx}^{(V)} + \varepsilon_{yx}^{(W)} & \varepsilon_{yy}^{(V)} + \varepsilon_{yy}^{(W)} & \varepsilon_{yz}^{(V)} + \varepsilon_{yz}^{(W)} \\ \varepsilon_{zx}^{(V)} + \varepsilon_{zx}^{(W)} & \varepsilon_{zy}^{(V)} + \varepsilon_{zy}^{(W)} & \varepsilon_{zz}^{(V)} + \varepsilon_{zz}^{(W)} \end{pmatrix} + kT_\Sigma. \end{gathered} \quad (40)$$

From Expression (40), it is seen that the tear part σ_n of the tensor σ is the sum of the tear parts of the tensors at the volumetric strain $\sigma_n^{(V)}$ and the surface load (friction) $\sigma_n^{(W)}$. The shear part σ_τ is the sum of the shear parts $\sigma_\tau^{(V)}$ and $\sigma_\tau^{(W)}$. This is the fundamental difference of the generalized approach to constructing a criterion for the MTD system limiting state.

We will distinguish the effective part of total energy (Expression (40)) according to items V and VIII and References [2,3]. To do this, we will introduce the coefficients $A_n(V)$, $A_\tau(V)$, and $A_T(V)$ of the corresponding dimension. The latter determine the absorbed energy fraction:

$$du_\Sigma^{eff} = \Lambda_{M\backslash T}(V)\left\{\Lambda_{n\backslash\tau}(V)\left[A_n(V)\sigma_n \cdot\cdot d\varepsilon_{ij} + A_\tau(V)\sigma_\tau \cdot\cdot d\varepsilon_{ij}\right] + A_T(V)kdT_\Sigma\right\} \quad (41)$$

or

$$du_\Sigma^{eff} = \Lambda_{M\backslash T}(V)\left\{\Lambda_{\tau\backslash n}(V)[A_n(V)du_n + A_\tau(V)du_\tau] + A_T(V)du_T\right\} \quad (42)$$

where $\Lambda_{M\backslash T}(V)$ and $\Lambda_{\tau\backslash\sigma}(V)$ are the functions of interaction between different energies. The subscript $\tau\backslash\sigma$ means the function Λ responsible for the interaction between the shear (τ) and tear (σ) components of the effective energy and the subscript $M\backslash T$—the function Λ is responsible for the interaction between the mechanical (M) and thermal (T) parts of the effective energy. Generally speaking, the coefficients A can be different at different points of volume V. This fact allows one to take into account the environment inhomogeneity.

Taking into consideration Expression (42), criteria (27) can be specified, not considering the environment influence:

$$\Lambda_{M\backslash T}(V)\left\{\Lambda_{\tau\backslash n}(V)\left[du_n^{eff} + du_\tau^{eff}\right] + du_T^{eff}\right\} = u_0. \quad (43)$$

When there is a linear relationship between stresses and strains, Expressions (41) and (42) will be of the following form:

$$u_\Sigma^{eff} = \Lambda_{M\backslash T}(V)\left\{\Lambda_{\tau\backslash n}(V)\left[\frac{1}{2}A_n(V)\,\sigma_n \cdot\cdot \varepsilon_{ij} + \frac{1}{2}A_\tau(V)\,\sigma_\tau \cdot\cdot \varepsilon_{ij}\right] + A_T(V)\,kT_\Sigma\right\}, \quad (44)$$

or

$$u_{\Sigma}^{eff} = \Lambda_{M\backslash T}(V)\Big\{\Lambda_{n\backslash\tau}(V)\,[A_n(V)u_n(V) + A_\tau(V)u_\tau(V)] + A_T(V)u_n(V)\Big\} = \\ = \Lambda_{M\backslash T}(V)\Big\{\Lambda_{n\backslash\tau}(V)\left[u_n^{eff}(V) + u_\tau^{eff}(V)\right] + u_T^{eff}(V)\Big\}. \quad (45)$$

With Expression (36) considered, criterion (43) can be presented as follows:

$$u_{\Sigma}^{eff} = \Big\{\left[u_n^{eff}(\sigma_n{}^{(V,\,W)}, \varepsilon_n{}^{(V,\,W)}) + u_\tau^{eff}(\sigma_\tau{}^{(V,\,W)}, \varepsilon_\tau{}^{(V,\,W)})\right]\Lambda_{n\backslash\tau} + u_T^{eff}\Big\}\Lambda_{T\backslash M} = u_0. \quad (46)$$

When time effects must be allowed for, criterion (46) will assume the form:

$$u_{\Sigma t}^{eff} = \int_0^t\Big\{\left[u_n^{eff}(\sigma_n{}^{(V,\,W)}, \varepsilon_n{}^{(V,\,W)},\, t) + u_\tau^{eff}(\sigma_\tau{}^{(V,\,W)}, \varepsilon_\tau{}^{(V,\,W)},\, t)\right]\Lambda_{n\backslash\tau}(t) + \\ + u_T^{eff}(t)\Big\}\Lambda_{T\backslash M}(t)\,dt = u_0. \quad (47)$$

So, Expression (45) is the concretization of Equation (25) and Equation (46)—the concretization of criterion (27) when the environment influence is not taken into account.

Criterion (27) in the forms of Expressions (46) and (47) states: when the sum of interacting effective energy components at force, frictional, and thermal (thermodynamic) loads reaches a critical (limiting) quantity u_0, the limiting (or critical) state of the MTD system (both as individual elements and the system as a whole) is implemented. Physically, it is attributed to many and different damage mechanisms.

Above, we noted the fundamental character of the parameter u_0. Based on References [66–78], we will understand parameter u_0 as the initial activation energy of the disintegration process. u_0 approximately means both sublimation heat for metals and crystals with ionic bonds and thermal destruction activation energy for polymers:

$$u_0 \approx u_T.$$

On the other hand, the quantity u_0 is determined as the activation energy for mechanical fracture:

$$u_0 \approx u_M.$$

Thus, the energy u_0 can be a constant of a material:

$$u_0 \approx u_M \approx u_T = \text{const}. \quad (48)$$

With the physical-mechanical and thermodynamic presentations of the damageability and fracture processes [67,68,70] taken into account, we write Expression (48) in the following form:

$$u_M = s_k\frac{\sigma_{th}}{E}\frac{C_a}{\alpha_V} = u_0 = kT_S\ln\frac{k\theta_D}{h} = u_T, \quad (49)$$

where, s_k is the reduction coefficient, σ_{th} the theoretical strength, E the elasticity modulus, C_a the atom heat capacity, α_V the thermal expansion of the volume, k the Boltzmann constant, T_S the melting point, θ_D the Debye temperature, and h is the Planck constant. According to Expression (49), we can approximately assume [67]:

$$u_0 \approx \varepsilon_*\frac{C_a}{\alpha_V}, \quad (50)$$

where $\varepsilon_* \approx 0.6$ is the limiting strain of the interatomic bond. Calculations according to Expression (50) are not difficult. The methods for experimental determination of u_0 have also been developed [68].

Equation (49) shows that u_0 is the activation energy of a given material and is by the order of magnitude equal to 1 … 10 eV per one particle or molecule (~10^2 … 10^3 kJ/mol), i.e., it is close to the energy of interatomic bond rupture in the solid [71] and does not depend on a way of reaching rupture: mechanically, thermally, or by their simultaneous action. Reference [68] contains the tables of the u_0 values for various materials.

Equation (49) gives a thermomechanical constant of a material [2]:

$$\frac{\sigma_{th}}{T_S} = E\frac{\alpha_V k}{C_a} \ln \frac{k\theta_D}{h} = \theta_\sigma. \tag{51}$$

The constant θ_σ is the strength loss per *1 K*.

Criterion (46) is written in absolute values of physical parameters: effective and critical energy components. We can make this criterion dimensionless: it must be by divided by the quantity u_0. Criterion (46) is presented in terms of irreversible (effective) damage:

$$\psi_u^{eff} = \frac{u_\Sigma^{eff}}{u_0} = 1. \tag{52}$$

The local (at the point) energy damageability measure ψ_u^{eff} is within the range:

$$0 \leq \psi_u^{eff} \leq 1, \tag{53}$$

or in expanded form:

$$\begin{gathered} 0 \leq \psi_u^{eff} = \\ = \frac{\Lambda_{T\backslash M}}{u_0}\left\{\left[u_n^{eff}\left(\sigma_n{}^{(V,\,W)}, \varepsilon_n{}^{(V,\,W)}\right) + u_\tau^{eff}\left(\sigma_\tau{}^{(V,\,W)}, \varepsilon_\tau{}^{(V,\,W)}\right)\right] \Lambda_{n\backslash\tau} + u_T^{eff}\right\} \leq 1. \end{gathered} \tag{54}$$

According to Expression (54), we can determine particular energy damageability measures

$$0 \leq \psi_n^{eff} = \frac{u_n^{eff}\left(\sigma_n^{(V,\,W)},\ \varepsilon_n^{(V,\,W)}\right)}{u_0} \leq 1, \tag{55}$$

$$0 \leq \psi_\tau^{eff} = \frac{u_\tau^{eff}\left(\sigma_\tau^{(V,\,W)},\ \varepsilon_\tau^{(V,\,W)}\right)}{u_0} \leq 1, \tag{56}$$

$$0 \leq \psi_T^{eff} = \frac{u_T^{eff}}{u_0} \leq 1, \tag{57}$$

at effective different energies determined by force (the subscript n), frictional (the subscript τ), and thermodynamic (the subscript T) loads, respectively. We can now write criterion (52) in dimensionless form:

$$\psi_u^{eff} = \left[\left(\psi_n^{eff} + \psi_\tau^{eff}\right)\Lambda_{n\backslash\tau} + \psi_T^{eff}\right]\Lambda_{N\backslash T} = 1. \tag{58}$$

Based on Expression (58), we can reach the MTD system limiting state at the sum of interacting damages ($0 < \psi < 1$) for mechanical and thermodynamic loads equal to 1. Criterion (46) in the form of Expression (58) finds convenient use because all damageability measures are dimensionless and are within $0 \leq \psi \leq 1$.

Since we cannot describe and predict exactly numerous and innumerable interactions between physical damages of many-type dislocation, vacancy, non-elastic deformation, etc., the analysis of the MTD system must use the concept of interaction between dangerous volumes [2] that contain a real complex of damages (defects as a result of the action of the corresponding stress/strain fields). By the statistical model of a deformable solid with a dangerous volume [71,72], such a volume of a solid must depend on its geometric parameters responsible for the working volume V_0, on the parameters of the distribution functions of $p(\sigma_{-1})$ and $p(\sigma)$ of the durability limit σ_{-1} and the effective stresses σ, considering both the effective stress probabilities P and γ_0, as well as the effective stress gradients G_σ:

$$V_{P\gamma} = F_V[p(\sigma_{-1}),\ p(\sigma),\ G_\sigma,\ V_0,\ P,\ \gamma_0,\ \vartheta_V]. \tag{59}$$

where, ϑ_V describes the influence of the shape of a body on the durability limit and the schemes of its loading in fatigue tests.

The dangerous volume can then be taken as the equivalent of the damage complex, since its value is proportional, in particular to the value of effective stresses, and hence, to the number (concentrations) of defects (damages).

As follows from Expression (59), the boundary between dangerous and safe volumes is generally blurred and probabilistic in nature. By increasing the damage probability P of the solid, the dangerous volume V_{P_γ} grows. At a given P value, the volume can vary depending on the confidence probability γ_0. It means that at P = const:

$$V_{P_{\gamma\ \min}} \leq V_{P_\gamma} \leq V_{P_{\gamma\ \max}}, \tag{60}$$

if $\gamma_{\min} \leq \gamma_0 \leq \gamma_{\max}$. Here, $\gamma_{\min}, \gamma_{\max}$ form a permissible range. If γ_0 = const, then the dangerous volume will have a single value associated with the damage probability P.

Not only the so-called smooth bodies, but also the elements with structural stress concentrators [71], are characterized by scattered damage within the dangerous volume. Figure 3 demonstrates several microcracks on the sharp cut (the rounding radius r = 0.5 mm, the theoretical stress concentration factor α_n= 8, in Figure 3a) and on the flat cut (r = 2 mm, α_n= 2.55, in Figure 3b) and also two fatigue cracks at a distance of 25 mm from each other at a fillet connection from the crankshaft journal to its web (r = 18 mm, α_n= 3.2, in Figure 3c). The crankshaft journal diameter is 360 mm.

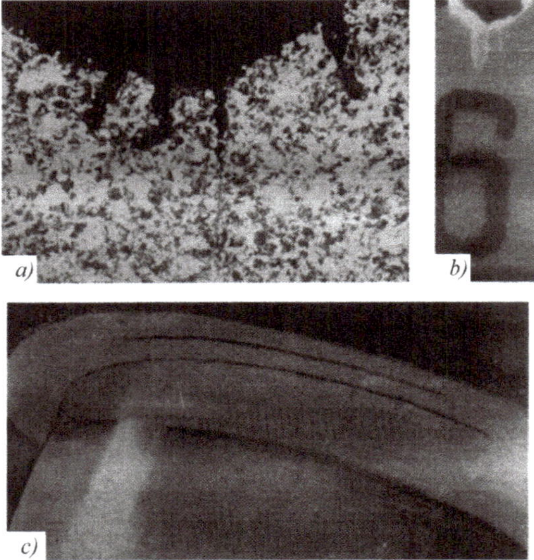

Figure 3. Microcracks in the zones of stress concentrators. (**a**) rounding-off radius r = 0.5 mm; (**b**) r = 2 mm;(**c**) r = 18 mm.

So, if in the case of the uniaxial stress state, the distribution of the stresses σ (x, y, z) in the x, y, z coordinates is known, then the dangerous volume is calculated by the formula:

$$V_{P\gamma} = \iiint\limits_{\sigma(x,y,z)>\sigma_{-1\min}} dxdydz, \tag{61}$$

where $\sigma_{-1\min}$–the lower boundary of the solid durability limit σ_{-1} is such that if $\sigma_{-1} < \sigma_{-1\min}$, then $P = 0$.

Expression (61) yields the generalized condition for fatigue fracture in of the form:

$$V_{P\gamma} > 0 \tag{62}$$

with some probability P at the confidence probability γ_0.

If

$$V_{P\gamma} = 0, \tag{63}$$

then fatigue fracture cannot occur physically (because in this case, $\sigma > \sigma_{-1\min}$); hence, Expression (63) is the generalized condition of non-fracture.

The methods to calculate dangerous volumes V_{ij} for friction pairs and tribo-fatigue systems are developed similar to Expression (59)

$$V_{ij} = V_{ij}\left(\sigma_n^{(V,\,W)},\ \sigma_\tau^{(V,\,W)},\ \sigma_{\lim}^{(V,\,W)},\ G_{\sigma_{ij}},\ V_0,\ P,\ \gamma_0\right) \tag{64}$$

and outlined in References [4,71–75]. Here, $\sigma_{\lim}^{(V,\,W)}$ is the limiting stress by the assigned criterion of damage and fracture.

Further, we can introduce the following dimensionless characteristics of damageability: integral energy damageability within the dangerous volume:

$$\Psi_u^{eff}(V) = \iiint\limits_{\psi_u^{eff}(dV)\geq 1} \frac{u_\Sigma^{eff}}{u_0} dV \tag{65}$$

and the average energy damageability (at each point of the dangerous volume):

$$\overline{\Psi}_u^{eff}(V) = \frac{1}{V_u} \iiint\limits_{\psi_u^{eff}(dV)\geq 1} \frac{u_\Sigma^{eff}}{u_0} dV. \tag{66}$$

The time accumulation of energy damageability within the dangerous volume is governed by the formulas:

$$\Psi_u^{eff}(V,t) = \int\limits_t \iiint\limits_{\psi_u^{eff}(dV)\geq 1} \frac{u_\Sigma^{eff}}{u_0} dV dt \tag{67}$$

$$\overline{\Psi}_u^{eff}(V,t) = \int\limits_t \frac{1}{V_u} \iiint\limits_{\psi_u^{eff}(dV)\geq 1} \frac{u_\Sigma^{eff}}{u_0} dV dt. \tag{68}$$

Based on Expressions (63)–(68), the MTD system damageability can be described and analyzed with the adoption of the most general concepts—the energy concepts allowing for the influence of numerous and different factors taken into account by Expression (59), including the scale effect, i.e., the changes in the size and shape (mass) of system elements.

In References [2,77], the function $\Lambda_{k\backslash l\backslash n}$ for damage interactions in the MTD system is determined by the effective energy ratio parameters:

$$\Lambda_{n\backslash k\backslash l} = \Lambda_{n\backslash k\backslash l}\left(\rho_{M\backslash T}, \rho_{n\backslash \tau}\right) \gtreqless 1, \tag{69}$$

$$\rho_{n\backslash\tau} = u_\tau^{eff}/u_n^{eff},\ \rho_{M\backslash T} = u_M^{eff}/u_T^{eff}. \tag{70}$$

The quantities Λ calculated by Expression (69) describe how the load parameter ratio affects the character and direction of interaction of irreversible damages [2–4]. If $\Lambda > 1$, then the system is self-softening, since, when hardening–softening phenomena are in balance, softening processes are dominant. If $\Lambda < 1$, then the system is self-hardening, since, when hardening–softening phenomena are in balance, hardening processes are dominant. At $\Lambda = 1$, the system is stable. The spontaneous hardening–softening phenomena are in balance. A particular article will deal with a general analysis of damage interactions in MTD systems because of its fundamental importance.

After criterion (27) has been basically formalized, the action of electrochemical loads (damages) should be taken into consideration in accordance with item VII. We must immediately emphasize that in the strict mechanothermodynamical statement, it is difficult to do this: when the environment interacts with a deformable solid, electrochemical reactions are very diverse, complex and insufficiently studied. That is why the approach proposed in References [2,3] was adopted: we introduced the simplification, according to which the damage of solids in the environment is determined by corrosion–electrochemical processes. In addition, the hypothesis is put forward, following which, the effective energy of corrosion–electrochemical damage is proportional to the square of the corrosion speed, i.e.,

$$u_{ch}^{eff} \sim v_{ch}^{2}. \tag{71}$$

If, in accordance with item VII, $0 \le D_{ch} \le 1$ is the parameter of corrosion–electrochemical damage of the solid, then from References [2,4,76], criterion (26) considering its shape will be of the form:

$$\Lambda_{M\backslash T}\left[\left(\frac{u_n^{eff}\left(\sigma_n^{(V,W)},\ \varepsilon_n^{(V,W)}\right)}{u_0(1-D_n)}+\frac{u_\tau^{eff}\left(\sigma_\tau^{(V,W)},\ \varepsilon_\tau^{(V,W)}\right)}{u_0(1-D_\tau)}\right)\Lambda_{n\backslash\tau}+\frac{u_T^{eff}}{u_0(1-D_T)}\right]=1,\ \Lambda \gtreqless 1 \tag{72}$$

where

$$0 \le \frac{u_n^{eff}\left(\sigma_n^{(V,W)},\ \varepsilon_n^{(V,W)}\right)}{u_0(1-D_n)} = \psi_{n(ch)}^{eff} \le 1, \tag{73}$$

$$0 \le \frac{u_\tau^{eff}\left(\sigma_\tau^{(V,W)},\ \varepsilon_\tau^{(V,W)}\right)}{u_0(1-D_\tau)} = \psi_{\tau(ch)}^{eff} \le 1, \tag{74}$$

$$0 \le \frac{u_T^{eff}}{u_0(1-D_T)} = \psi_{T(ch)}^{eff} \le 1, \tag{75}$$

$$1-D_T = b_{e(T)}\left(\frac{v_{ch}}{v_{ch(T)}}\right)^{m_{v(T)}};\ 1-D_n = b_{e(n)}\left(\frac{v_{ch}}{v_{ch(n)}}\right)^{m_{v(n)}}; \tag{76}$$

where v_{ch} is the corrosion speed in this environment, $v_{ch(T)}, v_{ch(\sigma)}, v_{ch(\tau)}$ is the corrosion speed in the same environment at thermal, force, and friction loads respectively, the b_e's are the coefficients responsible for corrosive erosion processes, the $M_{V(\bullet)}$'s are the parameters responsible for the electrochemical activity of materials at force (the subscript σ), friction (the subscript τ), and thermodynamic (the subscript T) loads, wherein $M_{V(\bullet)} = 2/A_{ch}$ and the parameter $A_{ch} \gtreqless 1$.

In Reference [76], we can find other methods for assessment of the parameter D_{ch}.

As seen, Equation (72) is the specification of criterion (27). According to this criterion, the limiting state of the MTD system is reached when the sum of dialectically interacting irreversible damages at force, friction, and thermodynamic loads (including electrochemical damage when acted upon by stress, friction, temperature) becomes equal to unity.

We consider the particular case: in Expression (46), it is assumed that $A_\sigma(V) = A_\sigma$ = const, $A_\tau(V)$ = A_τ = const, $A_T(V) = A_T$ = const, $A_{\tau\backslash\sigma}(V) = A_{\tau\backslash\sigma}$ = const, and $A_{M\backslash T}(V) = A_{M\backslash T}$ = const.

Firstly, the stress state is induced by volume deformation, for which all stress tensor components, with the exception of one component σ (one-dimensional tension–compression, pure bending), can be neglected. Secondly, the stress state is induced by surface friction, for which all stress tensor components, with the exception of one component τ_w, can be neglected. Expression (40) then assumes the form:

$$\Lambda_{M\backslash T}\left[\Lambda_{\tau\backslash n}\left(A_\sigma\sigma^2 + A_\tau\tau^2\right) + A_T T_\Sigma\right] = u_0,$$

or in accordance with Expression (72):

$$\Lambda_{M\setminus T}\left[\frac{a_T}{1-D_T}T_\Sigma + \Lambda_{n\setminus\tau}\left(\frac{a_n}{1-D_n}\sigma^2 + \frac{a_\tau}{1-D_\tau}\tau_w^2\right)\right] = u_0,\ \Lambda \gtreqless 1 \quad (77)$$

where $\frac{a_\sigma}{1-D_\sigma} = A_\sigma$, $\frac{a_\tau}{1-D_\tau} = A_\tau$, $\frac{a_T}{1-D_T} = A_n$.

Equation (77) is thus the simplest form of the energy criterion of the limiting state. Nevertheless, it is of great practical importance [2].

If the electrochemical influence of the environment is absent ($D_{ch} = 0$), then:

$$u_\Sigma^{eff} = \Lambda_{M\setminus T}\left[a_T T_\Sigma + \Lambda_{\tau\setminus n}\left(a_\sigma\sigma^2 + a_\tau\tau_w^2\right)\right] = u_0\ ,\ \ \Lambda \gtreqless 1. \quad (78)$$

Equation (78) is the simplest form of the energy criterion of the limiting state and is of great practical importance [2,76–81]. In particular, it is used to develop methods for assessment of a_T, a_σ, a_τ. In fact, at $\Lambda_{M\setminus T} = \Lambda_{\tau\setminus n} = 1$, the boundary conditions are:

$$\left.\begin{array}{ll} T_\Sigma = 0\ ,\tau_w = 0: & a_n\sigma_d^2 = u_0\ ,\ a_n = u_0/\sigma_d^2\ ; \\ T_\Sigma = 0\ ,\sigma = 0: & a_\tau\tau_d^2 = u_0\ ,\ a_\tau = u_0/\tau_d^2\ ; \\ \sigma = 0\ ,\tau_w = 0: & a_T\sigma_d = u_0\ ,\ a_T = u_0/T_d\ , \end{array}\right\} \quad (79)$$

where σ_d, τ_d are the force and friction limiting stresses as $T \to 0$ and are called the (mechanical) destruction limits, and T_d is the destruction temperature (when σ = 0, τ_w = 0) or the thermal destruction limit.

The effective ("dangerous") part of total strain energy can also be determined from the following physical considerations. The strain energy flux u, generated in the material sample at its cyclic strain ($\varepsilon = \varepsilon_{\max}\sin\omega t$) in the homogeneous (linear) stress state, is assumed to be similar to the light flux. In fact, it is continuously excited when the loading cycle is repeated with the speed $\omega 1/\lambda$. It can be considered as a wave of length λ. Some part of the energy u generated in such a way can be absorbed by material atoms and structural formations, which results in material damage. We denote the absorbed part of the energy by u^{eff}. The generated energy u is then equal to:

$$u = u^{eff} + u_{cons} \quad (80)$$

where, u_{cons} is the non-absorbed part of the generated energy u. In this case, it is called the conservative part.

If the analogy of light and energy strain is valid, then the strain absorption law may be similar to Bouguer's light absorption law. Consequently, the equation, linking the energy u_{cons} passed through the deformed material volume V and the generated energy u, has the form:

$$u_{cons} = u\exp(-\chi_\varepsilon V), \quad (81)$$

or, by Lambert, in differential form:

$$\frac{du}{u} = -\chi_\varepsilon V. \quad (82)$$

Here, as in Bourguer–Lambert's equation, the coefficient χ_ε independent of u is the energy absorption parameter.

Taking into account Equations (81) and (80), we obtain the strain energy absorption law:

$$u^{eff} = u[1 - \exp(-\chi_\varepsilon V)], \quad (83)$$

and hence, if $u = 0$ or $V = 0$, then $u^{eff} = 0$. If $V \to \infty$, it appears that according to Equation (81), $u_{cons} = u$, i.e., all input energy is dissipated within such a volume.

Physically, the strain energy absorption process occurs due to many phenomena:

- Electron transition in absorbing atoms from lower to higher energy levels (quantum theory).
- Generation and development of dislocation structures (dislocation theory).
- Emergence of II and III order residual strains (stresses) (elasticity theory).
- Formation and development of any imperfections (defects) of material composition and structure: point, planar, and spatial (physical materials science).
- Hardening–softening phenomena (including strain aging) developing in time (fatigue theory).
- Changes in (internal) tribo-fatigue entropy (wear-fatigue damage mechanics [2]).

It should be noted that approach (83) can also be extended to friction, since any indenter drives a strain wave upstream in the thin surface layer of the solid. The indenter is pressed to the solid. Here, χ_γ is the energy absorption parameter and the subscript γ denotes the shear strain. Similarly, heat absorption in the deformable solid can also be considered. Finally, by introducing the dangerous volume $V = V_{P\gamma}$ into Equations (81)–(83), we can easily solve the problem of strain energy absorption in the non-uniform (including complex) stress state.

It should be noted that, although criterion (78) is special, it is fundamental and general in nature. Its general nature follows from the fact that this case takes into consideration all four particular phenomena responsible for the state of the MTD system (although simplified by the statement of the stress-strain state) in accordance with item III. Its fundamental nature is that here, as in the complete solution of Expression (46), $\Lambda_{n\backslash\tau}$ takes into account the interaction of effective mechanical energy components due to friction τ_w and normal σ stresses, whereas $\Lambda_{M\backslash T}$ allows for the interaction of thermal and mechanical components of effective energy. The thermal component of the effective energy is determined by varying the total temperature $T_\Sigma = T_2 - T_1$ in the force contact zone induced by all heat sources, including the heat released during mechanical (spatial and surface) strain, structural changes, etc.

5. Mechanothermodynamical States

Within the framework of mechanothermodynamics, a special approach is being developed to assess the entropy in terms of a generalized energy state. Following this approach and Formula (77), the effective part of total energy (specific at some particular loads–force, temperature, etc.) directly spent for the damage production is defined by the experimentally found coefficients A_l in Formulas (41), (42) and (77) [2,51,76].

$$u_l^{eff} = A_l u_l, \tag{84}$$

where the u_l's are the specific internal energies at tear (u_n), shear (u_τ), and thermal action (u_T).

The total specific energy of an elementary volume and a rate of its change are then given as:

$$u = \sum_l \left[(1 - A_l) u_l + u_l^{eff} \right]; \tag{85}$$

$$\frac{du}{dt} = \sum_l \left[(1 - A_l) \frac{du_l}{dt} + \frac{du_l^{eff}}{dt} \right] \tag{86}$$

In addition, the Λ-functions are used to take account of a complex (non-additive) character of interactions between effective energies of different nature, expressed by Formula (42). This allows the total effective energy of the system to be assessed:

$$u_\Sigma^{eff} = \Lambda_\alpha\left(u_l^{eff}\right) = \Lambda_{M\backslash T}\left(\Lambda_{\tau\backslash n}, A_l u_l\right) = \Lambda_{M\backslash T}\left\{\Lambda_{\tau\backslash n}[A_n u_n + A_\tau u_\tau] + A_T u_T\right\}, \tag{87}$$

where the Λ_α's are the possible combinations of interaction of effective energies (irreversible damages).

The specific feature of Λ-functions is such that:

$$u_{\Sigma}^{eff} \gtreqless u_l^{eff} \quad (88)$$

and hence,

$$u_{\Sigma}^{eff} \gtreqless \Sigma u \quad (89)$$

By using coefficients A_l and Λ-functions, the energy interaction at different-nature loads can be found. Such interaction can give rise both to a sharp increase and a substantial decrease in the effective energy, resulting in damages and limiting states, in comparison to the energy calculated by the ordinary additivity model of type (17):

$$u_{\Sigma} = \sum A_l u_l. \quad (90)$$

By taking account of Formula (87), the total effective energy of volume V and its accumulation in time have the form:

$$U_{\Sigma}^{eff} = \int\limits_V \rho u_{\Sigma}^{eff}(V)dV \quad (91)$$

and

$$U_{\Sigma}^{eff}(t) = \int\limits_t \int\limits_V \rho u_{\Sigma}^{eff}(V,t)dVdt. \quad (92)$$

The principal moment of the mechanothermodynamical model is the account of the limiting state (limits of plasticity, strength, fatigue, etc.) according to item XIII (Section 3):

$$u_{\Sigma}^{eff} = u_0, \quad (93)$$

where u_0 is the limiting density of the internal energy treated as the initial activation energy of the disintegration process.

A relationship between the current state (mechanical, thermomechanical, energy) of an elementary volume of a solid (medium) and its limiting state enables one to construct the parameter of local energy damageability: dimensionless:

$$\psi_u^{eff} = \frac{u_{\Sigma}^{eff}}{u_0} \quad (94)$$

or dimensional:

$$\psi_{u*}^{eff} = u_{\Sigma}^{eff} - u_0. \quad (95)$$

Local energy damageability (Equation (94) or (95)) is most general among the damageability parameters constructed in terms of different mechanical (thermomechanical) states φ [2,51,76]:

$$\psi_q = \varphi_q / \varphi_q^{(*\lim)}, \quad (96)$$

where $\phi = \sigma, \varepsilon, u$; the σ's are the stresses, the ε's are the strains, u is the density of internal energy, the $\varphi_q^{(*\lim)}$'s are the limiting values of the state φ $q \in \left\{eqv, ij, i, S, \overset{D}{ij}, n, \tau, \text{int}, u, \overset{n}{u}, \overset{\tau}{u}, \overset{eff}{u}\right\}$, eqv is the equivalent mechanical state, the ij's are the components of the tensor ϕ, the i's are the main components of the tensor ϕ, S and $\overset{D}{ij}$ are the sphere and deviator parts of the tensor ϕ, n and τ are the normal and tangential components of the tensor ϕ, int is the intensity of ϕ, and u is the specific potential strain energy (internal energy density). The indices at u mean: $\overset{n}{u}$ and $\overset{\tau}{u}$ are the specific potential strain energy at tension–compression and shear, and $\overset{eff}{u}$ is the effective specific potential strain energy.

We can build integral damageability measures on the basis of local measures (Equation (96)) using the model of a deformable solid with a dangerous volume (Equations (64)–(68)) [4,76].

The dangerous volume is called the spatial region of a loaded solid. At each point of a solid, the local damageability value is smaller than the limiting one [4,51,76]:

$$V_q = \left\{ dV / \varphi_q \geq \varphi_q^{(*\lim)}, dV \subset V_k \right\}, \tag{97}$$

or

$$V_q = \left\{ dV / \psi_q \geq 1, dV \subset V_k \right\}.$$

Dangerous volumes are calculated by the following general formula:

$$V_q = \iiint\limits_{\psi_q(V) \geq 1} dV. \tag{98}$$

The integral condition of damageability of a solid or a system can be written in the form:

$$0 < \omega_q = \frac{V_q}{V_0} < 1, \tag{99}$$

where V_0 is the working volume of the solid.

To analyze, at a time, dangerous volumes and local damageability distributed within them, we introduce the function of damageability of unit volume:

$$d\Psi_q = \psi_q(V) dV. \tag{100}$$

The function of damageability of the entire volume V will then be as follows:

$$\Psi_q = \int\limits_{\psi_q \geq 1} \psi_q(V) dV. \tag{101}$$

The simplest functions of damageability accumulation in time for unit volume and total volume will be have the following form, respectively:

$$d\Psi_q^{(t)} = \int\limits_t \psi_q(t) dt; \tag{102}$$

$$\Psi_q^{(t)} = \int\limits_{\psi_q \geq 1} \int\limits_t \psi_q(V,t) dt dV. \tag{103}$$

The indices of volume-mean damageability

$$\overline{\Psi}_q^{(V)} = \frac{1}{V_q} \int\limits_{\psi_q \geq 1} \psi_q(V) dV \tag{104}$$

and its accumulation in time can be used

$$\overline{\Psi}_q^{(V,t)} = \frac{1}{V_q} \int\limits_t \int\limits_{\psi_q \geq 1} \psi_q(V,t) dV dt. \tag{105}$$

The analysis of Formulas (94), (100) and (102) leads to the conclusion: conceptually, they are related to the entropy concept as a difference (or relations) between two states (configurations) of a system, the degree of its organization (chaotic state). In relation to damageability, such states are current and limiting.

By using local energy damageability (Equation (94)), we construct specific (per unit mass) tribo-fatigue entropy (accurate constant):

$$s_{TF} = \psi_u^{eff}\left(\Lambda_\alpha, A_l, \sigma_{ij}, T\right) = \lim_{\Delta m -> 0} A_\psi \frac{u_\Sigma^{eff}(\Delta m)}{u_0 \Delta m}, \tag{106}$$

or

$$s_{TF} = s_{TF*} = \frac{\psi_{u*}^{eff}\left(\Lambda_\alpha, A_l, \sigma_{ij}, T\right)}{T} = \frac{u_\Sigma^{eff} - u_0}{T}. \tag{107}$$

where A_ψ is the dimensional parameter (J·mol^{-1}·K^{-1}).

On the basis of Expression (18) for entropy and Formulas (85) and (86), the local entropy and the rate of its change within an elementary volume will be:

$$s = \frac{1}{T}\sum_l [(1 - A_l)u_l] + s_{TF} \mathrm{d}\psi_q = \psi_q(V)dV \tag{108}$$

and

$$\frac{ds}{dt} = \frac{1}{T}\sum_l \left[(1 - A_l)\frac{du_l}{dt}\right] + \frac{ds_{TF}}{dt}. \tag{109}$$

Formulas (108) and (109) show that unlike the thermomechanical model, the state indicators of the mechanothermodynamical system u and s are not equivalent. This is due to the fact that the calculation of the tribo-fatigue entropy s_{TF} by Formula (106) is supplemented by the limiting state in the form of the limiting density of the internal energy u_0.

The tribo-fatigue entropy S_{TF} is calculated not within the total volume V, but only within its damaged part, i.e., within the energy effective dangerous volume V_u^{eff}:

$$V_u^{eff} = \left\{ dV / u_\Sigma^{eff} \geq u_0, dV \subset V_k \right\}. \tag{110}$$

Based on Formulas (11), (106) and (110), the tribo-fatigue entropy of volume V will be:

$$S_{TF} = \int\limits_{u_\Sigma^{eff}(V) \geq u_0} \rho s_{TF}(V)dV = \int\limits_{u_\Sigma^{eff}(V) \geq u_0} \rho \psi_u^{eff}(V)dV, \tag{111}$$

where,

$$\psi_u^{eff}(V) = \frac{u_\Sigma^{eff}(V)}{u_0} \text{ or } \psi_u^{eff}(V) = \frac{\psi_{u*}^{eff}(V)}{T} = \frac{u_\Sigma^{eff}(V) - u_0}{T(V)}, \tag{112}$$

and its accumulation will be:

$$S_{TF}(t) = \int\limits_t \int\limits_{u_\Sigma^{eff}(V,t) \geq u_0} \rho s_{TF}(V,t)dVdt = \int\limits_t \int\limits_{u_\Sigma^{eff}(V,t) \geq u_0} \rho \psi_u^{eff}(V,t)dVdt,. \tag{113}$$

where,

$$\psi_u^{eff}(V,t) = \frac{u_\Sigma^{eff}(V,t)}{u_0} \text{ or } \psi_u^{eff}(V,t) = \frac{\psi_{u*}^{eff}(V,t)}{T(V,t)} = \frac{u_\Sigma^{eff}(V,t) - u_0}{T(V,t)}. \tag{114}$$

We should emphasize the fundamental feature of tribo-fatigue total S_{TF} and specific s_{TF} entropies. So, a difference between two states can be assessed not only quantitatively (as thermomechanical entropy), but also qualitatively, because s_{TF} is calculated through the limiting density of the internal energy u_0. So, s_{TF} and S_{TF} allow us to answer how much the current state of a solid or a system is dangerous in comparison to limiting states.

The total entropy and the rate of its change for a system solid with regard to Equations (111) and (113) assume the form:

$$S = \int\limits_V \frac{1}{T(V)} \sum_l \rho[(1 - A_l(V))u_l(V)]\, dV + S_{TF} \tag{115}$$

and

$$\frac{dS}{dt} = \int\limits_V \frac{1}{T(V)} \sum_l \rho\left[(1 - A_l(V))\frac{du_l(V)}{dt}\right] dV + \frac{dS_{TF}}{dt} \tag{116}$$

Based on Formulas (106)–(116), we can build the function of total entropy accumulation in time:

$$\begin{gathered} S(t) = \int\limits_t \int\limits_V \sum\limits_l \rho s_l(V,t)dVdt + \int\limits_t \int\limits_{u_\Sigma^{eff}(V,t) \geq u_0} \rho s_{TF}(V,t)dVdt = \\ = \int\limits_t \int\limits_V \frac{1}{T(V,t)} \sum\limits_l \rho\left[(1 - A_l(V,t))\frac{du_l(V,t)}{dt}\right] dVdt + \int\limits_t \int\limits_{u_\Sigma^{eff}(V,t) \geq u_0} \rho\psi_u^{eff}(V,t)dVdt. \end{gathered} \tag{117}$$

Practically, bearing in mind the limiting states of a solid or a system, models (115)–(117) can answer whether the current state is a qualitative jump in the system, i.e., whether the current state is close to the limiting (critical fatigue fracture entropy) one. A similar (dialectical as a matter of fact) qualitative transition differs from the bifurcation point in the ability to predict the system behavior after a transition on the basis of the analysis of s_{TF} and S_{TF}. Particular limiting states (limit of strength, mechanical or contact fatigue, etc.) enable for predicting the situation after passing the given point: principal changes in the system properties and behavior or the formation of a new system based on the previous one.

An example can be non-linear deformation or generation of microcracks in the solid (or the system) that changes its strength and fatigue properties, and hence, its response to loads. In turn, formed macrocracks lead to local continuum violation—formation of new free surfaces (possibly, of new solids—destruction products), i.e., a new system.

It should be noted that models (115)–(117) were built using a traditional concept of entropy additivity (Equation (10)), although with the consideration of significant refinements. These models also contain reversible processes described by the entropy components s_l, not yielding primary damages, and hence, the limiting states: the points of qualitative change in the system.

The assessment of the entropy state on the basis of the mechanothermodynamical model of a solid, which uses only tribo-fatigue entropy, is more advisable for a qualitative and quantitative analysis of evolution of systems passing through the states traditionally defined as bifurcation branches. In this case, Formulas (111)–(113) for entropy and their accumulation will be of the form:

$$S = S_{TF} = \int\limits_{u_\Sigma^{eff}(V,t) \geq u_0} \rho s_{TF}(V)dV = \int\limits_{u_\Sigma^{eff}(V,t) \geq u_0} \rho\psi_u^{eff}(V)dV, \tag{118}$$

and

$$S(t) = S_{TF}(t) = \int\limits_t \int\limits_{u_\Sigma^{eff}(V,t) \geq u_0} \rho s_{TF}(V)dVdt = \int\limits_t \int\limits_{u_\Sigma^{eff}(V,t) \geq u_0} \rho\psi_u^{eff}(V,t)dVdt. \tag{119}$$

To identify the points of qualitative change in the limiting states of solids (systems), we can use the indices of relative integral entropy and its accumulation using the concept of the integral condition of solid damageability (Equation (99)):

$$\omega_S = \frac{S_{TF}}{V_0} = \frac{1}{V_0} \int\limits_{u_\Sigma^{eff}(V,t) \geq u_0} \rho s_{TF}(V)dV; \tag{120}$$

$$\omega_S(t) = \frac{S_{TF}(t)}{V_0} = \frac{1}{V_0}\int_t \int_{u_\Sigma^{eff}(V,t)\geq u_0} \rho s_{TF}(V) dV dt. \tag{121}$$

The values of S_{TF}, S_{TF} (t), ω_S, $\omega_S(t)$ can grow infinitely, allowing for not only describing the limiting states of type (93), but also different transmitting states. In essence, they "provide" a quantitative description of the entropy increase.

Now, based on Formulas (24), (115), (117) and (119), we construct generalized expressions for entropy, a rate of its change, and its accumulation in the MTD system consisting of a liquid (gas) medium of volume V and a solid of volume V_ψ:

$$\begin{aligned} S = &\int_V \rho s_T dV + \int_{V_\psi} \sum_l \rho s_l dV_\psi + \int_{u_\Sigma^{eff}\geq u_0} \rho s_{TF} dV_\psi = \int_V \tfrac{1}{T}\sigma_{ij}\varepsilon_{ij} dV + \int_V \tfrac{1}{T}\rho q dV + \\ &+ \int_V \tfrac{1}{T}\rho \sum_k \mu_k n_k \, dV + \int_{V_\psi} \tfrac{1}{T}\sum_k \rho[(1-a_k)u_k] dV_\psi + \int_{u_\Sigma^{eff}\geq u_0} \rho \psi_u^{eff} dV_\psi; \end{aligned} \tag{122}$$

$$\begin{aligned} \tfrac{dS}{dt} = &\int_V \rho \tfrac{ds_T}{dt} dV + \int_{V_\psi} \sum_l \rho \tfrac{ds_l}{dt} dV_\psi + \int_{V_\psi} \rho \tfrac{ds_{TF}}{dt} dV_\psi = \int_V \tfrac{1}{T}\sigma_{ij}\tfrac{d\varepsilon_{ij}}{dt} dV + \\ &+ \int_V \tfrac{1}{T}\rho \tfrac{dq}{dt} dV + \int_V \tfrac{1}{T}\rho \sum_k \mu_k \tfrac{dn_k}{dt} \, dV + \\ &+ \int_{V_\psi} \tfrac{1}{T}\sum_k \rho\Big[(1-a_k)\tfrac{du_k}{dt}\Big] dV_\psi + \int_{u_\Sigma^{eff}\geq u_0} \rho \tfrac{d\psi_u^{eff}}{dt} dV_\psi; \end{aligned} \tag{123}$$

$$\begin{aligned} S(t) = &\int_t \int_V \rho s_T dV \, dt + \int_t \int_{V_\psi} \sum_l \rho s_l dV_\psi dt + \int_t \int_{u_\Sigma^{eff}\geq u_0} \rho s_{TF} dV_\psi dt = \int_t \int_V \tfrac{1}{T}\sigma_{ij}\varepsilon_{ij} dV dt + \\ &+ \int_t \int_V \tfrac{1}{T}\rho q dV dt + \int_t \int_V \tfrac{1}{T}\rho \sum_k \mu_k n_k \, dV dt + \int_t \int_{V\psi} \tfrac{1}{T}\sum_l \rho[(1-a_l)u_l] dV_\psi dt + \\ &+ \int_t \int_{u_\Sigma^{eff}\geq u_0} \rho \psi_u^{eff} dV_\psi dt \end{aligned} \tag{124}$$

Similarly, we can build entropy state values for a system consisting of many media.

It should be noted that in Formulas (122)–(125), the interaction (contact) of two media, which can be complex in nature, is taken into account only implicitly in terms of medium state parameters (stress, strain, temperature). It is obvious that this is only the first step to a comprehensive (generalized) solution of the problem stated.

The simplified writing of Expression (123) for the entropy increment of the mechanothermodynamical system consisting of finite volumes dV and dV_ψ was presented in Reference [51] as follows:

$$dS = (dS)_T + (d_i S)_{TF} = \frac{dU + \Delta p dV}{T} - \frac{1}{T}\sum_k \mu_k dN_k + \Psi_u^{eff} dV_\psi. \tag{125}$$

Expression (125) can also be presented in terms of specific quantities as:

$$dS = \int_V \frac{\rho du + \rho dp}{T} dV - \int_V \frac{1}{T}\rho \sum_k \mu_k dn_k \, dV + \int_{u_\Sigma^{eff}\geq u_0} \rho d\psi_u^{eff} dV_\psi \tag{126}$$

or on the basis of Expression (123):

$$\frac{dS}{dt} = \int_V \frac{\sigma_{ij} d\varepsilon_{ij} + \rho dq}{T dt} dV - \int_V \frac{1}{T}\rho \sum_k \mu_k \frac{dn_k}{dt} \, dV + \int_{u_\Sigma^{eff}\geq u_0} \rho \frac{d\psi_u^{eff}}{dt} dV_\psi. \tag{127}$$

In Formulas (111)–(113) for calculation of the tribo-fatigue entropy S_{TF} and its accumulation $S_{TF}(t)$, the specific entropy s_{TF} is assumed to be integrated in terms of the damageable region of the solid alone—the dangerous volume. However, the influence of undamageable regions can also be allowed for by integrating S_{TF} within the total volume:

$$S_{TF} = \int_V \rho s_{TF}(V)dV = \int_V \rho\psi_u^{eff}(V)dV; \tag{128}$$

$$S_{TF}(t) = \int_t \int_V \rho s_{TF}(V,t)dVdt = \int_t \int_V \rho\psi_u^{eff}(V,t)dVdt, \tag{129}$$

where,

$$\psi_u^{eff} = \begin{cases} \frac{u_\Sigma^{eff}(V,t)}{u_0} \geq 1, \text{ if } u_\Sigma^{eff} \geq u_0; \\ \frac{u_\Sigma^{eff}(V,t)}{u_0} < 1, \text{ if } u_\Sigma^{eff} < u_0, \end{cases} \tag{130}$$

or

$$\psi_u^{eff} = \frac{\psi_{u*}^{eff}(V,t)}{T(V,t)} = \begin{cases} \frac{u_\Sigma^{eff}(V,t)-u_0}{T(V,t)} \geq 0, \text{ if } u_\Sigma^{eff} \geq u_0; \\ \frac{u_\Sigma^{eff}(V,t)-u_0}{T(V,t)} < 0, \text{ if } u_\Sigma^{eff} < u_0. \end{cases} \tag{131}$$

Expression (131) shows that $\psi_u^{eff} < 0$ is observed outside the dangerous volume (at $u_\Sigma^{eff} < u_0$). This means that the specific tribo-fatigue entropy s_{TF} also appears to be negative (or less than unity for its alternative definition) outside the dangerous volume where the limiting state is not reached. Negative values of ψ_u^{eff} and s_{TF} can then be interpreted as the case where damageability is absent. In other words, the structure and/or properties of the solid are preserved.

The foregoing reports that the entropy additivity assumption is wrong in the general case for a system, consisting of a solid and a liquid (gas), where chemical reactions can occur. By analogy with the Λ-functions of interaction of different energies (Equation (179)), the functions of interaction of different entropies must be introduced by adding them to Expression (125) to determine total effective entropy:

$$\begin{aligned} dS_{total}^{eff} &= \Lambda_{T\backslash TF}^{(S)}(dS_T + d_iS_{TF}) = \Lambda_{T\backslash TF}^{(S)}\left[\Lambda_{Q\backslash Ch}^{(S)}\left(dS_T^Q + dS_{Ch}^Q\right) + d_iS_{TF}\right] = \\ &= \Lambda_{T\backslash TF}^{(S)}\left[\Lambda_{Q\backslash Ch}^{(S)}\left(\frac{dU+\Delta pdV}{T} - \frac{1}{T}\sum_k \mu_k dN_k\right) + \Psi_u^{eff}dV_\psi\right], \end{aligned} \tag{132}$$

or

$$\begin{aligned} dS_{total}^{eff} &= \Lambda_{T\backslash TF\backslash Ch}^{(S)}(dS_T + d_iS_{TF}) = \\ &= \Lambda_{T\backslash TF\backslash Ch}^{(S)}\left[\frac{dU+\Delta pdV}{T} - \frac{1}{T}\sum_k \mu_k dN_k + \Psi_u^{eff}dV_\psi\right], \end{aligned} \tag{133}$$

where the subscripts Q and Ch denote the thermodynamic and chemical entropy components.

Formulas (132)–(133) are supplemented by the generalized interaction functions $\Lambda_{T\backslash TF}^{(S)}$, $\Lambda_{Q\backslash Ch}^{(S)}$, and $\Lambda_{T\backslash TF\backslash Ch}^{(S)}$ in MTD systems. This means that the hypothesis about the thermodynamic and tribo-fatigue entropy additivity is not accepted. The corresponding interaction Λ-functions must be concretized and introduced into Equations (132)–(133).

6. Entropy Calculation under Simultaneous Contact and Non-Contact Loading

Consider the example of entropy calculation for the tribo-fatigue system consisting of friction pair with the elliptic contact of the ratio between smaller b and bigger a semi-axes $b/a = 0.574$. One of the elements of the friction pair is loaded by non-contact bending. An example of such an element is the shaft in the roller/shaft tribo-fatigue system.

In the case of the contact interaction over the elliptical area, the pressure is expressed as:

$$p^{(n)}(x,y) = p_0^{(c)}\sqrt{(1-x^2/a^2-x^2/b^2)},$$

where $p_0^{(c)}$ is the maximum contact stress under the action of force F_c.

The entropy calculation system was based on the following initial data:

$$\begin{aligned} p_0^{(c)} &= \sigma_{zz}^{(n)}(F_c)\Big|_{x=0,y=0,z=0} = 2960\text{ MPa},\\ p_0^{(c,\lim)} &= p_0\left(F_c^{(\lim)}\right) = 888\text{ MPa} = 0.3p_0^{(c)} \end{aligned} \tag{134}$$

where $p_0^{(c,\lim)}$ is the contact fatigue limit (maximum contact stress under the action of the limiting force $F_c^{(\lim)}$ obtained in the course of mechano-rolling fatigue tests described in References [1–3]. The criterion of the limiting state in these tests was the limiting approach of the axes in the tribo-fatigue system (100 μm). The test base was equal to $3 \cdot 10^7$ cycles.

Calculations of the three-dimensional stress-strain state in the neighborhood of the elliptic contact for b/a = 0.574 [4] show that the maximum value of the strain energy u is related to the maximum contact pressure $p_0^{(c)}$ in the following way:

$$u = \max_{dV}[u(F_c,\, dV)] = 0.47p_0^{(c)}. \tag{135}$$

The limiting value of the strain energy $u^{(\lim)}$ under the action of the limiting force $F_c^{(\lim)}$ is:

$$u^{(\lim)} = \max_{dV}\left[u\left(F_c^{(\lim)},\, dV\right)\right] = 0.47p_0^{(c,\lim)}. \tag{136}$$

In the calculations performed, maximum stresses σ_a due to non-contact bending in the contact area were the following:

$$-0.34 \le \sigma_a/p_0^{(c)} \le 0.34.$$

Tangential surface forces (friction force is directed along the major semi-axis of the contact ellipse) are:

$$p^{(\tau)}(x,y) = -fp^{(n)}(x,y) = -fp_0^{(c)}\sqrt{(1-x^2/a^2-x^2/b^2)}.$$

The specific entropy distribution calculated according to Equation (112) shown in Figures 4–7 can be considered to be the characteristic of the probability of appearance of local damages (initial cracks). The higher the specific entropy at a point of a dangerous volume, the greater the probability of initiation of damage (crack) at this point. The values of dangerous volume and entropy are the integral damageability indices (including a possible number of cracks and their sizes) of a solid or a system.

From Figures 4–7 for $p_0 = p_0^{(c)}$ and the friction coefficient f = 0.2, the maximum specific entropy is in the center of the contact area.

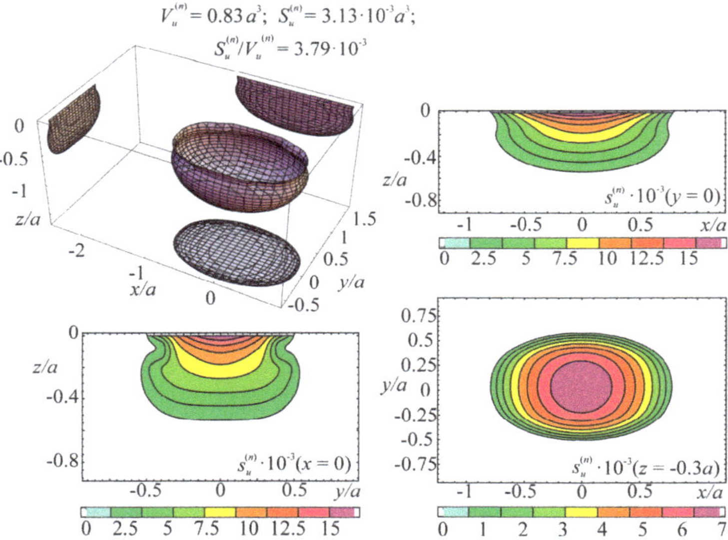

Figure 4. Energy dangerous volume and its sections, with specific entropy distributions for contact interaction without friction.

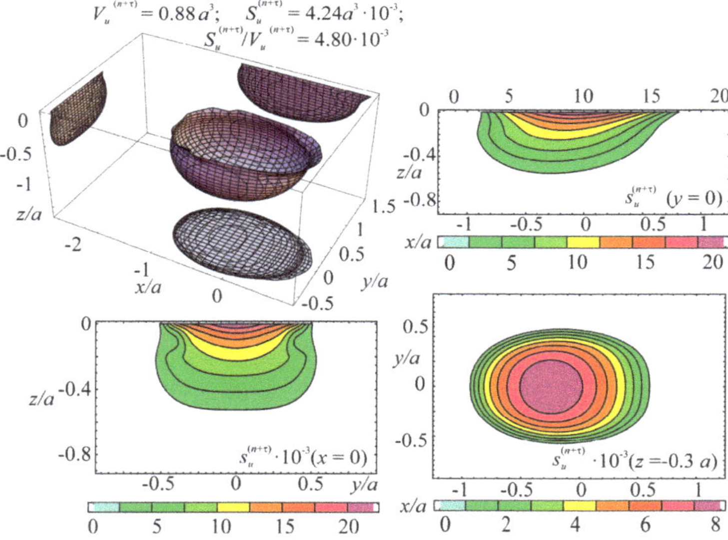

Figure 5. Energy dangerous volume and its sections, with specific entropy distributions for contact interaction with friction.

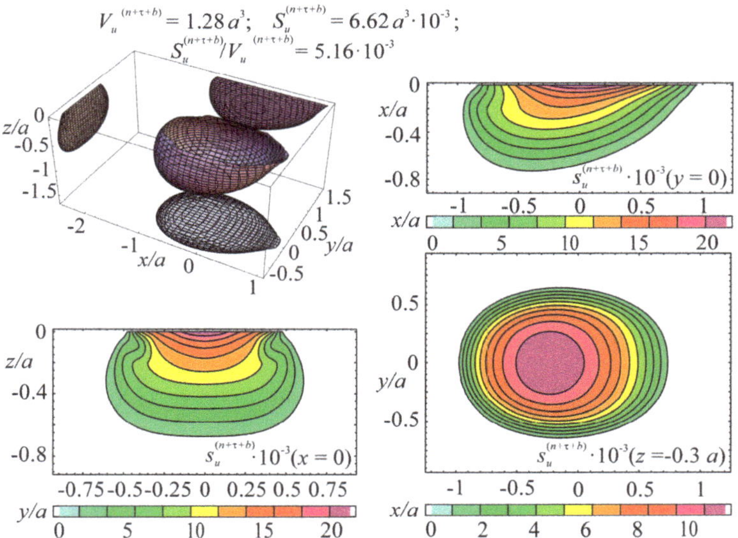

Figure 6. Energy dangerous volume and its sections, with specific entropy distributions for contact interaction with friction and tensile stresses $\sigma_a/p_0^{(c)} = 0.34$ in the contact area caused by non-contact bending.

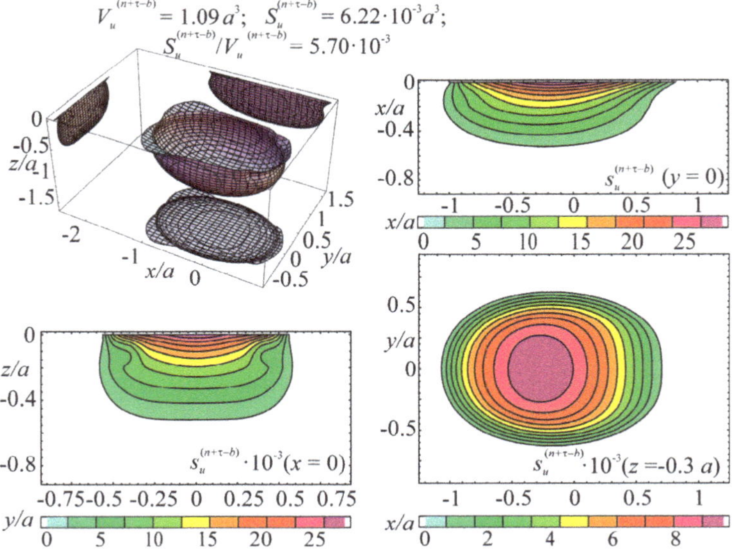

Figure 7. Energy dangerous volume and its sections, with specific entropy distributions for contact interaction with friction and compressive stresses $\sigma_a/p_0^{(c)} = -0.34$ in the contact area caused by non-contact bending.

Under the joint action of contact pressure and tangential surface forces (friction) $s_u^{(n+\tau)}$, the maximum specific entropy increases by about 30% in comparison with the maximum specific entropy $s_u^{(n)}$. The joint action of contact pressure, friction, and tension due to bending increases $s_u^{(n+\tau+b)}$ by about 30% in comparison with $s_u^{(n)}$. At compressive bending, $s_U^{(n+\tau-b)}$ increases by about 60% in comparison with $s_u^{(n)}$.

In case of frictional contact, the values of the dangerous volumes $V_u^{(n+\tau)}$, the entropy $S_u^{(n+\tau)}$, and the average entropy $S_u^{(n+\tau)}/V_u^{(n+\tau)}$, increase by about 6%, 35%, and 27%, as compared to $V_u^{(n)}$, $S_u^{(n)}$, and $S_u^{(n)}/V_u^{(n)}$, respectively.

A more detailed analysis of the considered effects might be done using Figure 8. It shows a significant growth of entropy with increasing contact pressure, friction coefficient, and stresses caused by non-contact loads. The entropy increases almost at the same level for the same absolute values of tensile and compressive non-contact stresses. This effect may be due to the fact that the energy u attains positive values.

The main conclusion of Figures 4–8 is that not only friction, but also non-contact forces significantly change entropy characteristics in the neighborhood of the contact area.

Note that according to Expressions (95), (97) and (112), calculations were performed for the simplest case when the energy applied to the system is fully absorbed. Similar calculations may be done for effective energies u^{eff}, determined by Expression (87).

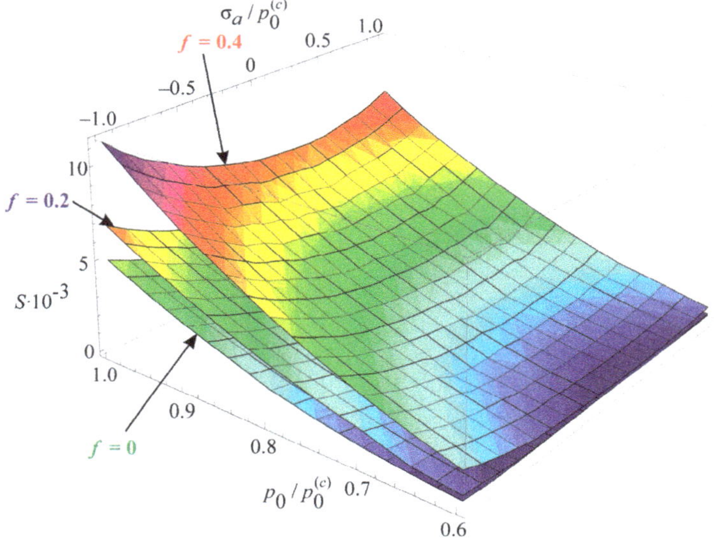

Figure 8. Entropy versus contact and non-contact stresses.

7. Translimiting States

The available information reports that the theory of translimiting states is still insufficiently developed [2]. Its elements will be set forth on the basis of solutions (72), (76) and (77).

Figure 9 analyzes the contribution of mechanical–chemical–thermal damage (parameters D) to reaching the limit state by the MTD system. Having analyzed Formulas (72), (76), and Figure 9, we concluded the following.

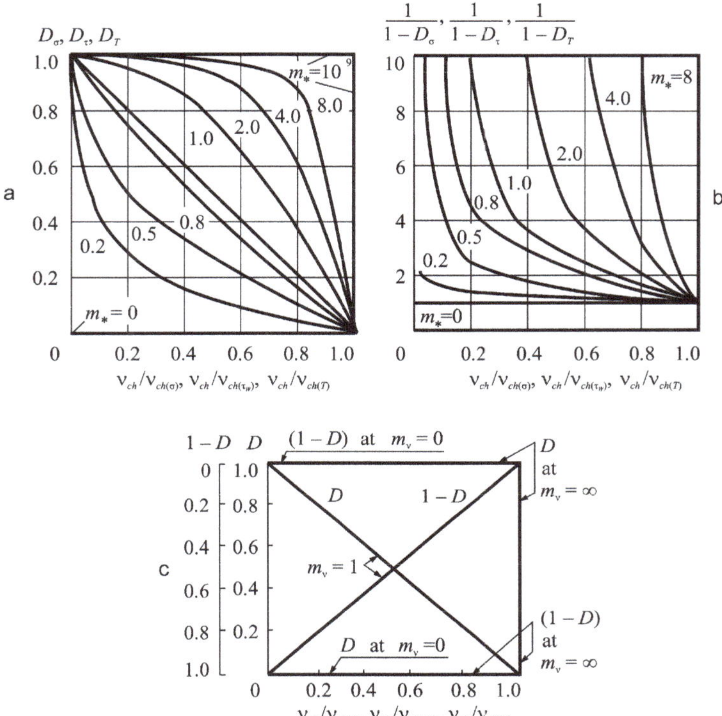

Figure 9. Effect of mechanical–chemical–thermal processes on the damage of a system (**a**) the dependence of parameters D on relative damage rate $v_{ch}/v_{ch(*)}$; (**b**)influence of parameter m_v; (**c**) specific cases analysis for D = 0,1.

1. The growth of parameters D means that the relative damage speed $v_{ch}/v_{ch(*)}$ decreases (Figure 9a). Mechanical–chemical–thermal damage speeds up the process of reaching the limiting state by the MTD system. It is faster for the greater magnitude of D parameter and/or speed $v_{ch(*)}$.

2. Parameter m_v greatly affects the system damage. The greater its effect, the larger this parameter is (Figure 9b). The MTD system is sensitive to mechanical load and temperature increase if electrochemical activity parameter $m_v > 5$. In this case, the translimiting state may occur. For a state, the damageability measure (Equation (53)) becomes greater than unity ($\psi_u^{eff} > 1$), while $\psi_u^{eff} = 1$ in Equation (52) is enough to obtain the limiting state.

The first specific case in Figure 9c is $D = 0$. Electrochemical corrosion does not affect wear-fatigue damage. However electrochemical corrosion may happen. According to Formula (76), when $D = 0$ for $m_v = 1$ we obtain:

$$1 - \frac{v_{ch}}{v_{ch(*)}} b_* = 0$$

Hence, the situation must be the following: $b_* = 1$ and $v_{ch}/v_{ch(*)} = 1$. In this case, the corrosion speed is not influenced by mechanical or frictional stresses. So, there are threshold values of σ^0, τ_w^0, and T^0 for a considered environment. The speed of corrosion for this environment according to Equation (77) stays the same at $\sigma \leq \sigma^0$, $\tau_w \leq \tau_w^0$, and $T_\Sigma \leq T^0$.

The second case is for $D = 1$, and hence, for $1/(1–D) \rightarrow \infty$. Damage of explosive type happens in a system if $\psi_u^{eff} \rightarrow \infty$. In this case, it should be:

$$\frac{v_{ch}}{v_{ch(*)}}b_* = 0$$

In case $v_{ch} = 0$ is an impossible event, then it may be assumed that $v_{ch(*)} \rightarrow \infty$. This is the condition for mechanical–chemical–thermal explosive event occurrence in a MTD system. This event is not just due to the environmental impact that is catastrophically increased by mechanical and temperature stress.

The damageability function of the MTD system (Equation (72)) can also be applied to the analysis of the system translimiting states. It can be done because of the possibility to take into account supercritical growth of frictional, mechanical, thermodynamic, and electrochemical loading by Equations (73)–(76), i.e.,

$$1 \leq \psi_u^{eff} = \Lambda_{T\backslash M}\left[\psi_{T(ch)} + \Lambda_{n\backslash\tau}\left(\psi_{n(ch)} + \psi_{\tau(ch)}\right)\right] \leq \infty \tag{137}$$

According to Equation (137), many translimiting states could be described by the $\psi_u^{eff} > 1$ condition. It may happen if the system limiting by damageability state occurs not only at one but at many points (elementary volumes) that constitute a dangerous volume. It could be assumed that there must exist many different types of such states.

Although the above criterion Equations (43), (47), (52), (58), (72) and (77) are constructed for the analysis of energy limiting state conditions, they could also be applied to the description of different translimiting states under supercritical loads (at fires, disasters, accidents, explosions, etc.).

A different general way to analyze the translimiting states uses a damage space defined by volume damageability measures according to Equations (59) and (64):

$$0 \leq \omega_{ij} = \frac{V_{ij}}{V_0} \leq 1 \tag{138}$$

On the basis of Equations (72)–(76), volume (space) damageability measures can be defined as:

$$\begin{aligned} \omega_{\sigma(ch)} &= \frac{V_{P\gamma}}{V_0(1-D_\sigma)} \\ \omega_{\tau(ch)} &= \frac{S_{P\gamma}}{S_0(1-D_\tau)} \\ \omega_{T(ch)} &= \frac{V_{T\gamma}}{V_0(1-D_T)} \end{aligned} \tag{139}$$

where, V_0, S_k are the working volumes. Criterion (77) can then be written with regard to (139):

$$\Lambda_{T\backslash M}\left[\frac{V_{T\gamma}}{V_0(1-D_T)} + \Lambda_{\sigma\backslash\tau}\left(\frac{V_{P\gamma}}{V_0(1-D_\sigma)} + \frac{S_{P\gamma}}{S_0(1-D_\tau)}\right)\right] = 1 \tag{140}$$

The advantage of Equation (140) is the following. Here, the interaction of dangerous volumes [2,4] at different loads when forming the limiting state of MTD systems is taken into account. Also, dangerous volumes are influenced by different metallurgical, technological, and structural factors as it is shown in Equation (59).

If interatomic bond ruptures are analyzed only at a dangerous section of a body at all its points (elementary surfaces) ($u_\Sigma^{eff} = u_0$), then it divides into two parts corresponding to $\omega_\Sigma = 1$, but if loads (mechanical, electrochemical, thermodynamic, etc.) are combined in such a way that "all" interatomic bonds undergo rupture over this section, then there occurs the process called the object disintegration. It corresponds to $\omega_\Sigma^* = \infty$:

$$1 \leq \omega_\Sigma^* = \Lambda_{T\backslash M}\left[\left(\omega_{\sigma(ch)} + \omega_{\tau(ch)}\right)\Lambda_{\sigma\backslash\tau} + \omega_{T(ch)}\right] \leq \infty \tag{141}$$

Naturally, Equation (141) is similar to (137). Their difference lies in the fact that condition (137) is formulated as energy damageability measures while condition (141) is formulated as volume (space) damageability measures.

Table 1 contains a classification of object states by volume damageability.

Table 1. Characteristics of the states of objects.

A-state	**Undamaged**	$\omega_\Sigma = 0$	
B-state	Damaged	$0 < \omega_\Sigma < 1$	*A*-evolution: characteristic states of a system (damage) ↓
C-state	Critical (limiting)	$\omega_\Sigma = 1 = \omega_c$	
D-state	Supercritical (translimiting)	$1 < \omega^*_\Sigma < \infty$	
E-state	Disintegration	$\omega^*_\Sigma = \infty$	

Irreversible damageability events in the MTD system can be interpreted using the failure probability.

If

$$0 \le P(\omega_\Sigma) \le 1 \tag{142}$$

is the traditional probability of failure by damageability ($0 \le \psi_\Sigma \le 1$) within the time interval (t_0, $T_\oplus$) (item XIV), then $P(\omega_\Sigma = \omega_c = 1) = 1$ is the reliable probability of unconditional functional failure. In case of supercritical states, the concept of reliable probabilities [79] can be formulated (see Figure 10):

$$1 < P_*\left(\omega^*_\Sigma\right) \le \infty \tag{143}$$

These supercritical damages $1 < \omega^*_\Sigma < \infty$ are consistent with numerous and innumerable shapes and sizes of particles forming during the system degradation (disintegration).

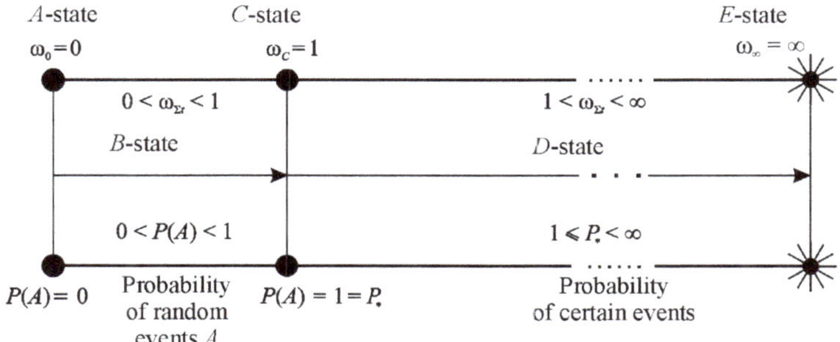

Figure 10. Connection between system damageability and event probability.

Data in Table 1 can be interpreted in the following way. If

$$\omega^*_\Sigma \to \infty. \tag{144}$$

then forming particles should have absolute size, according to Equation (32):

$$d^*_\omega \to 0. \tag{145}$$

To a first approximation, we assume a logarithmic relationship between d_ω and ω_Σ. Then,

$$d^*_\omega = e^{-\omega^*_\Sigma} \text{ or } \omega^*_\Sigma = -\ln d^*_\omega \tag{146}$$

As follows from the foregoing, all MTD system states (see Figure 11) caused by both continuous and discontinuous change of governing parameters are predicted by corresponding Equations (137) and/or (141). The law of MTD system decomposition (decay) can be formulated the in the following way:

$$\sum m_{V_{ijT}} = m_{V_0}. \tag{147}$$

Law (147) implies the conservation of mass of the system regardless of the conditions of its degradation and disintegration. The mass of disintegrated parts (particles) $\sum m_{V_{ijT}}$ (independently of their size) cannot exceed the initial system mass m_{V_0}.

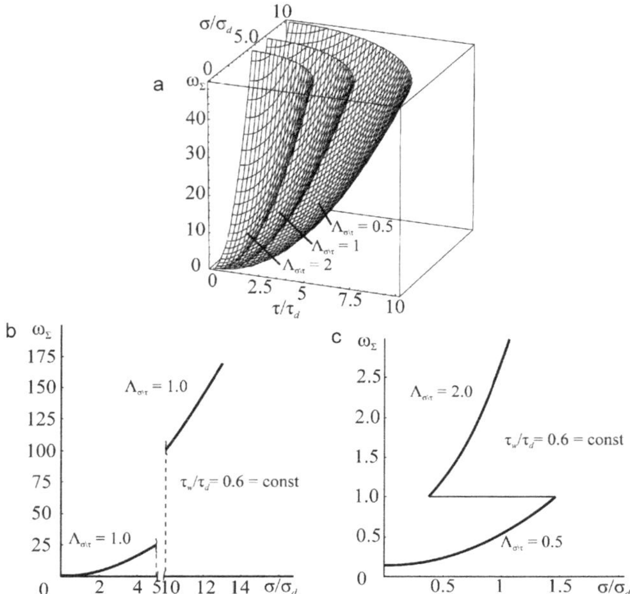

Figure 11. The surfaces of (**a**) damageability, (**b**) ω_Σ and (**c**) determined by parameters $\tau/\tau_d > 0$, $\sigma/\sigma_d > 0$, $\Lambda_{\sigma\backslash\tau} > 0$.

8. Analysis and Generalization of Experimental Data

It is extremely difficult to experimentally verify generalized criterion (72) of the MTD system limiting state due to the lack of experimental data. Below, we consider some particular cases of criterion (77) in the form of (78).

Let us obtain some applied formulas basing on criterion (78). Conditions of purely thermal at $\sigma = 0$ and $\tau_w = 0$ or purely mechanical damage at $T_\Sigma \to 0$ are the following:

$$a_T T_\Sigma = u_0; \tag{148}$$

$$\Lambda_{n\backslash\tau}\left(a_n\sigma^2 + a_\tau\tau_w^2\right) = u_0. \tag{149}$$

Isothermal mechanical fatigue at $\tau_w = 0$ could be described by:

$$\Lambda_{M\backslash T}\left(a_T T_\Sigma + a_n\sigma^2\right) = u_0, \tag{150}$$

and isothermal frictional fatigue at $\sigma = 0$:

$$\Lambda_{M\backslash T}\left(a_T T_\Sigma + a_\tau\tau_w^2\right) = u_0. \tag{151}$$

The analysis of these specific criteria drives us to the following conclusions.

(1) Increase of load parameters (σ, τ_w, T_Σ, D) yields the corresponding acceleration of reaching the limiting state (u_0).

(2) System limiting state can also be reached by increasing only one (any) of the load parameters (when the values of other parameters are invariable).

(3) If $\Lambda > 1$, the system damageability increases (i.e., the processes of its softening are dominant). If $\Lambda < 1$, damageability decreases (i.e., the processes of system hardening appear are dominant) in comparison to the damageability due to only a collective action of load parameters (when the dialectic interaction of irreversible damages is not allowed for).

The last conclusion also results from a fundamentally new approach to constructing the criterion of the limiting state of MTD systems [80]. According to this approach, not the mutual influence of the factors, but the interaction ($\Lambda \gtreqless 1$) of phenomena, is responsible for damageability processes in the MTD system [1,45–52,80]. In this regard, we synthesized the results of more than 600 diverse experimental data. This permitted the generalized MTD function of critical damageability states to be revealed.

We turn to a special case of criterion (78)—isothermal mechanical fatigue. From Equation (150) we have:

$$\log \sigma_{-1T} = \frac{1}{2}\log C_T; \qquad C_T = \left[u_0/\Lambda_{M\backslash T} - a_T T_\Sigma\right]\cdot\frac{1}{a_n} \tag{152}$$

Figure 12 convincingly confirms the dependence (Equation (152)) of σ_{-1T} on the parameter of thermomechanical resistance C_T for numerous steels of different grade tested for fatigue at different conditions [78,81,82]. The C_T magnitude changes by a factor of 100 or more and the value of fatigue limit σ_{-1T} by a factor of 10 or more. Testing temperature was thus varied from the helium temperature to 0.8 T_s (T_s is the melting point). As shown in Figure 12, Equation (152) adequately describes the results of more than 150 experiments.

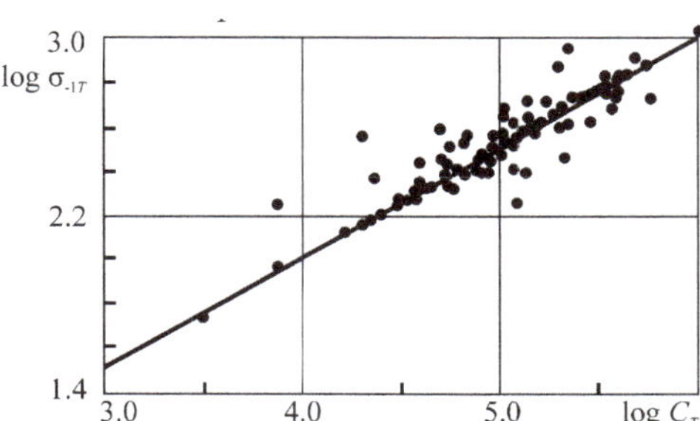

Figure 12. Fatigue limits of structural steels versus parameter of thermomechanical resistance C_T.

Equation (152) was also checked for different metals according to the results of fatigue test carried out by different authors (Figure 13a). In References [78,82], it is possible to find the list of references.

Figure 13b, analyzes the results of tensile tests under different temperatures (σ_{uT}—the strength limit). In Equation (152), $\sigma_{-1} = \sigma_{uT}$. The correlation coefficient is obviously very high even for the rare cases: $r = 0.722$. In most cases, the coefficient exceeds $r = 0.9$ for more than 300 test results that were analyzed. References [78,82] also contain other examples of successful experimental verification of criterion (152). We can hope that even more general criteria given by Equations (77) and (78) will be acceptable in applications.

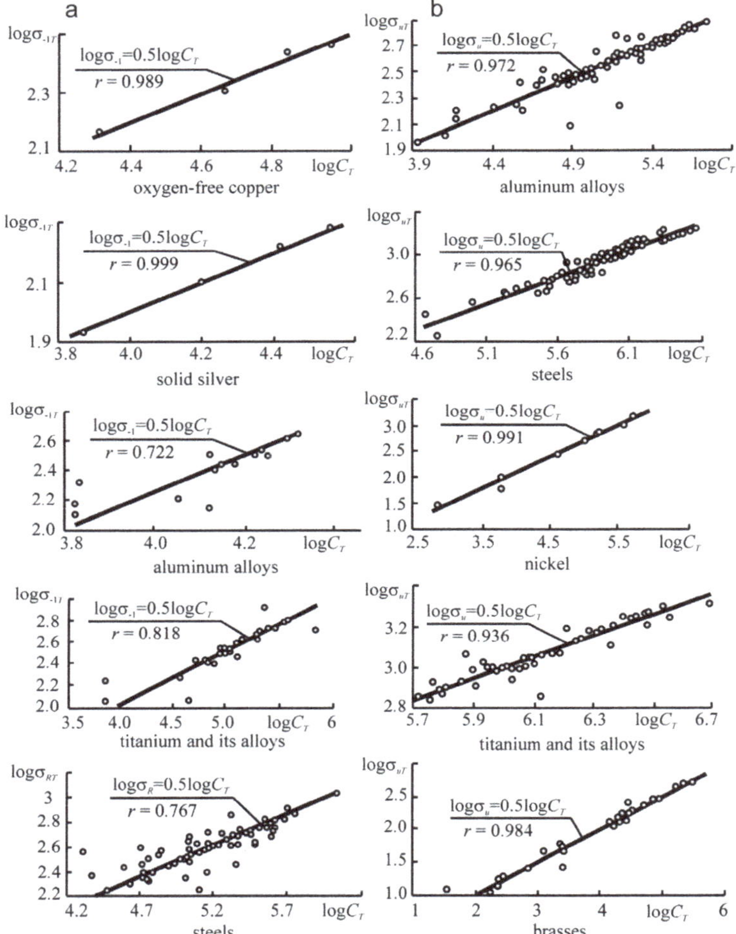

Figure 13. Dependences (**a**) σ_{-1} (C_T) and (**b**) σ_u (C_T) for different metals.

As said above, criterion Equation (149) is valid for $\sigma \leq \sigma_u$. For specific testing, the conditions τ_W can be treated as the largest contact pressure (p_0) at the contact zone center under rolling. It can also be treated as the sliding stress τ_w or as the nominal (average) pressure p_a at the contact area under sliding, or as the pressure (q) at fretting. If $\sigma = \sigma_{-1}$ is fixed, where $\sigma_{-1} << \sigma_u$, then Equation (28) can be presented in the form of the diagram of the limiting states of tribo-fatigue systems [2,81,82] (Figure 14).

Criterion Equation (149) clearly distinguishes the zones of realization of spontaneous hardening–softening processes (interaction function $\Lambda \gtreqless 1$). Figure 14 yields the above obvious conclusions: if $\Lambda < 1$, then the self-hardening system (during tests or during operation at these conditions) is considered. If $\Lambda > 1$, then the system turns to be self-softening. If $\Lambda < 1$ is found to convert into $\Lambda > 1$, then it implies that because of changing the determining operation conditions, hardening processes are replaced by softening ones.

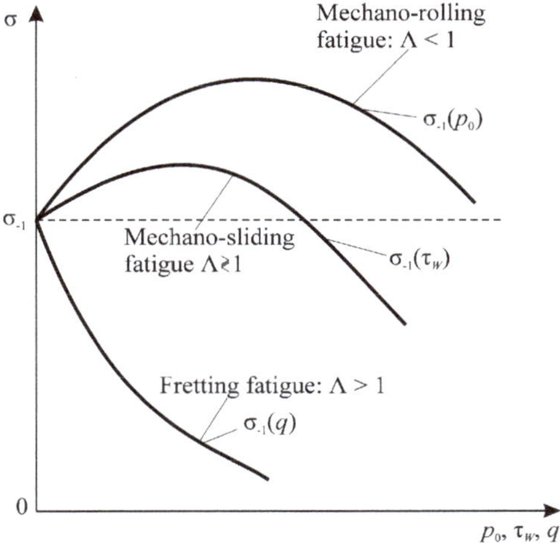

Figure 14. Main features of Λ-interactions in the tribo-fatigue system.

Figures 15–17, illustrate the additional experimental verification of these conditions. Note that for spontaneous hardening (for $\Lambda < 1$, Figures 14–16), the stress limit in wear-fatigue tests is higher than in routine fatigue tests. In these conditions, the friction and wear processes become "useful". Numerous works (see Reference [83]) illustrate that dosed wear in real tribo-fatigue systems (wheel/rail) causes an appropriate growth of their fatigue strength. When $\Lambda >> 1$ (Figure 14), they lead to a strong damageability growth: the fatigue limit decreases with increasing contact pressure q by a factor of 2 ... 3. In addition, there are many works (see Reference [84]), showing that the system wear suddenly decreases the fatigue strength.

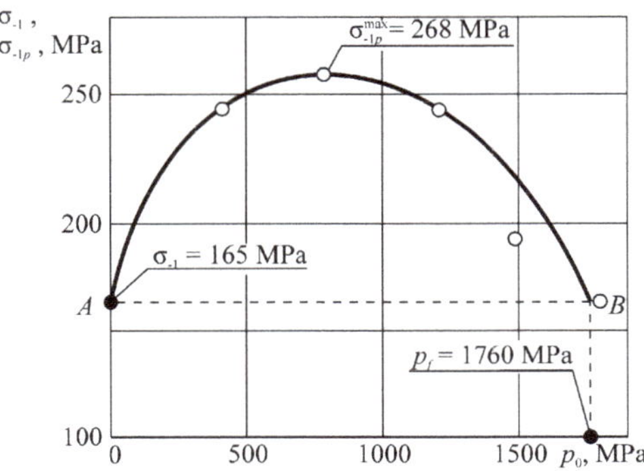

Figure 15. Influence of rolling friction on the resistance to mechano-rolling fatigue in the tests of the tribo-fatigue steel 45 (shaft)/steel 25 HGT (roller) system.

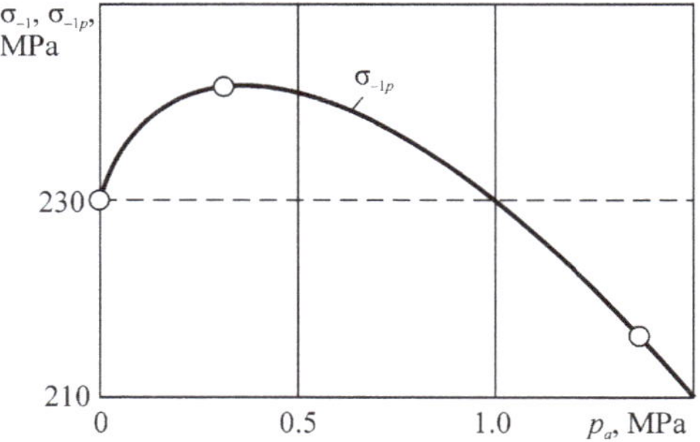

Figure 16. Limiting stresses versus the contact pressure for the tribo-fatigue steel 45 (shaft)/cast iron (partial bearing insert) system.

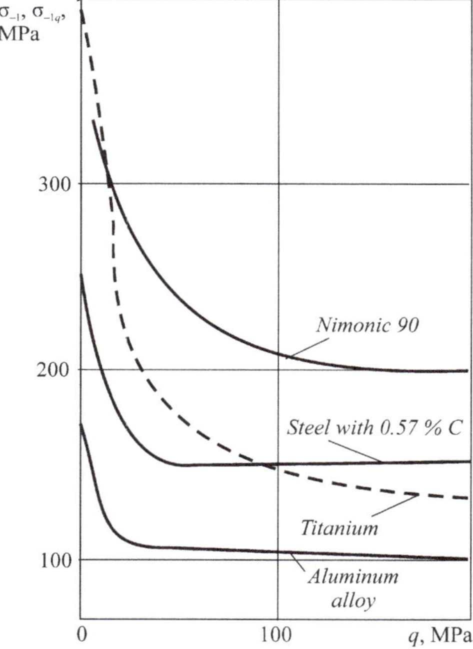

Figure 17. Contact pressure versus the fatigue limit under fretting fatigue.

Tables 2 and 3 summarize different physical signs (often encountered in practice) of the limiting state that can find use in relevant research areas.

As for the determination of the parameters $\Lambda_{M\backslash T}$ and $\Lambda_{n\backslash\tau}$, References [2,78] show that the parameter $\Lambda_{n\backslash\tau}$ is the function of the relative skewness coefficient of wear-fatigue damage:

$$\overline{\rho}_{n\backslash\tau} = \left(\frac{\tau_w}{\tau_f}\right)^2\left(\frac{\sigma_{-1}}{\sigma}\right)^2 \quad (153)$$

Hence $\overline{\rho}_{n\backslash\tau}$ depends not only on absolute values of effective (σ, τ_w) and limiting (σ_{-1}, τ_f) stresses, but also on their ratios: τ_w/σ, σ_{-1} / τ_f, σ_{-1} /σ, $\tau_w/\tau_f \gtreqless 1$. This means that very different patterns of accumulation of irreversible damages occur depending on the realization of inequalities $\sigma \gtreqless \sigma_{-1}$, $\tau_w \gtreqless \tau_f$. This conclusion is supported by the known experimental results and theoretical models. Figure 18 depicts the analysis of the possible dependences $\log \Lambda_{n\backslash\tau} - \log \overline{\rho}_{n\backslash\tau}$ based on References [2,78]. A more detailed analysis of the interdependences $\Lambda_{n\backslash\tau}\left(\overline{\rho}_{n\backslash\tau}\right)$ can be found in References [2,78].

Table 2. Main signs of the physical state.

Energy State		Condition of Reaching the Limiting (Critical) State
Symbol	**Physical State and Its Characteristic**	
N	Mechanical state σ_{ij}	$u_n^{eff} \overset{\sigma_{ij} \to \sigma_{\lim}}{\to} u_0$
T	Thermodynamic state T_Σ	$u_T^{eff} \overset{T_\Sigma \to T_S}{\to} u_0$
MTD	Mechanothermodynamical state σ_{ijT}, T_Σ	$u_\Sigma^{eff} \underset{T_\Sigma \to T_S}{\overset{\sigma_{ijT} \to \sigma_{\lim}(T)}{\Rightarrow}} u_0$
tMTD	Mechanothermodynamical state in time σ_{ijT}, T_Σ, t	$u_\Sigma^{eff} \underset{t \to t_{\lim}}{\overset{\sigma_{ijT} \to \sigma_{\lim}(T)}{\Rightarrow}} u_0$

Table 3. Specification of the characteristics and their physical signs of the limiting state.

Criterion	Condition of Reaching the Limiting State	Physical Sign
L1	$\sigma_{\lim} = \sigma_u$ σ_u–stress limit at tension	Static fracture
L2	$\sigma_{\lim} = \sigma_{-1}$ σ_{-1}–mechanical fatigue limit	Fatigue fracture (into parts)
L3	$\sigma_{\lim} = p_f$ p_f–limiting contact pressure at rolling	Pittings of critical density (critical depth), excessive wear
L4	$\sigma_{\lim} = \tau_f$ τ_f–limiting stress at sliding	Limiting wear
L5	$\sigma_{\lim} = \begin{cases} \sigma_{-1p} \\ \sigma_{-1\tau} \end{cases}$ σ_{-1p} $\sigma_{-1\tau}$–limiting stress during the direct effect implementation [2]	Fatigue fracture (into parts) depending on the contact pressure (subscript p) at rolling or depending on the friction stress (subscript τ) at sliding (direct effect in tribo-fatigue)
L6	$\sigma_{\lim} = \begin{cases} p_{f\sigma} \\ \tau_{f\sigma} \end{cases}$ $p_{f\sigma}$, $\tau_{f\sigma}$–limiting stresses during the inverse effect implementation [2]	Pittings of critical density (critical depth) or excessive wear at rolling or sliding depending on the level of cyclic stresses σ (subscript σ) (inverse effect in tribo-fatique)
L7	$\sigma_{\lim} = \sigma_{-1q}$ σ_{-1q}–fretting fatigue limit	Fatigue fracture at fretting corrosion and (or) fretting wear
L8	$\sigma_{\lim T} = \sigma_{-1T}$ σ_{-1T}_isothermal fatigue limit	Limiting state depending on temperature (isothermal fatigue)
L9	$T_{\lim} = T_s$ T_s_melting point	Thermal (thermodynamic) damage
L10	$t_{\lim} = t_c$ t_c_longevity	Time (physical) prior to the onset of the limiting state on the basis of any sign

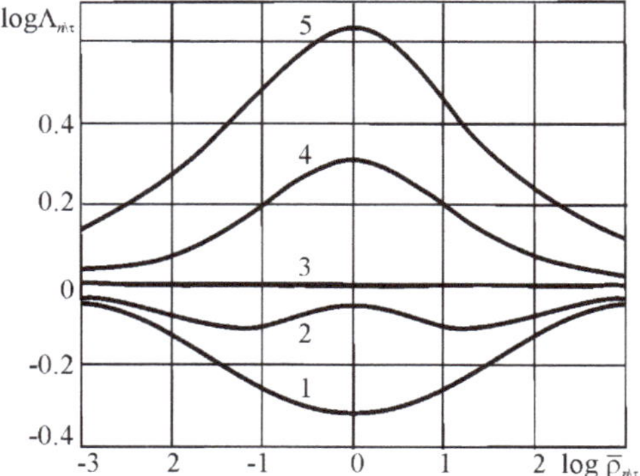

Figure 18. Typical plots of the character and direction of hardening–softening processes ($\Lambda \gtreqless 1$) versus the skewness coefficient of the damageability processes $\overline{\rho}$: 1, 2–mechano-rolling fatigue; 2, 3, 4–mechano-sliding fatigue; 4, 5–fretting-fatigue.

Here, $\sigma_{\lim}$ is the limiting stress, T_s is the melting point, $t_{\lim}$ is the longevity, σ_{ij} is the stress (strain) tensors, T_Σ is the temperature due to all heat sources, σ_{ijT} is the stress tensor in the isothermal (T_Σ = const) state, σ_{ijT} and T_Σ are the stress-strain state and the thermodynamic state, respectively, and σ_{ijT}, T_Σ, and t are the stress-strain state and the thermodynamic state in time, respectively

The plot of the $\Lambda_{T\backslash M}$ interactions versus the parameter $\overline{\rho}_{T\backslash M}$ can be analyzed in a similar way. Figure 19 illustrates the plots for steel, aluminum alloys, and nickel in the double logarithmic coordinates (according to the extensive experimental results [2,78]). The correlation coefficient r appears to be very high from 0.862 to 0.999. The plot of $\Lambda_{T\backslash M}(\overline{\rho}_{T\backslash M})$ suddenly changes for lg $\overline{\rho}_{T\backslash M} = 0$ ($\overline{\rho}_{T\backslash M}$=1) when thermal and stress damages turn to be in equilibrium (in comparison to the similar changes in the dependencies in Figure 18).

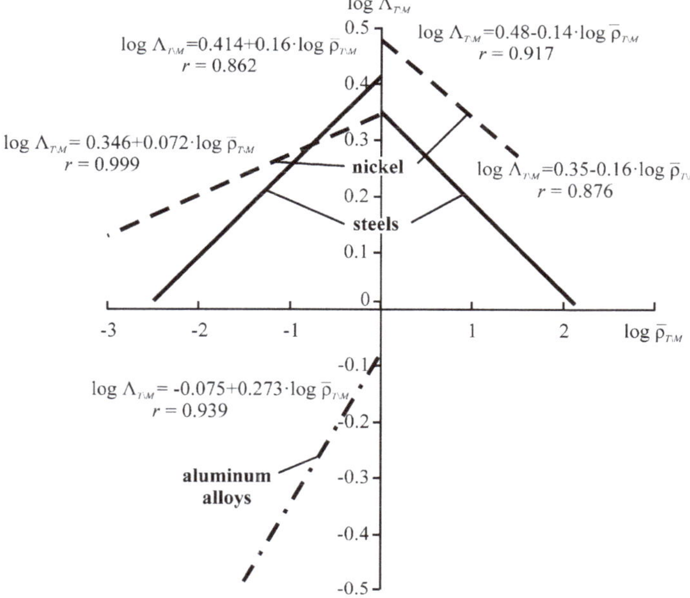

Figure 19. Logarithmic plots of $\Lambda_{T\backslash M}(\overline{\rho}_{T\backslash M})$ built on the basis of the experimental data.

For steels and nickel at $\overline{\rho}_{T\backslash M} < 1$, the direct dependence is found between $\Lambda_{T\backslash M}$ and $\overline{\rho}_{T\backslash M}$, and at $\overline{\rho}_{T\backslash M} > 1$ it becomes inverse. For aluminum alloys, the dependence $\Lambda_{T\backslash M}\left(\overline{\rho}_{T\backslash M}\right)$ is also direct, but located (at $\overline{\rho}_{T\backslash M} < 1$) in the III quadrant.

It is experimentally confirmed that the interaction parameter $\Lambda_{T\backslash M}$ is sensitive not only to the effective thermal-to-mechanical energy ratio, but also to the structure and composition (or nature) of metal materials. The last conclusion is also valid for the parameter $\Lambda_{n\backslash\tau}$: its numerical values appear to be significantly different, for example, for metal/metal and metal/polymer active systems even in the case when the ratios $\sigma\backslash\sigma_{-1}$ and $\tau_w\backslash\tau_f$ are identical for them.

In this section we briefly analyze the data of more than 600 tests of metals and their alloys (at isothermal conditions) obtained by many authors.

It was found that the thermodynamic dependence of limiting stresses can be presented in the $\log\sigma_{\lim} - \log C_T$ coordinates (Figures 12 and 13 and Formula (152)), where the function

$$C_T = C_T\left(T,\ u_0,\ a_n,\ a_T,\ \Lambda_{M\backslash T}\right) \tag{154}$$

is satisfactory at static tension ($\sigma_{\lim} = \sigma_u$) and fatigue fracture ($\sigma_{\lim} = \sigma_{-1}$) for numerous and various metal materials (steels; aluminum, titanium, alloys, etc.). In addition, interrelation (152) appears to be valid practically within the entire possible interval of temperature ($T \leq 0,8T_S$) and stress ($\sigma \leq \sigma_u$) varying with the correlation coefficient $r = 0.7$ in the specific cases and usually with $r > 0.9$. Model (152) then turns to be fundamental (Figure 20). The simplified model may seem dubious because in the known works (see Reference [85,86]), the explicit temperature dependence of limiting stresses is described by complex curves. This is attributed to the changes in the failure mechanisms of various materials at different testing conditions: normal, operating, and other temperatures.

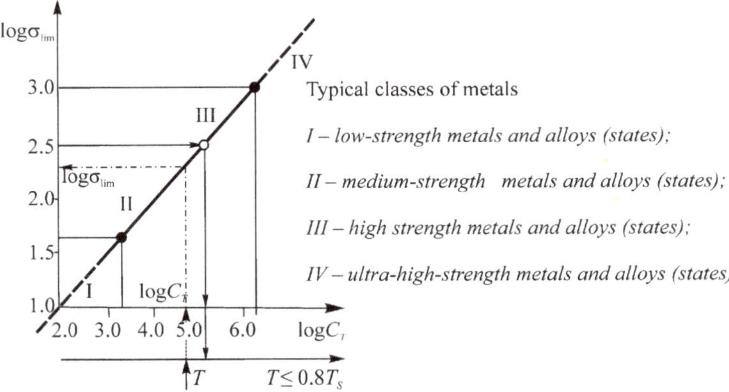

Figure 20. Generalized MTD function of the limiting states of metals and alloys ($\sigma_{\lim} \leq \sigma_u$; $T \leq 0,8T_S$), model (7).

Nevertheless, the fundamental nature of model (152) is supported experimentally (Figures 12 and 13).

From the theoretical standpoint, we can say the following in favor of model (152). It has four parameters (Formula (154)), one of them (u_0) is a fundamental constant of substance (Formulas (48) and (49) in Reference [80]), and the other two (a_T, a_n) are defined by the boundary conditions as the relations u_0 and physical constants σ_d and T_d of a given material [78]:

$$a_\sigma = u_0/\ \sigma_d^2,\ a_T = u_0/\sigma_d. \tag{155}$$

The methods to determine σ_d and T_d are outlined in References [2,78]. Here, we remind that material failure limit σ_d is obtained at tension fof $T_\Sigma \to 0$. Failure temperature T_d is obtained at the body heating for $\sigma = 0$. Therefore, in a general case, the accumulation of damages and failure due to mechanical stresses and thermal activation of these stresses in time is taken into account [67]. Finally, as it was briefly discussed above and given in References [4,76], the function 1 $\Lambda_{M\backslash T} \gtreqless 1$ takes into account damage interaction considering the change of ratio of $\sigma \gtreqless \sigma_{lim}$. Known studies [2,4,86] repeatedly and convincingly prove that this ratio determines the mechanism and character damage at different types of strain. The role of thermal fluctuations ($T_\Sigma < T_d$) is also studied in detail in References [67,68].

Further analysis of non-metalic (polymer) materials proves the fundamental nature of model (152). Table 4 and Figure 21 contain the analysis results of the polymer tests based on the experimental data [87]. It is obvious that model (152) is confirmed with the correlation coefficient $r = 0.917$. It should be noted that these test results are obtained not only for usual specimens (of ~5 mm diameter). Also, the results of tests of thin polymer films and threads are used not only under tensile deformation but also under torsion and bending. Large deviation of some points from the basic straight line could be explained by conventional accepting $\Lambda_{M\backslash T} = 1$ because of the lack of test data in order to estimate its actual value.

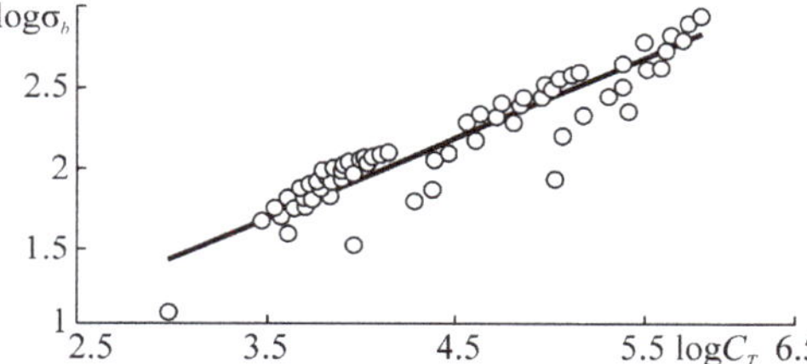

Figure 21. Dependence $\sigma_u(C_T)$ for polymer materials.

Table 4. Analysis of the main characteristic of polymer materials on the basis of the experimental data.

Material and Reference	u_0, $\frac{\text{kJ}}{\text{mol}}$	$\frac{a_T}{a_n}$, $\frac{\text{MPa}^2}{\text{K}}$ $\left(\frac{\text{kJ}}{\text{mol}\cdot\text{K}}/\frac{\text{kJ}}{\text{mol}\cdot\text{MPa}^2}\right)$	Tests Data	
			$\frac{\text{K}}{\sigma_b,\ \text{MPa}}$	Sample Size
Polyethylene high-density film (HDPF), grade 20806-024	108	$\frac{0.275}{2.94\cdot10^{-4}}$	$\frac{275...383}{32...386}$	5
Polypropylene film (PF) grade 03П10/005	119	$\frac{0.234}{1.70\cdot10^{-4}}$	$\frac{273...423}{150...570}$	5
Hardened staple fiber, polyvinyl alcohol (PVA) "Vinol MF"	111	$\frac{0.227}{7.62\cdot10^{-5}}$	$\frac{273...453}{80...802}$	5
Thread based on perchlorvinyl resin (PCV) grade "Chlorine"	114	$\frac{0.285}{2.56\cdot10^{-4}}$	$\frac{273...383}{60...376}$	5
Caprone thread (PCA) (GOST 7054067)	169	$\frac{0.282}{1.68\cdot10^{-4}}$	$\frac{275...453}{300...740}$	5
Polyethylene terephthalate film (PET) (TU 6-05-1597-72)	222	$\frac{0.342}{9.82\cdot10^{-4}}$	$\frac{279...498}{200...362}$	4
Polyamide film PM-1 (TU 6-05-1597-72)	202	$\frac{0.297}{2.1\cdot10^{-3}}$	$\frac{273...673}{12...240}$	7
Polystyrol (PS) at bending	281	$\frac{0.627}{2\cdot10^{-2}}$	$\frac{77...290}{56...108}$	10
Polymetalmethacrylate (PMMA) at bending	277	$\frac{0.558}{1.74\cdot10^{-2}}$	$\frac{77...290}{66...116}$	10
High-impact polystyrene (HIPS) at tension and torsion	277 252	$\frac{0.699}{2.53\cdot10^{-2}}$ $\frac{0...636}{1.84\cdot10^{-2}}$	$\frac{77...290}{48...94}$ $\frac{77...290}{50...105}$	10 10

Figure 22 illustrates the generalized experimentally verified MTD function of the limiting (by damageability) states. Figures 12 and 13 (compared to Figure 22) depict relatively large deviations of particular experimental points from the predicted ones. There are two reasons for that. The first one is that available references may have no data for a correct assessment of required parameters. The second reason is that the conducted experiments reveal significant errors, or they were not methodically correct.

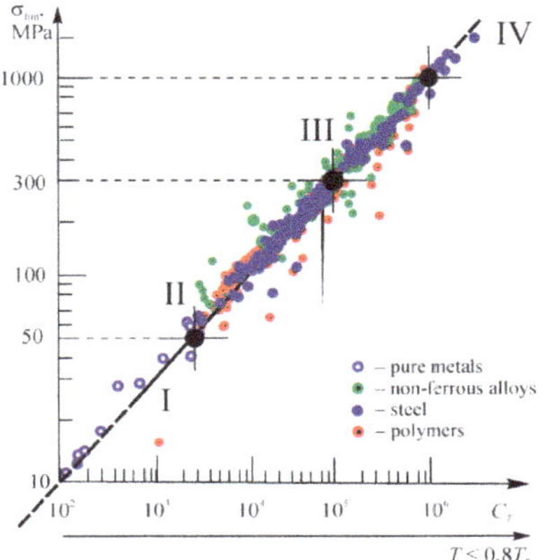

Figure 22. Experimentally verified MTD function of the critical by damageability states of metal and polymer materials.

Note that model (152) may seem to be non-fundamental because of its simplicity. However, we remind the classic dictum: the fundamental dependence cannot be complicated (or: every law is described by the simplest formula).

Model (152) can then serve for prediction of mechanical behavior of materials in the thermodynamic medium (shown by the arrows from T to $\sigma_{\lim}$ in Figure 20):

$$T \underset{\overline{\underset{a_n,\, a_T}{\uparrow}}}{\overset{\underline{\overset{u_0,\, \Lambda_{M\backslash T}}{\downarrow}}}{\rightarrow}} \log C_T \rightarrow \log \sigma_{\lim}\left(T, u_0, a_n, a_T, \Lambda_{M\backslash T}\right) \rightarrow \sigma_{\lim(T)}. \tag{156}$$

The parameters T, a_T, and $\Lambda_{M\backslash T}$ are responsible for the medium state in Equation (153).

Predictions by Equations (152) and (156) could be applied to the materials of different nature and structure. They are irrespective of damage and fracture mechanisms under static and cyclic loads.

Of course, because of the linearity of Equation (152), the reverse prediction could be possible and effective. In case a mechanical state of material (defined by the parameters u_0, $\sigma_{\lim(T)}$) is known, then the requirements can be formulated to the medium (defined by the parameters T, a_T, $\Lambda_{M\backslash T}$) where the system can work (the arrows from $\sigma_{\lim}$ to T in Figure 20):

$$\sigma_{\lim(T)} \rightarrow \log \sigma_{\lim(T)} \underset{\overline{\underset{a_n,\, a_T}{\uparrow}}}{\overset{\underline{\overset{u_0,\, \Lambda_{M\backslash T}}{\downarrow}}}{\rightarrow}} \log C_T \rightarrow C_T\left(T, u_0, a_n, a_\tau, \Lambda_{M\backslash T}\right) \rightarrow T. \tag{157}$$

Note that the attempts to construct an explicit temperature dependence of limiting stresses in uniform, semi-logarithmic, and logarithmic coordinates for various materials and different testing conditions are quite ineffective (Figure 23). We will further briefly analyze a more complex problem of the MTD system operation in the medium under the processes of thermal corrosion and corrosion at stress. From Equation (77), at $\tau_w = 0$ we have

$$\Lambda_{M\backslash T}\left(\frac{a_T}{1-D_T}T_\Sigma + \frac{a_n}{1-D_n}\sigma^2\right) = u_0 \tag{158}$$

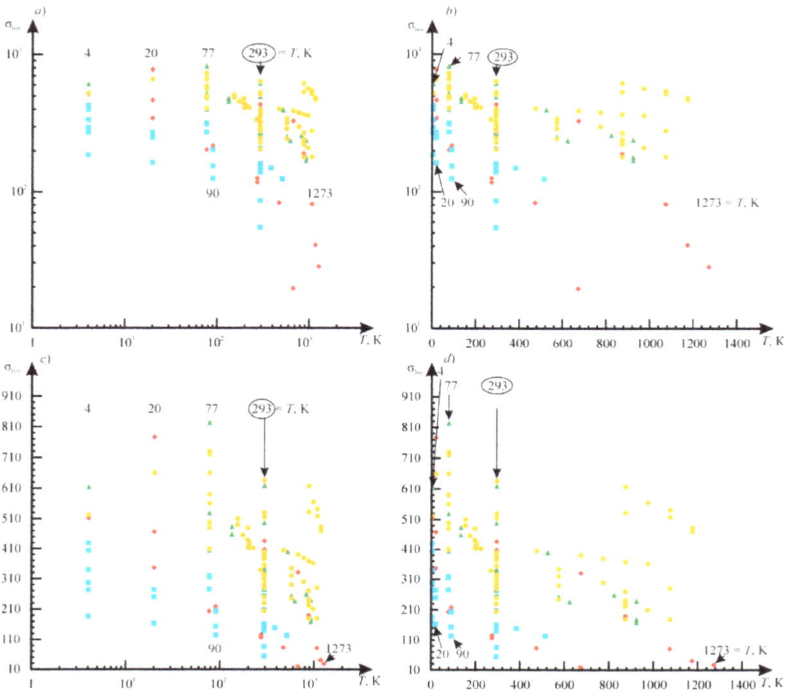

Figure 23. Dependencies of metals fatigue limit (according to 136 results of tests of many authors) on temperature.

Upon simple manipulations we obtain:

$$\sigma_{\lim(T,\,ch)} = \frac{1}{2}\log C_{T(ch)} \tag{159}$$

where the parameter of thermal resistance to corrosion at stress is:

$$C_{T(ch)} = C_{T(ch)}\Big(T,\ u_0,\ a_n,\ a_T,\ \Lambda_{M\backslash T},\ v_{ch},\ v_{ch(\sigma)},\ m_{v(\sigma)},\ v_{ch(T)},\ m_{v(T)}\Big). \tag{160}$$

It is seen that models (152) and (159) are fundamentally (and formally) identical. They differ because corresponding functions (154) and (160) use the parameters describing the damageability processes characteristic of the analyzed phenomena. In function (160), parameters v_{ch}, $v_{ch(\sigma)}$, $m_{v(\sigma)}$, $v_{ch(T)}$, $m_{v(T)}$ describe the processes of thermal corrosion at stress (Formula (76) in Reference [80]). Based on models (159) and (160), it is easy to develop prediction algorithms (type (156) and (157)) of resistance to thermal corrosion at stress.

A detailed analysis of models (157) and (160) is beyond the scope of the present work. It can be made in the future as applications to the novel results described in References [3,4,76].

It should be noted that solutions (77)–(151) can be analyzed in a similar way for other operating (or testing) conditions.

9. Discussion

The foregoing gives three main conclusions:

1. Damage is a fundamental physical property (and a functional duty) of any system and all its elements.

2. Damageability of each object (any existing one) inevitably grows up to its death–decomposition (disintegration) into a set of particles of arbitrarily small size, i.e., it is the unidirectional time process.

3. Evolution of the system by damageability is characteristic not only of the unity and struggle of opposites, but also of the directivity of various and complex physical hardening–softening processes (depending on the load and time level). It means that the Λ-function of interaction of different-nature damages can take three classes of values: (1) $\Lambda < 1$ when the hardening process is dominant, (2) $\Lambda > 1$ when the softening process is dominant, and (3) $\Lambda = 1$ when a stable hardening-to-softening process ratio is found.

So, the first law of mechanothermodynamics states that the evolution of any system inevitably needs a unidirectional process of its damage and disintegration, finally, into an infinitely large number of small components (fragments, atoms, elementary particles, etc.). In fact, it is equivalent to the recognition of the evolution endlessness, if it is taken into account that disintegration products of any system become a construction material for new systems. Thus, the evolution hysteresis is formed.

The second law of mechanothermodynamics states: interaction Λ-functions must take three classes of values ($\Lambda \gtreqless 1$) to describe not only the unity and struggle of opposites, but also the directivity of physical hardening–softening processes in the system, i.e., the system evolution by damageability [50,52,76].

Figure 24 generalizes the above results [1,2]. It is seen that the system state can be equivalently described in terms of either energy or entropy. The main drawback of such descriptions is the known unreality of energy, and hence, of entropy: physical energy carriers are not detected and, apparently, do not exist. As Feynman [88] said, figuratively, they cannot be touched. Damages are completely different: they are physically real, can be touched, and actually define any of the conceivable states of material bodies and systems. The kinetic process of their accumulation, as well as the time stream, is inevitable and unidirectional.

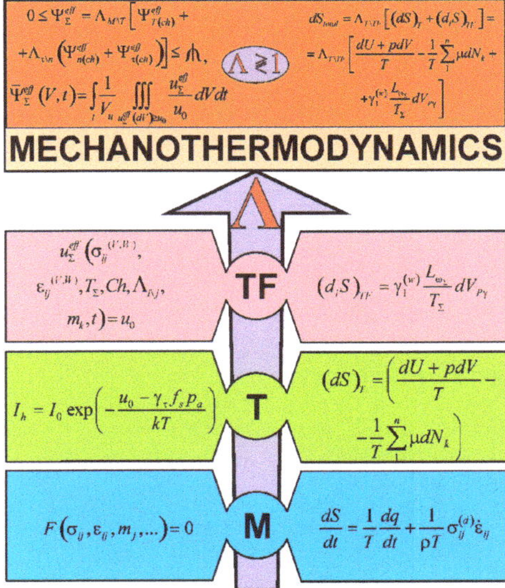

Figure 24. Energy (**left**) and entropy (**right**) approaches to developing mechanothermodynamics (M: mechanics, T: thermodynamics, TF: tribo-fatigue).

An attempt was made to formulate the basic principles of a new (or, better to say, integrated) physical discipline–mechanothermodynamics with the use of the energy principles. This discipline combines two branches of physics in order not to argue or not to compete with each other, but to take a fresh look at the MTD system evolution (Figure 25).

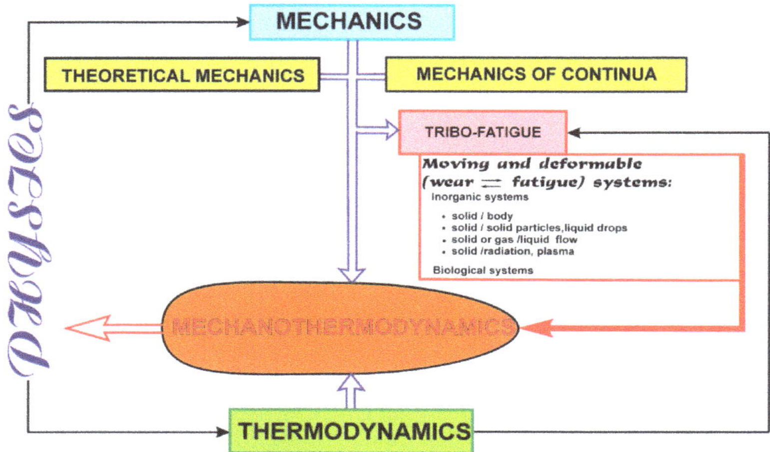

Figure 25. Ways towards mechanothermodynamics as a new branch of physics.

Figure 26 shows that the principles of mechanothermodynamics can be formulated in two ways: (1) mechanics → tribo-fatigue → mechanothermodynamics and (2) thermodynamics → tribo-fatigue → mechanothermodynamics. So, tribo-fatigue has become a bridge to pass from mechanics and thermodynamics to mechanothermodynamics.

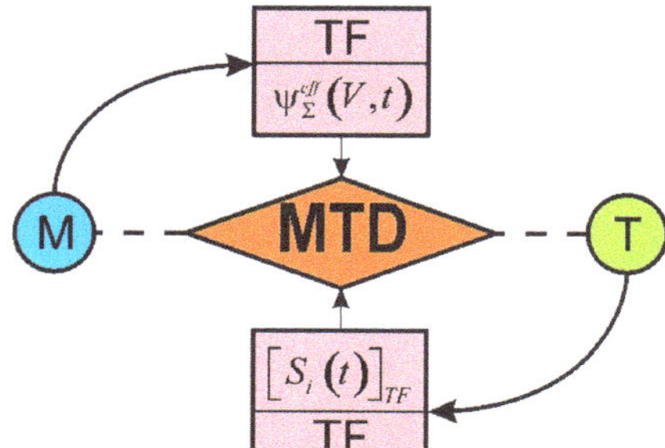

Figure 26. Tribo-fatigue bridges from mechanics (**M**) and thermodynamics (**T**) to mechanothermodynamics (MTD) are denoted by the solid lines with arrows and the unrealized ways (during more than 150 years, from M or T to MTD)—by the dashed lines.

The fact that both ways lead to one objective and, finally, yield the same (unified) result, means that the above-mentioned two methodologies of analysis are valid, correct, and do not contradict each other.

10. Conclusions

1. It is shown that a generalized physical discipline—mechanothermodynamics—can be created by making two main bridges. The first one is the tribo-fatigue entropy that allowed transfer from

thermodynamics to mechanics. The second one is the fundamental tribo-fatigue understanding of irreversibility of damage of everything that allowed transfer from mechanics to thermodynamics.

2. Main principles (I–XV) founding the general theory of evolution of MTD systems were formulated. The following models and theories were developed:

- Limiting state energy theory (Section 3).
- Damageability energy theory (Section 4).
- Fundamentals of electrochemical damageability theory (Section 5).
- Elements of MTD system transmitting state theory (Section 6).

3. Procedures and methods of calculation of effective (dangerous) energy expended for generation, accumulation, and motion of irreversible damages were developed (Formulas (79)–(83) and the text related to them).

4. Fundamentals of the theory of Λ-interaction between damages due to different loads of nature: thermodynamic, mechanical, etc. (Formulas (69), (70) and (155) and the text related to them) were outlined. This theory allows consideration of the effect of accidental hardening–softening processes on the limiting by damageability state of MTD.

5. The relationship between the damages of the system and the event probability (Figure 10) in the course of its evolution was analyzed. The idea of reliable damageability probability $1 < P_* < \infty$ at the stage of translimiting states was proposed.

6. The physical signs and specific characteristics of limiting states of objects and systems (Tables 2 and 3) were given. These may be of use for specialists in the relevant research areas.

7. Practically, a unified MTD function of critical by damageability (limiting) states of polymer and metal materials working under different and complex conditions (Formula (152) and Figure 15) was obtained in the present work. The analysis of more than 600 experimental results (Figures 7, 8, 14, 16 and 17) showed the fundamental nature of this function since it is applicable for high-, average-, and low-strength states of alloys, pure metals, and polymers. MTD function can be used for a wide range of medium temperatures (from 0.8 T_S where T_S is the melting point of material to temperature of helium), limiting values of mechanical stresses (up to the limit of strength under static loading), and the fatigue life of $10^6 \ldots 10^8$ cycles. This function can effectively predict the behavior of specific MTD systems at different working (testing) conditions (procedures (156) and (157)). Models (159) and (160) were proposed to describe the effects of corrosion at stress and thermal corrosion on the change in materials' limiting states.

In conclusion, it should be noted that the research in mechanothermodynamics is in its initial stage. Expanding and deepening the front of research in this promising new area of knowledge could be expected in the near future.

Author Contributions: Conceptualization, L.A.S. and S.S.S.; Formal analysis, S.S.S.; Investigation, L.A.S. and S.S.S.; Methodology, L.A.S. and S.S.S.; Validation, L.A.S.; Visualization, S.S.S.; Writing–original draft, L.A.S. and S.S.S.; Writing–review & editing, S.S.S.

Funding: This research received no external funding.

Conflicts of Interest: The authors declare no conflict of interest.

References

1. Sosnovskiy, L.A.; Sherbakov, S.S. Mechanothermodynamical system and its behavior. *Contin. Mech.* **2012**, *24*, 239–256. [CrossRef]
2. Sosnovskiy, L.A. *Mechanics of Wear-Fatigue Fracture*; Sosnovskiy, L.A., Ed.; BelSUT Press: Gomel, Belarus, 2007; p. 434. (In Russian)
3. Sosnovskiy, L.A. *Tribo-Fatigue: Wear-fatigue Damage and Its Prediction*; Springer: Berlin, Germany, 2005; p. 424.
4. Sherbakov, S.S.; Sosnovskiy, L.A. *Mechanics of Tribo-Fatigue Systems*; Sherbakov, S.S., Sosnovskiy, L.A., Eds.; BSU: Minsk, Belarus, 2010; p. 407. (In Russian)

5. Basaran, C.; Yan, C.Y. A thermodynamic framework for damage mechanics of solder joints. *J. Electron. Packag.* **1998**, *120*, 379–384. [CrossRef]
6. Basaran, C.; Nie, S. An irreversible thermodynamics theory for damage mechanics of solids. *Int. J. Damage Mech.* **2004**, *13*, 205–223. [CrossRef]
7. Basaran, C.; Li, S.; Abdulhamid, M.F. Thermomigration induced degradation in solder alloys. *J. Appl. Phys.* **2008**, *103*, 123520. [CrossRef]
8. Yao, W.; Basaran, C. Computational damage mechanics of electromigration and thermomigration. *J. Appl. Phys.* **2013**, *114*, 103708. [CrossRef]
9. Basaran, C.; Nie, S. A thermodynamics based damage mechanics model for particulate composites. *Int. J. Solids Struct.* **2007**, *44*, 1099–1114. [CrossRef]
10. Ye, H.; Basaran, C.; Hopkins, D.C. Deformation of solder joint under current stressing and numerical simulation. *Int. J. Solids Struct.* **2004**, *41*, 4939–4958. [CrossRef]
11. Gomez, J.; Basaran, C. Nanoindentation of Pb/Sn solder alloys; experimental and finite element simulation results. *Int. J. Solids Struct.* **2006**, *43*, 1505–1527. [CrossRef]
12. Basaran, C.; Zhao, Y.; Tang, H.; Gomez, J. A damage-mechanics-based constitutive model for solder joints. *J. Electron. Packag.* **2005**, *127*, 208–214. [CrossRef]
13. Temfack, T.; Basaran, C. Experimental verification of thermodynamic fatigue life prediction model using entropy as damage metric. *Mater. Sci. Technol.* **2015**, *31*, 1627–1632. [CrossRef]
14. Basaran, C.; Lin, M. Damage mechanics of electromigration in microelectronics copper interconnects. *Int. J. Mater. Struct. Integr.* **2007**, *1*, 16–39. [CrossRef]
15. Kondepudi, D. *Modern Thermodynamics. From Heat Engines to Dissipative Structures*; Kondepudi, D., Prigogine, I., Eds.; John Wiley & Sons: New York, NY, USA, 1998; p. 510.
16. Mase, G. *Theory and Problems of Continuum Mechanics*; McGraw-Hill: New York, NY, USA, 1970; p. 221.
17. Sedov, L.I. *A Course in Continuum Mechanics*; Wolters–Noordhoff: Groningen, The Netherlands, 1971; p. 242.
18. Truesdell, C.A. *A First Course in Rational Continuum Mechanics*; Academic Press: Cambridge, MA, USA, 1976; p. 295.
19. Yokobori, T. *An Interdisciplinary Approach to Fracture and Strength of Solids*; Wolters–Noordhoff: Groningen, The Netherlands, 1968; p. 335.
20. Kachanov, L.M. *Introduction to Continuum Damage Mechanics*; Kluwer Academic Publishers: Dordrecht, The Netherlands, 1986; p. 148.
21. Panin, V.E. *Physical Mesomechanics of Heterogeneous Media and Computer-Aided Design of Materials*; Int. Sci. Publishing: Cambridge, UK, 1998; p. 450.
22. Houlsby, G.T.; Puzrin, A.M. *Principles of Huperplasticity. An Approach to Plasticity Theory Based on Thermodynamic Principles*; Springer: London, UK, 2006; p. 363.
23. Lemaitre, J.; Desmorat, R. *Engineering Damage Mechanics: Ductile, Creep, Fatigue and Brittle Failures*; Springer: New York, NY, USA, 2007; p. 402.
24. Palmov, V.A. *Vibrations of Elasto-Plastic Bodies*; Springer: Berlin, Germany, 1998; p. 311.
25. Gomez, J.; Lin, M.; Basaran, C. Damage mechanics modeling of concurrent thermal and vibration loading on electronics packaging. *Multidiscip. Modeling Mater. Struct.* **2006**, *2*, 309–326. [CrossRef]
26. Besson, J. Continuum models of ductile fracture: A review. *Int. J. Damage Mech.* **2010**, *19*, 3–52. [CrossRef]
27. Krajcinovic, D. Damage mechanics: Accomplishments, trends and needs. *Int. J. Solids Struct.* **2000**, *37*, 267–277. [CrossRef]
28. Basaran, C.; Desai, C.S.; Kundu, T. Thermomechanical Finite Element Analysis of Problems in Electronic Packaging Using the Disturbed State Concept Part I: Theory and Formulation. *Trans. ASME J. Electron. Packag.* **1998**, *120*, 41–47. [CrossRef]
29. Butuc, M.C.; Gracio, J.J.; Barata da Rocha, A. An experimental and theoretical analysis on the application of stress-based forming limit criterion. *Int. J. Mech. Sci.* **2006**, *48*, 414–429. [CrossRef]
30. Collins, I.F.; Houlsby, I.F. Application of thermomechanical principles to the modeling of geotechnical materials. *Proc. R. Soc. Lond. Ser. A* **1997**, *453*, 1975–2001. [CrossRef]
31. Brunig, M. An anisotropic ductile damage model based on irreversible thermodynamics. *Int. J. Plast.* **2003**, *19*, 1679–1713. [CrossRef]
32. Brünig, M.; Chyra, O.; Albrecht, D.; Driemeier, L.; Alves, M. A ductile damage criterion at various stress triaxialities. *Int. J. Plast.* **2008**, *24*, 1731–1755. [CrossRef]

33. Voyiadjis, G.Z.; Dorgan, R.J. Framework using functional forms of hardening internal state variables in modeling elasto-plastic-damage behavior. *Int. J. Plast.* **2007**, *23*, 1826–1859. [CrossRef]
34. Chandaroy, R.; Basaran, C. Damage mechanics of surface mount technology solder joints under concurrent thermal and dynamic loading. *J. Electron. Packag.* **1999**, *121*, 61–68. [CrossRef]
35. Basaran, C.; Chandaroy, R. Using finite element analysis for simulation of reliability tests on solder joints in microelectronic packaging. *Comput. Struct.* **2000**, *74*, 215–231. [CrossRef]
36. Desai, C.S.; Basaran, C.; Dishongh, T.; Prince, J. Thermomechanical Analysis in Electronic Packaging with Unified Constitutive Models for Materials and Joints. *IEEE Trans. Comp. Pack. Manu. Tech. Part B Trans. Adv. Packag.* **1998**, *21*, 87–97. [CrossRef]
37. Tang, H.; Basaran, C. Influence of microstructure coarsening on thermomechanical fatigue behavior of Pb/Sn eutectic solder joints. *Int. J. Damage Mech.* **2001**, *10*, 235–255. [CrossRef]
38. Chen, B.; Li, W.-G.; Peng, X.-H. A microvoid evolution law involving change of void shape and micro/macroscopic analysis for damaged materials. *J. Mater. Process. Techn.* **2002**, *122*, 189–195. [CrossRef]
39. Maire, E.; Bordreuil, C.; Babout, L.; Boyer, J.-C. Damage initiation and growth in metals. Comparison between modelling and tomography experiments. *J. Mech. Phys. Solids.* **2005**, *53*, 2411–2434. [CrossRef]
40. Hammi, Y.; Horstemeyer, M.F. A physically motivated anisotropic tensorial representation of damage with separate functions for void nucleation, growth, and coalescence. *Int. J. Plast.* **2007**, *23*, 1641–1678. [CrossRef]
41. Weck, A.; Wilkinson, D.S.; Maire, E.; Toda, H. Visualization by X-ray tomography of void growth and coalescence leading to fracture in model materials. *Acta Materialia.* **2008**, *56*, 2919–2928. [CrossRef]
42. Tutyshkin, N.; Muller, W.H.; Wille, R.; Zapara, M. Strain-induced damage of metals under large plastic deformation: Theoretical framework and experiments; N. *Int. J. Plast.* **2014**, *59*, 133–151. [CrossRef]
43. Demir, I.; Zbib, H.M. A mesoscopic model for inelastic deformation and damage. *Int. J. Eng. Sci.* **2001**, *39*, 1597–1615. [CrossRef]
44. Einav, I.; Houlsby, G.T.; Nguyen, G.D. Coupled damage and plasticity models derived from energy and dissipation potentials. *Int. J. Solids Struct.* **2007**, *44*, 2487–2508. [CrossRef]
45. Hayat, T.; Khan, M.I.; Qayyum, S.; Alsaedi, A. Entropy generation in flow with silver and copper nanoparticles. *Colloids Surf. A* **2018**, *539*, 335–346. [CrossRef]
46. Hayat, T.; Qayyum, S.; Khan, M.I.; Alsaedi, A. Entropy generation in magnetohydrodynamic radiative flow due to rotating disk in presence of viscous dissipation and Joule heating. *Phys. Fluids* **2018**, *30*, 017101. [CrossRef]
47. Hayat, T.; Qayyum, S.; Khan, M.I.; Alsaedi, A.; Khan, M.I. New thermodynamics of entropy generation minimization with nonlinear thermal radiation and nanomaterials. *Phys. Lett. A* **2018**, *382*, 749–760. [CrossRef]
48. Khan, M.I.; Hayat, T.; Waqas, M.; Khan, M.I.; Alsaedi, A. Entropy generation minimization (EGM) in nonlinear mixed convective flow of nanomaterial with Joule heating and slip condition. *J. Mol. Liq.* **2018**, *256*, 108–120. [CrossRef]
49. Khan, M.I.; Qayyum, S.; Hayat, T.; Waqas, M.; Khan, M.I.; Alsaedi, A. Entropy generation minimization and binary chemical reaction with Arrhenius activation energy in MHD radiative flow of nanomaterial. *J. Mol. Liq.* **2018**, *259*, 274–283. [CrossRef]
50. Sosnovskiy, L.A.; Sherbakov, S.S. Possibility to Construct Mechanothermodynamics. *Nauka I Innov.* **2008**, *2*, 24–29.
51. Sosnovskiy, L.A.; Sherbakov, S.S. *Surprises of Tribo-Fatigue*; Magic book: Minsk, Belarus, 2009; p. 200.
52. Vysotskiy, M.S.; Vityaz, A.; Sosnovskiy, L.A. Mechanothermodynamical system as the new subject of research. *Mech. Mach. Mech. Mater.* **2011**, *2*, 5–10. (In Russian)
53. Hibbeler, R.C. *Mechanics of Materials*; Pearson Prentice Hall: Boston, MA, USA, 2016; p. 896.
54. Giurgiutiu, V.; Reifsnider, K.L. Development of strength theories for random fiber composites. *J. Compos. Technol. Res. JCTRER* **1994**, *16*, 103–114.
55. Pisarenko, G.S.; Lebedev, A.A. *Deformation and Strength of Materials in Complex Stress States*; Naukova dumka: Kiev, Ukraine, 1976; p. 412. (In Russian)
56. Troschenko, V.T. (Ed.) *Strength of Materials and Constructions*; Akademperiodika: Kiev, Ukraine, 2005; p. 1088. (In Russian)
57. Yu, M.-H. *Unified Strength Theory and Its Applications*; Springer: Berlin, Germany, 2004; p. 448.

58. Makhutov, N.A. *Structural Strength, Resource and Technological Safety*; Nauka: Novosibirsk, Russia, 2005; Volume 1, p. 494; Volume 2, p. 610. (In Russian)
59. Makhutov, N.A. *Deformation and Fracture Criteria and Durability Calculation of Structural Elements*; Mashinostroenie: Moscow, Russia, 1981; p. 272. (In Russian)
60. Zhuravkov, M.A.; Chumak, N.G. Analysis of stationary and progressive cracks under the dynamic loading in finite solid bodies. *Can. J. Mech. Sci. Eng.* **2011**, *2*, 127–136.
61. Zhuravkov, M.A. *Mathematical Modeling of Deformation Processes in Solid Deformable Media: (Using the Examples of Problems in Mechanics of Rocks and Massifs)*; BSU Press: Minsk, Belarus, 2002; p. 456. (In Russian)
62. Khonsari, M.M.; Amiri, M. *An Introduction to Thermodynamics of Mechanical Fatigue*; CRC Press, Taylor & Francis Group: Boca Raton, FL, USA, 2013; p. 145.
63. Naderi, M.; Khonsari, M.M. A thermodynamic approach to fatigue damage accumulation under variable loading. *J. Mater. Sci. Eng.* **2010**, *527*, 6133–6139. [CrossRef]
64. Amiri, M.; Naderi, M.; Khonsari, M.M. An experimental approach to evaluate the critical damage. *Int. J. Damage Mech.* **2011**, *20*, 89–112. [CrossRef]
65. Zhurkov, S.N. Kinetic concept of the strength of solids. *Vestn. Akad. Nauk SSSR.* **1968**, *3*, 46–52. [CrossRef]
66. Zhurkov, S.N. Dilaton Mechanism of the Strength of Solids. *Sov. Phys Solid State* **1983**, *20*, 1797–1800.
67. Regel, V.R.; Slutsker, A.I.; Tomashevskii, E.E. *Kinetic Nature of the Strength of Solids*; Nauka: Moscow, Russia, 1974; p. 560.
68. Gabar, I.G. Analysis of the failure of BCC-and FCC-metals and the concept of interrelation of fatigue curve parameters. *Strength Mater.* **1989**, *11*, 61–64.
69. Ivanova, V.S.; Terentiev, V.F. *Nature of Metals Fatigue*; Metallurgiya: Moscow, Russia, 1975; p. 456.
70. Cherepanov, G. Fracture Mechanics and kinetic theory of strength. *Strength Mater.* **1989**, *11*, 3–8. [CrossRef]
71. Sosnovskiy, L.A. *Statistical Mechanics of Fatigue Damage*; Nauka i tekhnika: Minsk, Belarus, 1987; p. 288. (In Russian)
72. Sosnovskiy, L.A.; Sherbakov, S.S. Vibro-impact in rolling contact. J. *Sound Vib.* **2007**, *308*, 489–503. [CrossRef]
73. Shcherbakov, S.S. Modeling of the damaged state by the finite-element method on simultaneous action of contact and noncontact loads. *J. Eng. Phys. Thermophys.* **2012**, *85*, 472–477. [CrossRef]
74. Shcherbakov, S.S. State of volumetric damage of tribo-fatigue system. *Strength Mater.* **2013**, *45*, 171–178. [CrossRef]
75. Bogdanovich, A.V. *Prediction of Limiting State of Active Systems*; GrSU Press: Grodno, Belraus, 2008; p. 372. (In Russian)
76. Sosnovskiy, L.A.; Sherbakov, S.S. *Mechanothermodynamics*; Springer: Berlin, Germany, 2016; p. 155.
77. Makhutov, N.A.; Sosnovskiy, L.A.; Bogdanovich, A.V. Theory of Accumulation ofWear-Fatigue Damage. In *Tribo-Fatigue–98/99: Annual*; Sosnovskiy, L.A., Ed.; S&P Group TRIBOFATIGUE: Gomel, Belarus, 2000; p. 60. (In Russian)
78. Sosnovskiy, L.A. *L-Risk (Mechanothermodynamics of Irreversible Damages)*; BelSUT Press: Gomel, Belaus, 2004; p. 317. (In Russian)
79. Sosnovskiy, L.A. *Tribo-Fatigue: The Dialectics of Life*, 2nd ed.; BelSUT Press: Gomel, Belraus, 1999; p. 116. (In Russian)
80. Sosnovskiy, L.A. *Fundamentals of Tribo-Fatigue: A Manual for Students of Higher Technical Educational Institutions*; BelSUT Press: Gomel, Belraus, 2003; Volume 1 and 2, pp. 234, 246. (In Russian)
81. Sosnovskiy, L.A. Principal ways for improving operational reliability of tribo-fatigue system brake block-wheel-rail. In Proceedings of the VI Intern. symposium on Tribo-Fatigue ISTF 2010, Belarusian State University, Minsk, Belraus, 25 October–1 November 2010; BSU Press: Minsk, Belraus, 2010; Volume 2, pp. 187–201. (In Russian).
82. Oleynik, N.V. *Fatigue Strength of Materials and Machines Components in Corrosive Media*; Oleynik, N.V., Magdenko, A.N., Sklar, S.P., Eds.; Naukova dumka: Kiev, Ukraine, 1987; p. 198.
83. Serensen, S.V. Problem of fatigue and wear resistance of details of machine components (brief overview); Increase of wear resistance and lifetime of machines. *Kiev Publ. Acad. Sci. Ukr. SSR.* **1960**, *1*, 10–14. (In Russian)
84. Forrest, G. *Fatigue of Metals*; Pergamon Press Ltd: Oxford, UK, 1962; p. 425.
85. Ponomarev, S.D. Strength calculations in mechanical engineering. *Mosc. State Sci. Tech. Publ. House Mach. Build. Lit.* **1958**, *2*, 974.

86. Prokopchuk, N.V. Temperature dependence of activation energy of mechanical failure of polymeric materials. *Strength Mater.* **1984**, *10*, 46–50. [CrossRef]
87. Sitamov, S.; Kartashov, E.M.; Khukmatov, A.I. Processes of failure of polymers in various types of loading in the brittle and quasibrittle state. *Strength Mater.* **1989**, *1*, 37–40. [CrossRef]
88. Feynman, R. *Lectures on Physics*; Moscow: Mir Moscow, Russia, 1963; Volume 4, p. 261. (In Russian)

Article

An Entropy-Based Failure Prediction Model for the Creep and Fatigue of Metallic Materials

Jundong Wang [1] **and Yao Yao** [1,2,*]

[1] School of Mechanics, Civil Engineering and Architecture, Northwestern Polytechnical University, Xi'an 710072, China; wangjundong@nwpu.edu.cn

[2] School of Civil Engineering, Xi'an University of Architecture and Technology, Xi'an 710055, China

* Correspondence: yaoy@nwpu.edu.cn

Received: 29 September 2019; Accepted: 29 October 2019; Published: 12 November 2019

Abstract: It is well accepted that the second law of thermodynamics describes an irreversible process, which can be reflected by the entropy increase. Irreversible creep and fatigue damage can also be represented by a gradually increasing damage parameter. In the current study, an entropy-based failure prediction model for creep and fatigue is proposed based on the Boltzmann probabilistic entropy theory and continuum damage mechanics. A new method to determine the entropy increment rate for creep and fatigue processes is proposed. The relationship between entropy increase rate during creep process and normalized creep failure time is developed and compared with the experimental results. An empirical formula is proposed to describe the evolution law of entropy increase rate and normalized creep time. An entropy-based model is developed to predict the change of creep strain during the damage process. Experimental results of metals and alloys with different stresses and at different temperatures are adopted to verify the proposed model. It shows that the theoretical predictions agree well with experimental data.

Keywords: entropy increase rate; creep strain; damage mechanics; fatigue; metallic material

1. Introduction

In the past decades, fatigue of materials has been investigated extensively with respect to crack nucleation, propagation, and life prediction under cyclic loading. Numerous theoretical models have been proposed based on statistics or empirical methods. The models adopted in industry are usually empirical and the physical mechanism of fatigue damage and life prediction still requires further study.

Creep is time-dependent and can be accelerated by increasing of the stress and temperature. It is one of the common damage modes in engineering such as turbine blades, thermal plants, and thermonuclear installations, especially at high-temperature conditions. The creep deformation can emerge even when the applied stress is below the elastic limit, which is more pronounced when the ambient temperature approaching the melting point of materials. Generally, the creep deformation behavior is distinguished by three stages: The creep strain rate decreases constantly in the first stage; the creep strain rate keeps almost constant in the second stage; and in the third stage, the creep rate increases rapidly until failure. The potential physical mechanism of different stages can be explained by the dislocation theory for metallic materials [1–6]. The dislocation density changes during the creep process, micro-voids nucleate in the first stage, and the coalescence and propagation mechanism of micro-voids occur in the second and the third stages simultaneously, which leads to the final fracture.

The relationship between creep strain rate and creep life was widely investigated theoretically and experimentally for different engineering materials. Monkman and Grant [7] proposed a model to describe the evolution law of steady creep strain rate and creep life and it was successfully applied to metallic materials. This model was subsequently modified by considering the damage parameters to describe the creep behavior [8–11]. Dyson and Gibbons [12] related the normalized creep strain and time by considering the applied stress and damage variable during the creep process. The proposed model linked the strain with time by introducing a damage variable. One of the widely adopted creep life prediction models for carbon steel is proposed by Fields [13]. A power law is applied to relate the creep stress and time. The parameters in this model can be determined from experiments [14].

The applications of thermodynamic methodology to contact problems [15–17] introduced entropy into solid mechanics. The specific entropy was applied for the complex systems under mechanical fatigue, thermal loading, friction, and wear conditions [18–20]. The entropy increase rate was studied under the framework of Boltzmann probabilistic theory and continuum damage mechanics [21]. A low-cycle fatigue life prediction model was proposed with respect to the entropy increase rate [21]. Based on the second law of thermodynamics, creep damage process is also irreversible, which can be represented by the increasing of entropy in the entire creep life.

In the current study, the entropy increase rate model and its application in fatigue life prediction is reviewed. Then, the entropy increase rate model is applied to describe the creep behavior. The relationship between entropy increase rate and normalized creep time is investigated based on the experimental analysis. A unified entropy increase trend was observed for metallic materials under different experimental conditions. An empirical formula is then proposed to describe the entropy increase rate during creep process. An entropy-based creep life prediction model is obtained by solving an ordinary differential equation. Comparison with experimental data indicates that the proposed model can accurately predict creep behavior of metallic materials with different stresses and temperatures.

2. The Change Regulation of Entropy Increase Rate during Degeneration Process

For ideal gas system, Boltzmann [22] defined a precise relationship between the disorder state and entropy:

$$S = k_0 \ln(W) \tag{1}$$

where k_0 is the Boltzmann constant and W is the disorder state parameter, which represents the probability of the system to exist in the current state relative to all the possible states. Although it is difficult to determine the value of W, Equation (1) provides an approach to determine the disorder of molecular thermal motion in the system.

The relation between entropy per unit mass and the disorder parameter was improved by Basaran et al. [20]. The disorder state parameter W was defined as a function of the entropy S, Avogadro constant, and the specific mass m_s: $W = \exp(Sm_s/N_0)$. Basaran et al. [23–27] proposed a relation between the entropy per unit mass and disorder state parameter; a damage law is then developed, which links the damage parameter D and entropy S:

$$D = D_{cr}\frac{W - W_0}{W} = D_{cr}\left[1 - \exp\left(-\frac{m_s(S - S_0)}{N_0 k_0}\right)\right] \tag{2}$$

where W_0 represents the disorder corresponding to the initial state of the continuous medium with entropy S_0 and D_{cr} is the critical value of damage approaching final failure.

The degeneration process, such as creep and fatigue, is not only a damage process, but also irreversible, which is consistent with the second law of thermodynamics. Although the damage variable is an artificially defined quantity in the viewpoint of solid mechanics, it has the same trend for entropy without decreasing. Determination of damage parameters requires different physical quantities including the elasticity modulus, micro-hardness, density, and electrical resistance etc. Variation of these physical quantities represents the corresponding microstructure change of material. The variation of entropy during the damage process represents the logarithm change of the molecular configurations [22]. Both quantities represent the microstructure change in different states, one for an outward manifestation and the other for the essential molecular configurations. In addition, the entropy and damage parameters are all monotonically changed during the degeneration process. Hence, it is possible to establish a connection between the damage variable and entropy.

A classical damage rate model was proposed by Bonora [28] based on the continuous damage mechanics and the plastic part of the Ranberg–Osgood power law:

$$\dot{D} = -\dot{\lambda}\frac{\partial \Phi}{\partial Y} = \frac{K^2}{2Ea_0}\frac{(D_{cr}-D)^{\frac{\alpha-1}{\alpha}}}{p}f\left(\frac{\sigma_m}{\sigma_{eq}}\right)\frac{\dot{\lambda}}{1-D} \tag{3}$$

where K and a_0 are the material constants, $\dot{\lambda}$ is plastic multiplier, α is the damage exponent and can be obtained by determining the change of elastic modulus during the damage process, $f\left(\frac{\sigma_m}{\sigma_{eq}}\right)$ is a factor, and for uniaxial loading $f\left(\frac{\sigma_m}{\sigma_{eq}}\right) = 1$. By applying the relation between plastic multiplier $\dot{\lambda}$ and cumulative plastic strain rate $\dot{p}$: $\dot{\lambda} = \dot{p}\cdot(1-D)$, Equation (3) can be written in another form:

$$\dot{D} = \frac{K^2}{2Ea_0}\frac{(D_{cr}-D)^{\frac{\alpha-1}{\alpha}}}{p}f\left(\frac{\sigma_m}{\sigma_{eq}}\right)\frac{\dot{p}}{p} \tag{4}$$

The damage variable D has two threshold values, D_0 and D_{cr}. The threshold D_0 represents the initial value of damage variable presented in material microstructure or the value at the beginning of creep or fatigue damage accumulation. The threshold D_{cr} is the critical value of damage variable when creep or fatigue failure occurs. The corresponding cumulative plastic strain for $D = D_0$ and $D = D_{cr}$ are p_{th} and p_{cr}, respectively. Integrating Equation (4) between $[D, D_{cr}]$ and $[p, p_{cr}]$ gives:

$$(D - D_{cr})^{1/\alpha} = \frac{K^2}{2Ea_0\alpha}ln\left(\frac{p_{cr}}{p}\right)f\left(\frac{\sigma_m}{\sigma_{eq}}\right). \tag{5}$$

Based on the plasticity damage dissipation potentials in Equation (2), an entropy increasing rate model for uniaxial state was proposed (detailed derivation can be found in [21]):

$$\dot{S} = \frac{N_0 k_0 \alpha}{m_s \ln(p_{cr}/p) f(\sigma_m/\sigma_{eq})}\left[\dot{f}\left(\frac{\sigma_m}{\sigma_{eq}}\right)ln\left(\frac{p_{cr}}{p}\right) - f\left(\frac{\sigma_m}{\sigma_{eq}}\right)\frac{\dot{p}}{p}\right] \tag{6}$$

For the uniaxial loading case, Equation (6) can be written as:

$$\dot{S} = \frac{N_0 k_0 \alpha}{m_s \ln(p_{cr}/p)}\cdot\frac{\dot{p}}{p} \tag{7}$$

Equation (7) describes the entropy increase rate for general mechanical process and the proposed model was successfully applied to the low-cycle fatigue life prediction of metallic materials [21]:

$$\frac{N_{in}}{N_f} = -\frac{N_0 k_0 \alpha}{s_{cr} m_s}\left[\ln\left(\ln\left(\frac{p_{cr}}{p}\right)\right) - \ln\left(\ln\left(\frac{p_{cr}}{p_{th}}\right)\right)\right] \tag{8}$$

All the parameters in Equation (8) have clear physical meaning, where N_0 and k_0 are physical constants; α, m_s, s_{cr}, p_{th}, and p_{cr} are parameters related to the material properties and can be obtained from experiments. It should be note that Equation (8) can also be applied in the accelerated fatigue test; the fatigue life can be obtained by the same initial cycles of fatigue with a well-determined database.

3. The Relation of Entropy Increase Rate and Normalized Creep Time

As the creep process is also an irreversible degradation process, Equation (7) can be applied in the creep process.

To investigate the increase rate of entropy in the creep process, a wide variety of creep experimental data for metals and alloys from literature were adopted [29–38]. Detailed experimental data sources are listed in Table 1 and summarized as follows: Creep tests for 9Cr–1Mo steel were performed at different temperatures (500 °C, 550 °C, 600 °C, and 650 °C) under various stress levels from 80 MPa to 320 Mpa by using a uniaxial-load creep test frame [29,30]. Creep tests for 9Cr–3W–3Co–1CuVNbB were performed at different temperatures (625 °C, 650 °C, and 675 °C) and stress levels (120–220 MPa) by using creep machines (RDJ 50 CRIMS) [31]. Creep tests for aluminum alloys [32,33] were adopted to verify the proposed model. Creep tests for Bar 257 were performed at 650 °C with a stress range from 70 MPa to 100 MPa [34]. Creep tests of Ni-base superalloy were performed at different directions and heat treatments [35]. Creep tests of Q460 steel were carried out at nine temperatures in the range of 300–900 °C and at various stress levels ranging from 13 MPa to 509 MPa [36]. Constant load creep tests of Co–Cr–Mo alloy were conducted at a temperature range of 650–800 °C and a stress range of 240–330 MPa [37]. The creep samples of DZ125 were machined such that the applied stress is along the [001] orientation [38]. The creep entropy increase rate was determined by Equation (7). The cumulative plastic strain rate was obtained by taking the slope of figures, which contains two coordinate axes: Creep strain and creep time. Other parameters were determined in the previous research [21]. The negative entropy increase rate (−dS/dt) with normalized creep time (t/tf) for materials at different temperatures and applied creep stresses are shown in Figure 1.

Table 1. The experimental data adopted in the current study.

Materials	Experimental Data Sources
9Cr–1Mo steel	[29,30]
9Cr–3W–3Co–1CuVNbB martensite ferritic steel	[31]
Aluminium alloy at 150 °C	[32]
Al 2024-T3	[33]
Al 2124	[33]
Bar 257 steel at 650 °C	[34]
Ni-base superalloy	[35]
Q460 steel	[36]
Co–Cr–Mo alloy	[37]
DZ125 super alloy	[38]

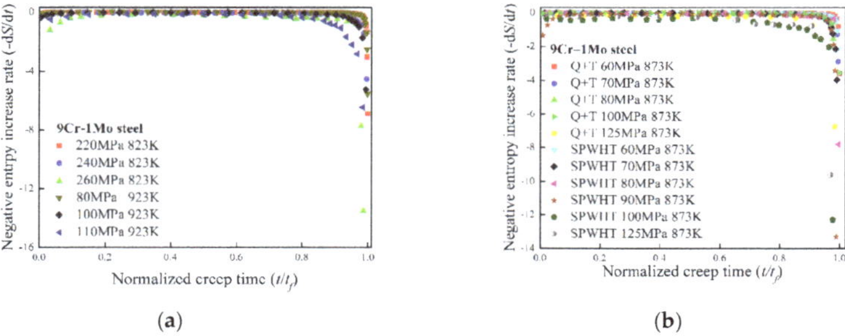

(a) (b)

Figure 1. *Cont.*

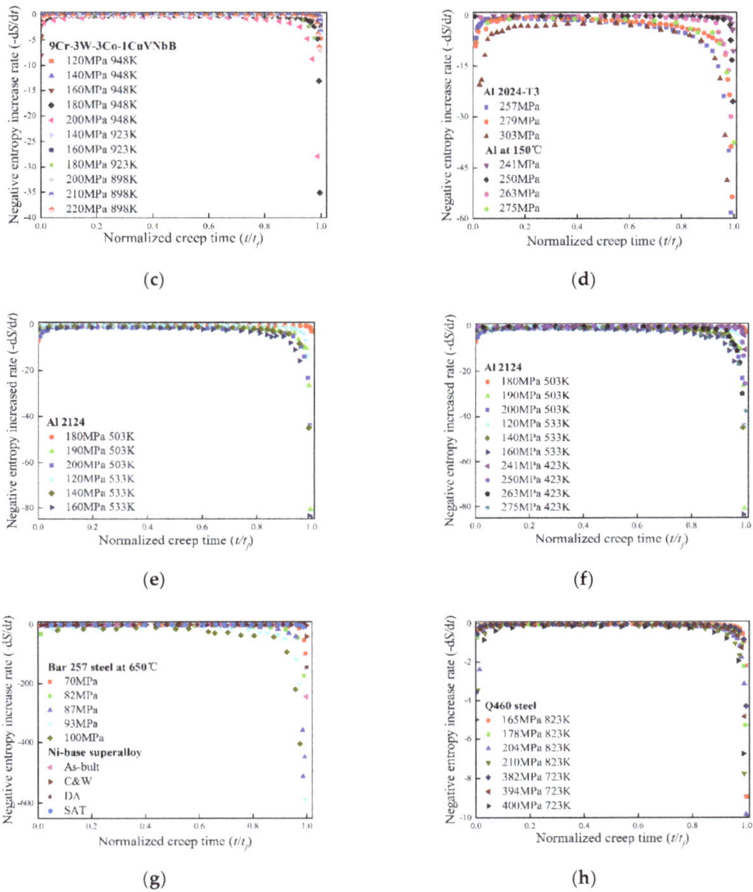

Figure 1. Change of the entropy increase rate with normalized creep time in the whole creep process: (**a**) 9Cr–1Mo steel at different temperatures; (**b**) 9Cr–1Mo steel under different treatment conditions (Q + T and SPWHT); (**c**) 9Cr–3W–3Co–1CuVNbB steel at different temperatures; (**d**–**f**) aluminum alloy at different temperatures; (**g**) Bar 257 steel and Ni-base super alloy at different temperatures; (**h**) Q460 steel at different temperatures.

As shown in Figure 1, the variation trend of entropy increase rate in creep process remains almost identical for different materials. In the early stage of creep, the dislocation multiplication and continuous movement lead to hardening of material. The entropy increasing rate decreases rapidly with time and then reaches a balance state. In the second stage, the creep strain rate achieves the minimum value and the entropy of the system increases at a fixed rate. In the last stage, the creep strain rate increases rapidly until the final fracture. The massive point defect separates out quickly at the grain boundary. The vacancy defect accelerates the creep strain rate and the final fracture. The entropy increasing rate of the system also increases rapidly in the last creep stage, which corresponds to increasing of the disorder degree of microstructure; the entropy increase rate attains infinity large when the final fracture occurs.

It should be noted that the entropy increase rate always keeps positive although its value reduces first and then increases; this phenomenon is consistent with the second law of thermodynamics.

Comparison with experimental data shows that degeneration process during creep can be well represented by Equation (7).

To describe the variation trend of entropy increase rate with normalized creep time, an empirical formula is proposed based on the boundary features and the characteristics of Figure 1:

$$\dot{S} = A \cdot \frac{1}{t} \cdot \frac{1}{\ln\left(t/t_f\right)} \tag{9}$$

where A is a parameter related to the constant creep stress, environment temperature, and material properties; t_f is the time when final creep fracture occurs. The curve shape of Equation (9) shows sufficient similarity to ensure the characterization of entropy increase rate by continuous functions.

Experimental data for different metallic materials have been adopted to verify Equation (9). For simplification, six group of experimental data for different material are shown in Figure 2. In general, the theoretical predictions agree well with the experimental data.

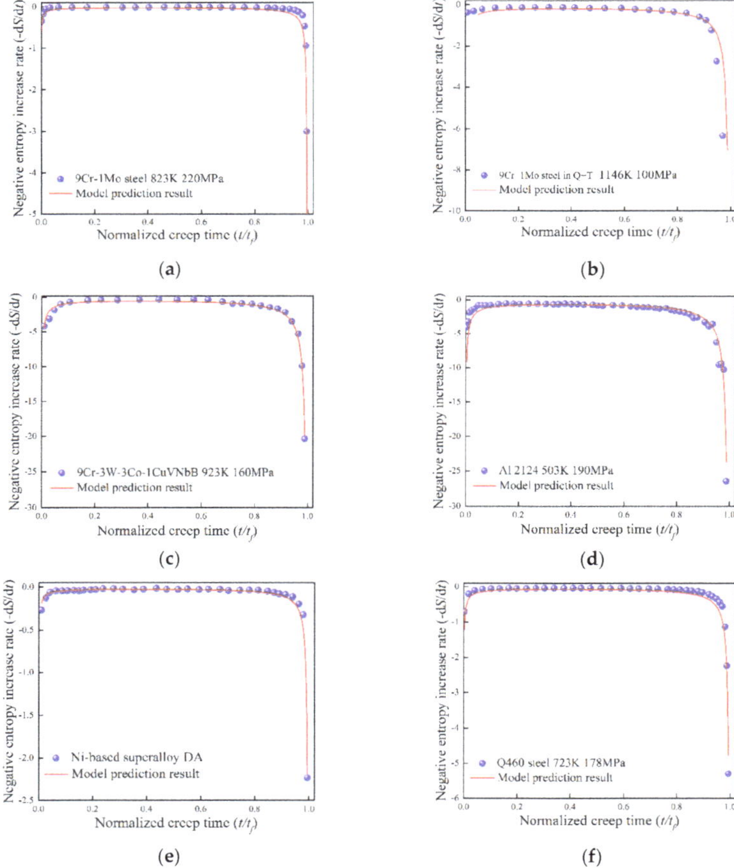

Figure 2. Verification of the proposed model compared with experimental data: (**a**) 9Cr–1Mo steel at 823 K with creep stress = 220 MPa [26]; (**b**) 9Cr–1Mo steel in quenched and tempered (Q + T) at 1146 K with creep stress = 100 MPa [27]; (**c**) 9Cr–3W–3Co–1CuVNbB at 923 K with creep stress = 160 MPa [28]; (**d**) Al 2124 at 503 K with creep stress = 190 MPa [30]; (**e**) Ni-based super alloy direct aging treatment at 718 °C/8 h/FC + 621 °C/10 h/AC [32]; (**f**) Q460 at 723 K with creep stress = 178 MPa [33].

As shown in Figure 2, the entropy increase rate can be well described by Equation (9). The slight fluctuation of some experimental data points in the stable creep stage may originate from the deviation of data recording form reference. Entropy approaches infinity when the final fracture occurs, thus the change of entropy increase rate in the first creep stage is smaller than that of the third stage. While in the first stage, the entropy is limited although the entropy increase rate is discontinuous at the moment of applying stress. In the stable creep stage, the entropy increase rate approaches zero and this phenomenon is consistent for different experimental conditions. Thus, the increasing of entropy during creep process is related to the change of dislocation, which corresponds to the microstructure changes in the thermodynamic level.

4. The Entropy-Based Creep Strain Prediction Model

The change regulation of entropy increase rate during creep process is investigated. From Equations (7) and (9), an entropy-based creep strain rate prediction equation can be obtained:

$$\dot{p} = \frac{Am_s p}{N_0 k_0 \alpha} \cdot \frac{1}{t ln\left(t/t_f\right)} \tag{10}$$

Equation (10) can be regarded as an ordinary differential equation. After adjustment of Equation (10) and integral on both sides, the creep strain can be obtained:

$$p = p_{cr} / \exp\left(\exp\left(\ln\left(\ln\left(\frac{p_{cr}}{p_{th}}\right)\right) - B\left[\ln\left(\ln\left(\frac{t_f}{t}\right)\right) - \ln\left(\ln\left(\frac{t_f}{t_{th}}\right)\right)\right]\right)\right) \tag{11}$$

$$B = \frac{Am_s}{N_0 k_0 \alpha} \tag{12}$$

where p_{th} is the initial value of cumulative plasticity in the microstructure of material, which represents the value at the beginning of creep damage accumulation; p_{cr} is the threshold value of cumulative plastic variable when creep failure occurs. The corresponding creep time when $p = p_{th}$ and $p = p_{cr}$ are t_{th} and t_f, respectively. The parameter B is related to the applied stress, temperature, and material properties. Equation (11) can be used to predict the creep strain during the creep damage process.

To verify the developed model, experimental data for different metals and alloys were adopt for comparison. The main parameter in Equation (11) is B, which is related to the applied stress, temperature, and material properties such as elastic modulus. It can be obtained by fitting the experimental data through Equation (11). The threshold value of cumulative plastic strain and threshold time can be obtained from experiments. The initial value of cumulative plastic strain and creep time is determined by taking the first group of experimental data point and make the iterative operations. The predictions of the developed model are compared with experimental data for different materials in the following sections.

4.1. Ni-Base Super Alloy

The creep properties of Ni-base super alloy at different build directions and heat treatments were studied experimentally by Kuo et al. [35]. Comparison of theoretical prediction with experimental data is shown in Figure 3. In general, the predictions agree well with experimental results. The applied stress (550 MPa) and temperature (923 K) remain unchanged for different test conditions. The main difference comes from the microstructure, which is reflected by different values of B in the developed model. However, it is difficult to effectively quantify microstructure for Ni-base super alloy at different build directions and heat treatments and establish the relation between B and microstructure.

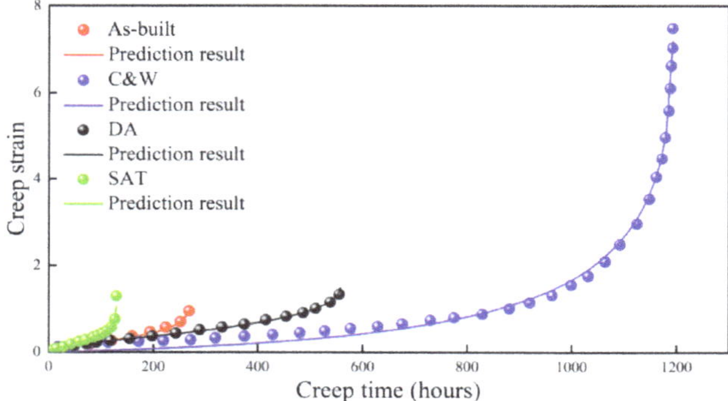

Figure 3. Prediction of Ni-base super alloy at different build directions and heat treatments. C&W: Cast-and-wrought; STA: Solution treatment and aging treatment; DA: Direct-aging treatment.

4.2. Q460 Steel

The creep property of high-strength Q460 steel at different temperatures was studied by Wang et al. [36]. Decreasing of the maximum creep strain with the stress level was considered because sufficient plasticity can be developed with longer duration [36]. The maximum creep strain shows no obvious relation with creep stress for most of the cases. As shown in Figure 4, the prediction results agree well with the experimental data for both 723 K and 823 K cases. The absolute value of B is obtained by fitting the experimental data, which increases with the applied stress. The detailed discussion of B is given in the next section.

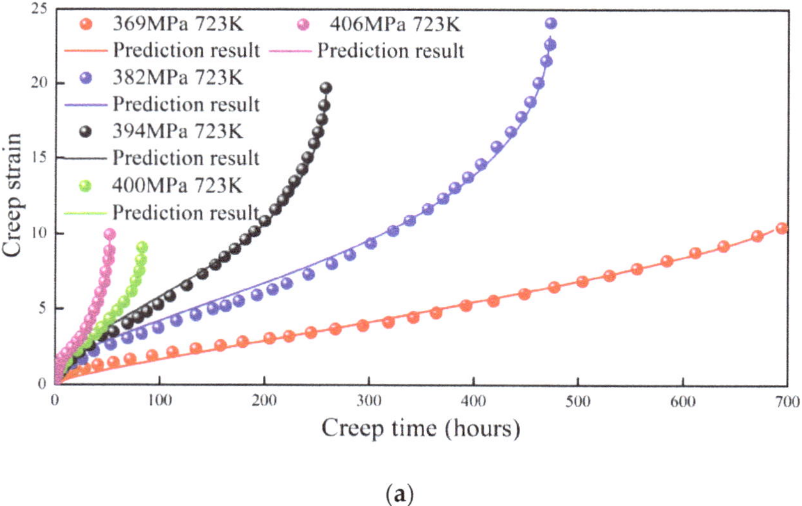

(a)

Figure 4. *Cont.*

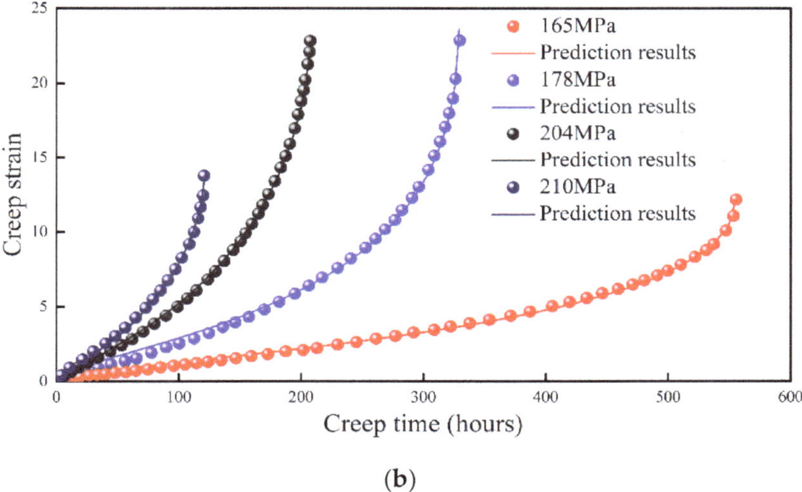

(b)

Figure 4. Prediction result of high strength Q460 steel at different temperatures; (**a**) 723 K; (**b**) 823 K.

4.3. Bar 257 Steel at 923 K

As shown in Figure 5, the predictions of proposed model agree well with the experimental data for Bar 257 steel at 923 K [34], except for the applied stress of 70 MPa. The predicted creep strain is slight larger around the final fracture region. The absolute value of B also increases with the applied stress. The prediction with an applied stress of 87 MPa does not cover the last two data points as the prediction is composed of 1000 data points; this problem can be solved by making the data points denser (for instance, 3000 data points).

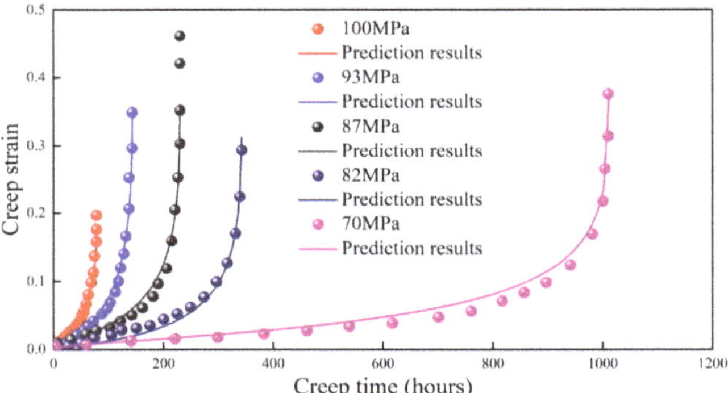

Figure 5. Prediction result of Bar 257 steel at 923 K.

4.4. Al 2124 at 503 K and 533 K

The high-temperature creep behaviors of Al 2124 at 473 K, 503 K, and 533 K were studied by Li et al. [33]. The theoretical prediction results are shown in Figure 6. Because the discontinuity of experimental data at 473 K (235 MPa) during the creep damage process, only the experimental data of 503 K and 533 K were selected to benchmark the proposed model. It shows that the prediction results agree well with the experimental data. With increasing of the applied stress, the absolute value of B increases as well.

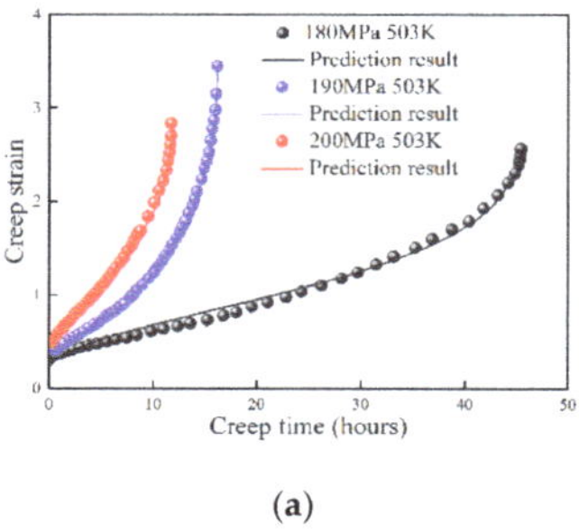

(a)

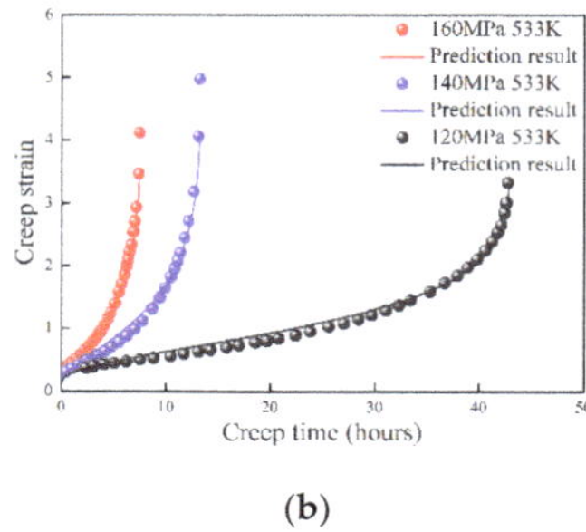

(b)

Figure 6. Prediction results of Al 2124 at (**a**) 503 K and (**b**) 533 K.

4.5. Cr–1Mo Steel at 823 K and 923 K

The creep behavior of modified 9Cr–1Mo steel at 823 K and 923 K were studied by Zhang et al. [30]. Comparison of the experimental data and prediction results are shown in Figure 7. The maximum creep strain approaches at 923 K for different stresses. However, this phenomenon was not observed for most of the other metals and alloys. Thus, it is hard to take the maximum plastic strain to evaluate the creep life [30]. The creep strain was well predicted with creep time by the proposed model. The absolute value of B increases with the applied stress.

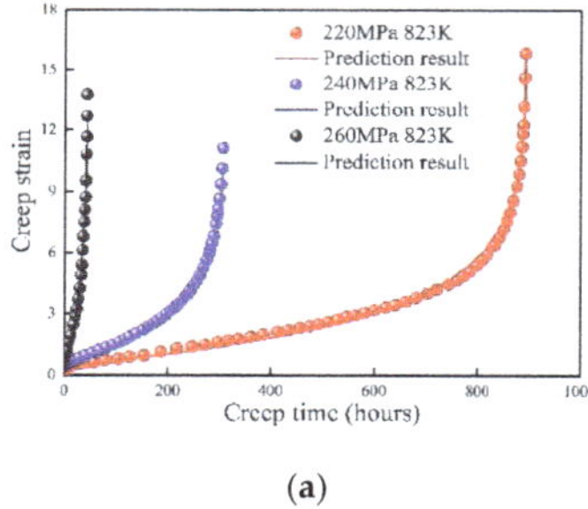

(a)

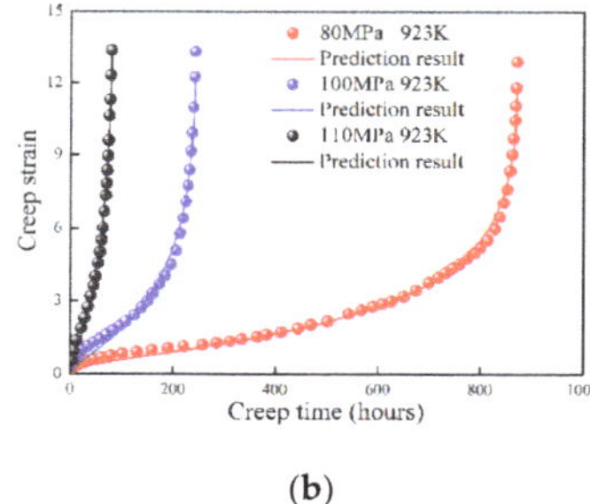

(b)

Figure 7. Prediction results of 9Cr–1Mo steel at (**a**) 823 K and (**b**) 923 K.

4.6. Cr–3W–3Co–1CuVNbB Martensite Ferritic at 898 K, 923 K, and 948 K

Xiao et al. [31] systematically investigated the creep behavior of 9Cr–3W–3Co–1CuVNbB martensite ferritic steel for a temperature range of 898 K to 948 K under uniaxial tensile stress from 120 to 220 MPa. As shown in Figure 8, predictions of the proposed model agree well with the experimental data. For the creep behavior of 9Cr–3W–3Co–1CuVNbB steel under 898 K, the first creep stage has a relatively longer time compared with other temperatures. It is difficult for the traditional models to predict the creep strain under these circumstances (it only maintains a certain precision in the first or third stages) [32]. The predictions of proposed model agree well with the experimental data except under the applied stress of 120 MPa at 948 K. Considering that the proposed model is single-parameter, the accuracy of prediction is acceptable.

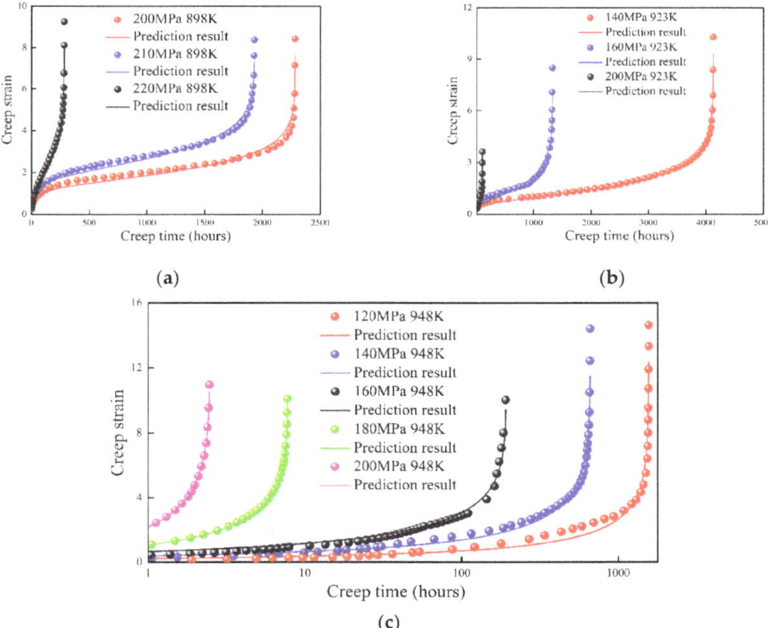

Figure 8. Prediction result of 9Cr–3W–3Co–1CuVNbB martensite ferritic steel at (**a**) 898 K, (**b**) 923 K, and (**c**) 948 K.

4.7. Cr–1Mo Ferritic Steel in Quenched and Tempered (Q+T) and Simulated Post Weld Heat Treatment (Spwht) Conditions

The creep behavior of 9Cr–1Mo ferritic steel under SPWHT and Q + T treatment was studied by Choudhary [30]. The experimental data are compared with theoretical prediction, as shown in Figure 9. The predictions of proposed model agree well with the experimental data except under Q + T and SPWHT condition with the applied stress of 90 MPa. The second case is because the estimated p_{th} is not converged. The prediction is obtained by setting p_{th} as the first experimental data point. At the same time, the creep experiments usually accompanied with a certain degree of discreteness.

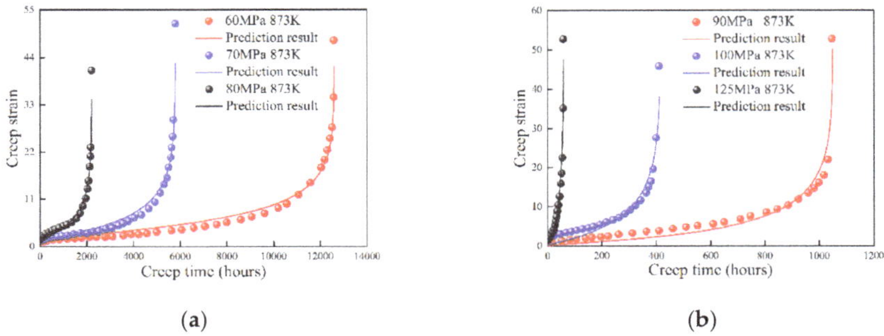

Figure 9. *Cont.*

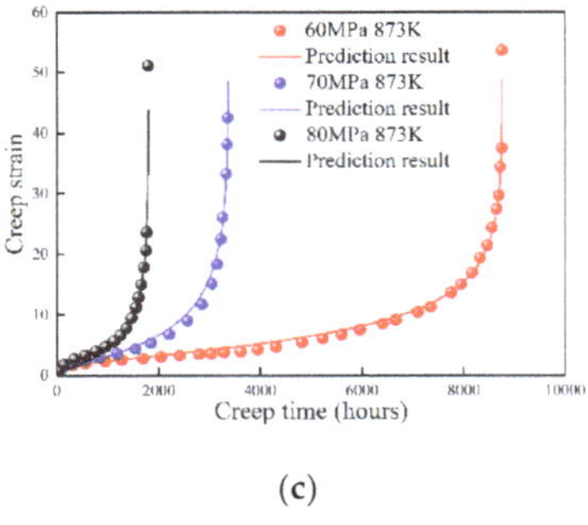

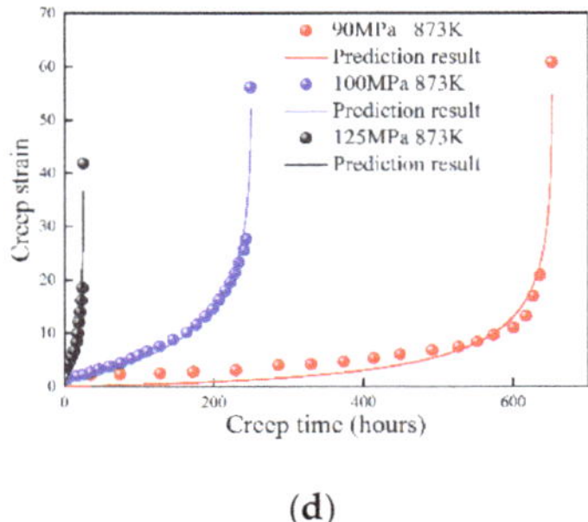

(c) (d)

Figure 9. Prediction result of 9Cr−1Mo ferritic steel in quenched and tempered (Q + T) and simulated post weld heat treatment (SPWHT) conditions: (**a**) Q + T 60~80 MPa; (**b**) Q + T 90~125 MPa; (**c**) SPWHT 60~80 MPa; and (**d**) SPWHT 90~125 MPa.

4.8. Co–Cr–Mo Alloy

The creep behavior of Co−Cr−Mo alloy at different temperatures (923 K, 973 K, 1023 K, 1073 K) was investigated by Sun et al. [37]. The experiments were carried out at a constant applied stress (240 MPa). The theoretical prediction is compared with experimental data, as shown in Figure 10. The transverse axes is taken as logarithmic coordinates to make the difference of experimental data more obvious. Generally, the prediction results agree well with the experimental data.

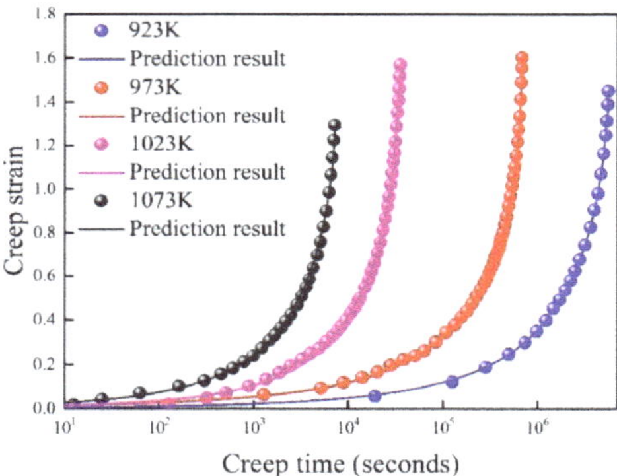

Figure 10. Prediction result of Co–Cr–Mo alloy at different ambient temperature.

4.9. DZ125 Super Alloy

The creep behavior of DZ125 super alloy at different creep stresses and temperatures was investigated by Fu et al. [38]. The theoretical prediction is compared with experimental data, as shown in Figure 11. The theoretical predictions agree well with the all five groups of experimental data.

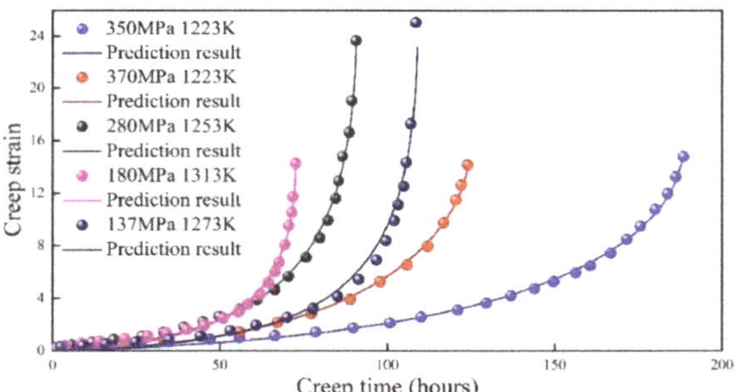

Figure 11. Prediction result of DZ125 super alloy at different creep stresses and temperatures.

5. Parametric Analysis of the Proposed Model

As is well known, the creep strain is strongly influenced by the applied stress and temperature. The creep strain will increase with higher applied stress and temperature. These effects cannot be ignored in the creep analysis. Therefore, the parameter B in Equation (11) should be associated with the applied stress and temperature. Assuming that these two effects are irrelevant, which is commonly accepted in the traditional models [39], the parameter B should have a form as follows:

$$B = M \cdot f(\sigma) \cdot g(T) \tag{13}$$

where M is a coefficient that contains α, m_s, k_0, and N_0. The relation between B and applied stress are verified by the experimental data in Section 4. The values of B and its linear fitting with applied stress are shown in Figure 12:

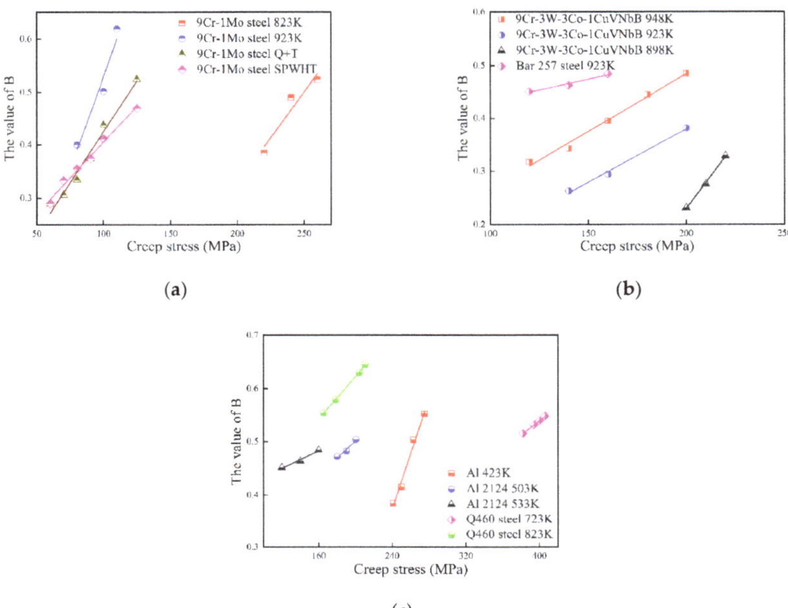

Figure 12. The relation between the parameter B and applied stress. (**a**) 9Cr-1Mo steel; (**b**) 9Cr-3W-3Co-1CuVNbB and Bar 257 steel; (**c**) Al and Q460 steel.

As shown in Figure 12, with increasing of the applied stress, the value of B also shows a near-linear increasing trend for different materials. In the current study, the value of B has a relative concentration range of 0.2~0.7 (maximum = 0.645 for Q460 steel at 210 MPa 823 K; minimum = 0.231 for 9Cr–3W–3Co–1CuVNbB at 210 MPa 898 K).

The relationship between *B* and temperature is shown in Figure 13. With the increase in temperature, the value of B also increases for 9Cr–3W–3Co–1CuVNbB and Co–Cr–Mo alloy. Although the relationship between *B* and temperature approaches linear for Co–Cr–Mo alloy, determination of an accurate relationship still requires more experimental data.

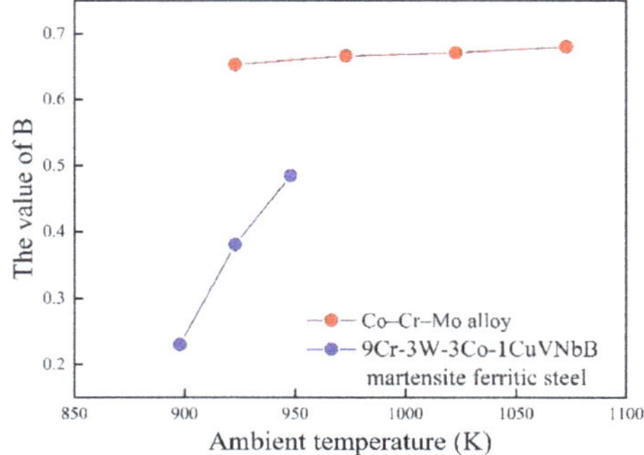

Figure 13. The relation between the parameter *B* and temperature.

6. Conclusions

Based on the continuum damage mechanics and the statistical definition of entropy, the entropy increasing rate during creep process is investigated. The conclusions are summarized as follows:

1. The entropy increasing rate for creep is investigated with experimental data for different metallic materials. Similar entropy increasing trend is observed with normalized creep time. A theoretical model is proposed to describe the relationship based on the characteristic of boundary conditions. Comparison with experimental data shows that the developed model gives reasonably accurate estimation of entropy increasing rate in the creep process.
2. An entropy-based creep strain prediction model is proposed with respect to the entropy increasing rate. Predictions of the proposed model agree well with the experimental data for different metallic materials.
3. The single parameter B in the proposed model is associated with the applied stress and temperature. In general, the parameter increases linearly with the applied stress.

Author Contributions: Conceptualization, Y.Y.; supervision, Y.Y.; writing-original draft, J.W.

Funding: This research was funded by National Natural Science Foundation of China: 11572249,11772257.

Acknowledgments: The authors acknowledge the financial support by National Natural Science Foundations of China (11572249 and 11772257) and Fundamental Research Funds for the Central Universities (No. G2019KY05212).

Conflicts of Interest: The authors declare no conflict of interest.

References

1. Burke, P.M.; Sherby, O.D. Mechanical Behavior of Crystalline Solids at Elevated Temperature. *Progr. Mater. Sci.* **1968**, *13*, 323–390.
2. Brehm, H.; Daehn, G.S. A framework for modeling creep in pure metals. *Metall. Mater. Trans. A* **2002**, *33*, 363–371. [CrossRef]

3. Kassner, M.E.; PérezPrado, M.T. Five-power-law creep in single phase metals and alloys. *Progr. Mater. Sci.* **2000**, *45*, 1–102. [CrossRef]
4. Yao, Y.; Wang, J.D.; Keer, L.M. A phase transformation based method to predict fatigue crack nucleation and propagation in metals and alloys. *Acta Mater.* **2017**, *127*, 244–251. [CrossRef]
5. Yao, Y.; Long, X.; Keer, L.M. A review of recent works on mechanical behavior of lead-free solder materials. *Appl. Mech. Rev.* **2017**, *69*, 040802. [CrossRef]
6. Yao, Y.; Fine, M.E.; Keer, L.M. An energy approach to predict fatigue crack propagation in metals and alloys. *Int. J. Fract.* **2007**, *146*, 149–158. [CrossRef]
7. Monkman, F.C.; Grant, N.J. An empirical relationship between rupture life and minimum creep rate in creep rupture tests. *Proc. Am. Soc. Test. Mater.* **1956**, *56*, 593–620.
8. Dobes, F.; Milicka, K. The relation between minimum creep rate and time to fracture. *Met. Sci.* **1976**, *10*, 382–384. [CrossRef]
9. Rabotnov, Y.N.; Leckie, F.A.; Prager, W. Creep Problems in Structural Members. *J. Appl. Mech.* **1970**, *37*, 249. [CrossRef]
10. Leckie, F.A.; Hayhurst, D.R. Constitutive equations for creep rupture. *Acta Metall.* **1977**, *25*, 1059–1070. [CrossRef]
11. Ashby, M.F.; Dyson, B.F. *Advances in Fracture Research*; Pergamon Press: Oxford, UK, 1984.
12. Dyson, B.F.; Gibbons, T.B. Tertiary creep in nickel-base superalloys: Analysis of experimental data and theoretical synthesis. *Acta Metall.* **1987**, *35*, 2355–2369. [CrossRef]
13. Fields, B.A.; Fields, R.J. *Elevated Temperature Deformation of Structural Steel*; Report NISTIR 88-3899; NIST: Gaithersburg, MA, USA, 1989.
14. Cowan, M.; Khandelwal, K. Modeling of high temperature creep in ASTM A992 structural steels. *Eng. Struct.* **2004**, *80*, 426–434. [CrossRef]
15. Kozyrev, Y.P.; Sedakova, E.B. Application of a thermodynamic model for analysis of wear resistance of materials. *J. Mach. Manuf. Reliab.* **2008**, *37*, 60–62.
16. Bryant, M.D. Entropy and Dissipative Processes of friction and Wear. *FME Trans.* **2009**, *37*, 55–60.
17. Amiri, M.; Khonsari, M.M. On the Thermodynamics of Friction and Wear—A Review. *Entropy* **2010**, *12*, 1021–1049. [CrossRef]
18. Sosnovskiy, L.A.; Sherbakov, S.S. A Model of Mechanothermodynamic Entropy in Tribology. *Entropy* **2016**, *19*, 115. [CrossRef]
19. Sosnovskiy, L.A.; Sherbakov, S.S. Mechanothermodynamical system and its behavior. *Contin. Mech. Thermodyn.* **2012**, *24*, 239–256. [CrossRef]
20. Sosnovskiy, L.A.; Sherbakov, S.S. *Mechanothermodynamics*; Springer: Berlin/Heidelberg, Germany, 2016.
21. Wang, J.D.; Yao, Y. An Entropy Based Low-Cycle Fatigue Life Prediction Model for Solder Materials. *Entropy* **2017**, *19*, 503. [CrossRef]
22. Boltzmann, L. *Lectures on Gas Theory*; University of California Press: Berkeley, CA, USA, 1964.
23. Basaran, C.; Gomez, J.; Lin, M. Damage Mechanics Modeling of Concurrent Thermal and Vibration Loading on Electronics Packaging. *Multidiscip. Model. Mater. Struct.* **2013**, *2*, 309–326.
24. Basaran, C.; Yan, C.Y. A Thermodynamic Framework for Damage Mechanics of Solder Joints. *J. Electron. Packag.* **1998**, *120*, 379–384. [CrossRef]
25. Basaran, C.; Tang, H. Implementation of a Thermodynamic Framework for Damage Mechanics of Solder Interconnects in Microelectronic Packaging. *Int. J. Damage Mech.* **2002**, *11*, 87–108. [CrossRef]
26. Basaran, C.; Lin, M.; Ye, H. A thermodynamic model for electrical current induced damage. *Int. J. Solids Struct.* **2003**, *40*, 7315–7327. [CrossRef]
27. Basaran, C.; Nie, S. An Irreversible Thermodynamics Theory for Damage Mechanics of Solids. *Int. J. Damage Mech.* **2004**, *13*, 205–223. [CrossRef]
28. Bonora, N. A nonlinear CDM model for ductile failure. *Eng. Fract. Mech.* **1997**, *58*, 11–28. [CrossRef]
29. Shrestha, T.; Basirat, M.; Charit, I.; Potirniche, G.P.; Rink, K.K.; Sahaym, U. Creep deformation mechanisms in modified 9Cr–1Mo steel. *J. Nucl. Mater.* **2012**, *423*, 110–119. [CrossRef]
30. Choudhary, B.K. Tertiary creep behaviour of 9Cr–1Mo ferritic steel. *Mater. Sci. Eng. A* **2013**, *585*, 1–9. [CrossRef]

31. Xiao, B.; Xu, L.; Zhao, L.; Jing, H.; Han, Y.; Zhang, Y. Creep properties, creep deformation behavior, and microstructural evolution of 9Cr–3W–3Co–1CuVNbB martensite ferritic steel. *Mater. Sci. Eng. A* **2017**, *711*, 434–447. [CrossRef]
32. Li, B.; Lin, J.; Yao, X. A novel evolutionary algorithm for determining uniÿed creep damage constitutive equations. *Int. J. Mech. Sci.* **2002**, *44*, 987–1002. [CrossRef]
33. Li, L.T.; Lin, Y.C.; Zhou, H.M.; Jiang, Y.Q. Modeling the high-temperature creep behaviors of 7075 and 2124 aluminum alloys by continuum damage mechanics model. *Comput. Mater. Sci.* **2013**, *73*, 72–78. [CrossRef]
34. Hyde, T.H.; Becker, A.A.; Sun, W.; Williams, J.A. Finite-element creep damage analyses of P91 pipes. *Int. J. Press. Vessel. Pip.* **2006**, *83*, 853–863. [CrossRef]
35. Kuo, Y.L.; Horikawa, S.; Kakehi, K. Effects of build direction and heat treatment on creep properties of Ni-base superalloy built up by additive manufacturing. *Scr. Mater.* **2017**, *129*, 74–78. [CrossRef]
36. Wang, W.; Yan, S.; Liu, J. Studies on temperature induced creep in high strength Q460 steel. *Mater. Struct.* **2017**, *50*, 68. [CrossRef]
37. Sun, S.H.; Koizumi, Y.; Kurosu, S.; Lib, Y.P.; Chiba, A. Phase and grain size inhomogeneity and their influences on creep behavior of Co–Cr–Mo alloy additive manufactured by electron beam melting. *Acta Mater.* **2015**, *86*, 305–318. [CrossRef]
38. Fu, C.; Chen, Y.D.; Yuan, X.F.; Tin, S.; Antonov, S.; Yagi, K. A modified θ projection model for constant load creep curves-II Application of creep life prediction. *J. Mater. Sci. Technol.* **2019**, *35*, 687–694. [CrossRef]
39. Kraus, H.; Saunders, H. Creep Analysis. *J. Mech. Des.* **1982**, *104*, 530. [CrossRef]

Article

Intelligent Analysis Algorithm for Satellite Health under Time-Varying and Extremely High Thermal Loads

En-Hui Li [1], Yun-Ze Li [1,2,3,*], Tian-Tian Li [1], Jia-Xin Li [1], Zhuang-Zhuang Zhai [4] and Tong Li [5]

1 School of Aeronautic Science and Engineering, Beihang University, Beijing 100191, China; lienhui@buaa.edu.cn (E.-H.L.); litiantian@buaa.edu.cn (T.-T.L.); jxin.lee@buaa.edu.cn (J.-X.L.)
2 Institute of Engineering Thermophysics, North China University of Water Resources and Electric Power, Henan 450045, China
3 Advanced Research Center of Thermal and New Energy Technologies, Xingtai Polytechnic College, Hebei 054035, China
4 School of Automation Science and Electrical Engineering, Beihang University, Beijing 100191, China; zhaizz@buaa.edu.cn
5 Chengyi Academy of PKUHS, Peking University, Beijing 100080, China; litong@i.pkuschool.edu.cn
* Correspondence: liyunze@buaa.edu.cn; Tel.: +86-10-82338778; Fax: +86-10-82315350

Received: 12 September 2019; Accepted: 8 October 2019; Published: 10 October 2019

Abstract: This paper presents a dynamic health intelligent evaluation model proposed to analyze the health deterioration of satellites under time-varying and extreme thermal loads. New definitions such as health degree and failure factor and new topological system considering the reliability relationship are proposed to characterize the dynamic performance of health deterioration. The dynamic health intelligent evaluation model used the thermal network method (TNM) and fuzzy reasoning to solve the problem of model missing and non-quantization between temperature and failure probability, and it can quickly evaluate and analyze the dynamic health of satellite through the collaborative processing of continuous event and discrete event. In addition, the temperature controller in the thermal control subsystem (TCM) is the target of thermal damage, and the effects of different heat load amplitude, duty ratio, and cycle on its health deterioration are compared and analyzed.

Keywords: satellite; dynamic health evaluation; fuzzy reasoning

1. Introduction

With the development of satellite space missions, satellite health is facing severe challenges due to the drastic changes of thermal environment and internal thermal load. Due to the change of satellite orbit or transfer [1] or the different working modes of satellite components, time-varying thermal load will be caused. In addition, due to the highly integrated package of electronic equipment [2,3], the use of high heat flux density components [4] and wireless energy transmission, the transient thermal load of the satellite is extremely large. Such as lasers, electronic chips, and advanced propulsion devices, are expected to involve high heat fluxes (above 100 W/cm^2) [5,6]. Time variations and extremely high thermal loads can affect satellite health and even lead to satellite system failures [7,8]. Therefore, the rapid and effective dynamic health assessment of satellites is of great significance.

According to data from insurance analysis, from 25% to 28% [9,10] of satellite failures on orbit, are related to the electrical power system. The main failure mode of electronic equipment is thermal failure, and its failure rate increases exponentially with increasing temperature [11]. The traditional researches on the influence of temperature on the failure rate of components are based on empirical models, such as the exponential model of Arrhenius [12]. Therefore, there is a lack of accurate quantitative relationship between temperature and failure rate.

The past research on satellite health is to calculate the lifetime and reliability of satellites through reliability analysis. Traditional reliability analysis is based on probabilistic statistical analysis of large amounts of in-orbit data. Therefore, to study the satellite fault caused by thermal failure requires necessary data to determine various parameters of the model. Unfortunately, limited empirical data and statistical analysis of satellite reliability exist in the technical literature [13]. Many methodologies such as failure mode effect analysis (FMEA) and the fault tree analysis (FTA) are used in the reliability analysis [14,15]. However, this model has some limitations in reliability analysis. It is not easy for these models to conduct further quantitative analysis automatically due to the lack of effective means of mathematical expression [16]. In addition, the traditional reliability analysis model only considers failure death or survival and lacks analysis methods for sub-health.

In order to analyze the satellite thermal health response under time-varying and extreme thermal loads, an intelligent evaluation model based on fuzzy logic is proposed. In addition, this paper presents an evaluation index that can represent satellite sub-health state. Fuzzy reasoning is used to solve the problem of missing models and non-quantization which are caused by temperature. The continuous process simulation and discrete event simulation of thermal load change complete efficient collaborative computation by thermal health assessment algorithm. Finally, the effects of typical thermal loads on satellite health are analyzed.

2. Dynamic Health Intelligent Evaluation Model

In order to analyze the health deterioration of satellites under time-varying and extreme thermal loads, the dynamic health intelligent evaluation model (DHIEM) is presented in this paper. Meanwhile, new definitions such as health degree and failure factor and new topological system (satellite-subsystems-components) considering the reliability relationship are proposed to characterize the dynamic performance of health deterioration. This section introduces the principle and algorithm of the model in detail.

2.1. Principle Description and Evaluation Index

The DHIEM is combined with the TNM. TNM solves the dynamic temperature distribution of the satellite based on thermal load, and DHIEM can dynamically evaluate the thermal health index of the satellite based on the output of TNM. So, this subsection first introduces the three definitions used in the model.

The failure probability of the component is obtained according to its dependence on temperature, as shown in formula 1. Then, according to the failure probability of the component, the health degree and failure factor of the subsystems can be obtained. Similarly, according to the health degree of the subsystems, the health degree and failure factor of the satellite can be calculated, as shown in Equations (2)–(3).

2.1.1. Definition of Evaluation Index

- Failure Probability $\xi(t)$:

The failure probability refers to the instantaneous probability of failure of the components when the components run to a certain time. The failure probability value is affected by temperature and satisfies the equation:

$$\xi(t) = f(T, dT/dt) \tag{1}$$

where, $\xi(t)$ is the failure probability value at a certain time; T and dT/dt are the temperature and its difference of the moment respectively; Function relation f is fuzzy inference.

- Health Degree $H(t)$:

Health degree is to evaluate the health of the whole system through the statistics of the damage of each component of the system. Health degree refers to the percentage of the number of units in the system that are not damaged at the current moment when the system runs to a certain time t.

$$H(t) = 1 - \frac{N_{failure}}{N_{total}} \tag{2}$$

where, N_{total} is the total number of the underlying component units constituting the system, and $N_{failure}$ is the number of failure failures of the underlying component units in the system at the current moment.

- Failure Factor $F(t)$:

Failure factor describes the deterioration speed of system failure. Failure factor refers to the ratio of the increment of failure sub-units in the system within unit time t to the number of healthy component units at time t.

$$F(t) = \frac{N_{failure}(t + \Delta t) - N_{failure}(t)}{\Delta t \cdot [N_{total} - N_{failure}(t)]} \tag{3}$$

where, $N_{failure}(t + \Delta t) - N_{failure}(t)$ is the number of newly added fault failure unit in time; $N_{total} - N_{failure}(t)$ is the number of remaining underlying component units, that is, the number of component units that have not failed by time t; Δt is the time interval taken.

2.1.2. Principle Description

The principle of satellite dynamic health intelligent evaluation model is shown in Figure 1. First, the dynamic temperature $T_n(t)$ of satellite component n under different thermal load conditions can be calculated by using the TNM. Considering the influence of temperature and its difference on the failure probability of satellite components, fuzzy reasoning is used to quantitatively analyze the failure probability $\xi_n(t)$ of components. Then, the health degree $H_k(t)$ and failure factor $F_k(t)$ of subsystem k are obtained by using the thermal health evaluation I to cooperative solve continuous process simulation and discrete event simulation. Finally, the thermal health evaluation II is used to calculate the satellite's health degree $H(t)$ and failure factor $F(t)$ according to the subsystem's health degree $H_k(t)$.

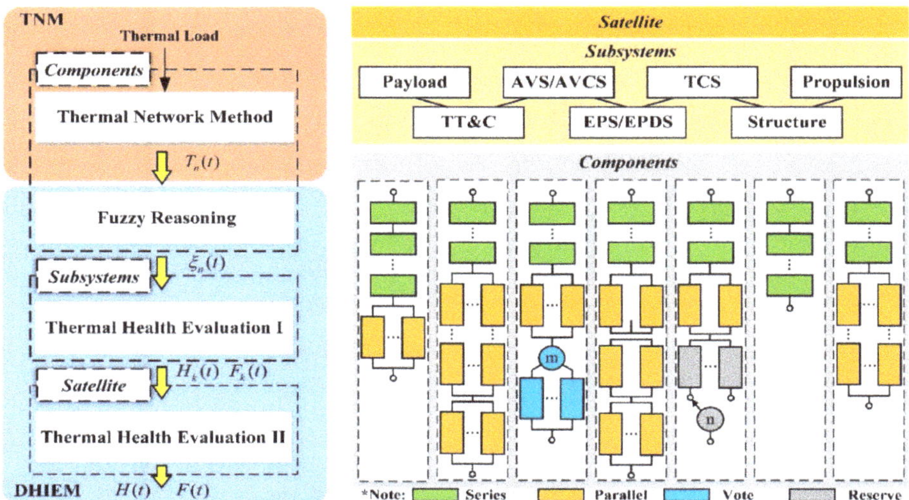

Figure 1. The principle of satellite dynamic health intelligent evaluation model: (**a**) Model schematic; (**b**) Topological system (satellite-subsystems-components).

The model divides the satellite into satellite-subsystems-components topological system according to its functional composition. In general, a satellite consists of a payload and the common subsystems supporting the payload. Typical public subsystems include structural subsystem, thermal control subsystem (TCS), energy/distribution subsystem, telemetry tracking and command subsystem (TT&C), attitude/orbit control subsystem and propulsion subsystem [17]. Each subsystem is composed of different functional components. The topological system takes into account the series, parallel, vote and reserve relations of various components of the satellite, so as to reflect the influence of component failure on subsystems or satellite health more accurately.

Section 2.2 uses the thermal network model to calculate the dynamic temperature distribution of components. In Section 2.3, the relationship between the failure probability of the component and the temperature of the component is analyzed quantitatively by fuzzy inference. Section 2.4 introduces a dynamic evaluation algorithm that calculates the health degree and failure factor of subsystems and satellite by analyzing the failure of components.

2.2. Component Temperature Dynamic Modeling

In order to study the effects of thermal damage on satellite health, it is necessary to calculate the dynamic distribution of satellite temperature. The thermal network model is used to solve the temperature distribution of satellite. In the satellite thermal network model, each equipment component is treated as an isothermal body and a node, and the node temperature represents the average temperature of the isothermal body. It is in itself an approximate method because of the discretization needed to solve the heat transfer differential equation. However, due to simplicity and agility, lumped parametric models are more and more widely used in satellite thermal analysis.

The TNM is described in detail elsewhere [18,19]. For any satellite component (node) j, the non-linear algebraic transient heat exchange equations that are obtained and that have to be solved is expressed by Equation (4).

$$(cm)_j\frac{dT_j}{dt} = Q_{sj} + Q_{pj} + \sum_i \left(\frac{A_c\lambda}{\delta}\right)_{i,j}(T_j - T_i) + \sum_i (\varepsilon A_r)_{i,j}\sigma(T_j^4 - T_i^4) \tag{4}$$

where, Q_{sj} represents the external heat absorbed by node j; Q_{pj} represents the heat that it is directly produced in the j node itself. Any satellite component (node) has heat exchange with other nodes through heat conduction and heat radiation. $\sum\limits_i (\frac{A_c\lambda}{\delta})_{i,j}(T_j - T_i)$ represents the thermal conduction heat transfer between node j and the rest of the nodes of the model. A_c is the thermal conductivity area, λ is the thermal conductivity coefficient, and δ is the thermal conductivity thickness. The subscript indicates that the three parameters corresponding to the heat conduction and exchange between different nodes have different values.$\sum\limits_i (\varepsilon A_r)_{i,j}\sigma(T_j^4 - T_i^4)$ represents the radiation heat transfer between node j and the rest of the nodes of the model.ε is the emissivity of the node, A_r is the radiation area of the node, and σ is the Stefan-Boltzmann constant. The relative positions of different nodes must also be considered in calculating the radiative heat transfer between them. $(cm)_j$ is the heat capacity of node j. The external heat flux Q_{sj} varies with the position and spatial attitude of the satellite in orbit, which is a function of time. The thermal power Q_{pj} may also be a function of time, depending on what the component needs to accomplish.

2.3. Component Failure Probability Fuzzy Modeling

In order to quantitatively analyze the influence of component temperature and change on its failure probability, an intelligent calculation model based on fuzzy reasoning is proposed. The failure probability analysis based on fuzzy logic solves the law analysis of the influence of temperature factor on the failure probability of components and completes the qualitative to quantitative analysis and calculation. As a kind of "grey box" system, fuzzy reasoning table has the characteristics of short development cycle, nonlinearity and no need to establish mathematical model, but its difficulty lies in the acquisition of fuzzy rules [20,21].

2.3.1. Fuzzy Reasoning

Fuzzy reasoning goes through three basic processes: fuzziness, fuzzy reasoning, and clearness. Fuzzification—compare input variables and membership functions to obtain membership values of each language identifier; Fuzzy reasoning—performs union operation (usually multiplication or minimization) on the membership function of the initial part to obtain the activation right of each rule, and relies on the activation right to produce the effective result of each rule; Clarity—overlay all valid results to produce a clear output.

The failure probability of unit is related not only to temperature but also to the rate of change of unit time temperature. Therefore, the intelligent evaluation model algorithm proposed in this paper adopts the fuzzy reasoning decision structure with double input and single output. The double inputs are the temperature and the rate of change of temperature per unit time, and the output is the failure probability of component units. The fuzzy reasoning decision system is shown in Figure 2.

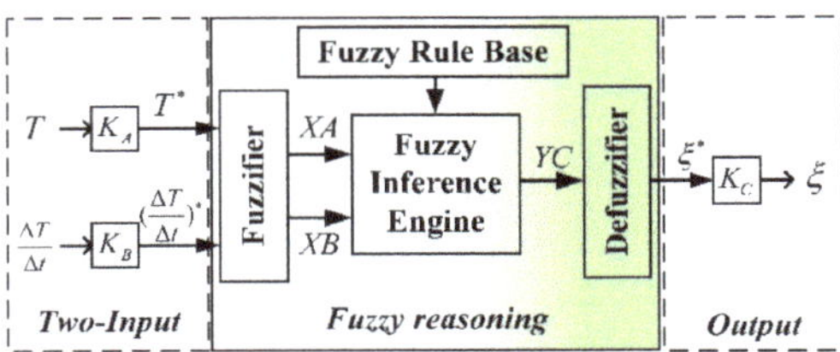

Figure 2. Fuzzy reasoning system.

Figure 2 shows the fuzzy reasoning system consisting of a fuzzifier, a fuzzy inference engine, a defuzzifier, and a fuzzy rule base. The double inputs to the fuzzifier are the temperature T^* and its difference $(\frac{\Delta T}{\Delta t})^*$ normalized by the factors K_A and K_B. Considering that the actual variation range of the theoretical domain of different component elements may be different, in order to normalize the model, we uniformly adopt the standardized theoretical domain [−1, 1] and use different transformation factors to transform the practical theoretical domain. Similarly, the output ξ scaled by the factor K_c. The transformation factor satisfies the following equation:

$$\begin{cases} T^* = K_A\left(T - \frac{T_{\max}+T_{\min}}{2}\right) \\ (dT/dt)^* = K_B\left[(dT/dt) - \frac{(dT/dt)_{\max}+(dT/dt)_{\min}}{2}\right] \end{cases}$$

$$\begin{cases} K_A = \frac{2}{T_{\max}-T_{\min}} \\ K_B = \frac{2}{(dT/dt)_{\max}-(dT/dt)_{\min}} \end{cases} \tag{5}$$

The input and output variables to the fuzzy reasoning system are characterized by the fuzzy sets, linguistic values, and associated analytical ranks shown in Table 1. In order to facilitate the formulation of fuzzy rule table, each fuzzy set is divided into 9 element levels. The subscript of the fuzzy set represents which element level the language value is related to. Each fuzzy set is defined by a Gaussian membership function shown in Figure 3. The membership functions have an overlap with each other to provide a smooth output transition between the regions. The input of fuzzy logic (temperature, temperature change rate per unit time) is not completely attributed to a certain class (fuzzy set). That is to say, there is no strict boundary between higher and higher temperatures, and fuzzy theory uses membership degree to measure the degree of attribution of variables to each set. The process of transforming the input of logic into the membership of each fuzzy set is called fuzzification.

Table 1. Fuzzy sets and linguistic values: (**a**) input; (**b**) output.

(a)			(b)		
Fuzzy Sets	**Ranks**	**Linguistic Values**	**Fuzzy Sets**	**Ranks**	**Linguistic Values**
NB	−4	Negative big	ZE	0	Zero
NM	−3	Negative medium	PZ1	1	Positive zero 1
NS	−2	Negative small	PZ2	2	Positive zero 2
NZ	−1	Negative zero	PS1	3	Positive small 1
ZE	0	Zero	PS2	4	Positive small 2
PZ	1	Positive zero	PM1	5	Positive medium 1
PS	2	Positive small	PM2	6	Positive medium 2
PM	3	Positive medium	PB1	7	Positive big 1
PB	4	Positive big	PB2	8	Positive big 2

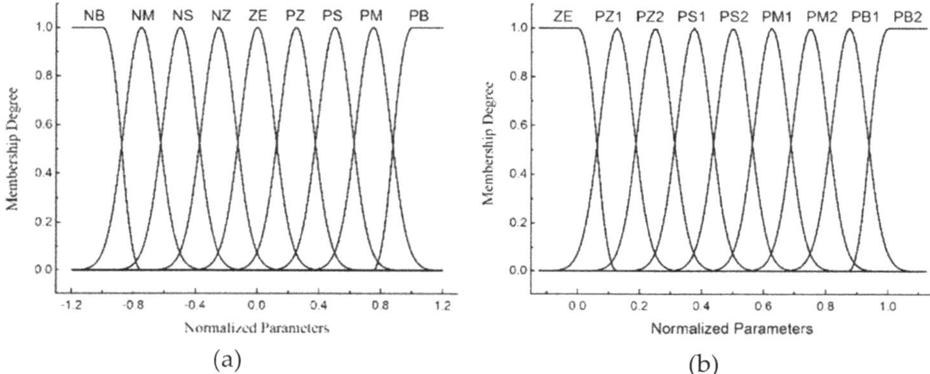

Figure 3. Membership degree functions: (**a**) input; (**b**) output.

The failure probability of a component is not only related to temperature, but also related to electrical stress, mechanical stress, fatigue and other factors. Some component failure probabilities are mainly affected by temperature, while some component temperatures have less influence on their failure probability. Therefore, the specific dependence of the failure probability of components of different functions on temperature determines the range of failure probability of the actual component. We also use the standardization domain for generalization transformation. Considering the non-negative nature of failure probability, the corresponding failure probability uses the standardization domain [0, 1], its fuzzy set and fuzzy language as Table 1(b).

The fuzzy reasoning output is determined using the linguistic rules in the following form:

$$\text{IF } T^* \text{ is } XA_i \text{ and } (\frac{\Delta T}{\Delta t})^* \text{ is } XB_j, \text{ THEN } \xi^* \text{ is } YC_{l(i,j)}$$

where XA_i, XB_j, and $YC_{l(i,j)}$ are the fuzzy sets reflecting the linguistic values of T^*, $(\frac{\Delta T}{\Delta t})^*$, and ξ^*, respectively, and the subscript variables i, j, and (i, j) denote the analytical rank associated with the linguistic values in a nine-element set defined in Table 1.

Defuzzification is the inverse of fuzziness. The function of clearness is to transform the fuzzy quantity obtained by fuzzy reasoning into the actual clearness and adopt the clearness method given by the following.

$$\xi^* = \frac{\sum_{i=1}^{9} \sum_{j=1}^{9} \xi^*{}_{l(i,j)} \mu_{l(i,j)}}{\sum_{i=1}^{9} \sum_{j=1}^{9} \mu_{l(i,j)}} \tag{6}$$

In (6), $\xi^*{}_{l(i,j)}$ and $\mu_{l(i,j)}$ are the discrete element and membership degree of the output fuzzy set $YC_{l(i,j)}$ representatively.

2.3.2. Fuzzy Rule Design

Realize the influence of temperature and unit time temperature on the failure probability of components. In the environment of drastic temperature change, the material and composite parts will be damaged such as fracture and avulsion. Temperature and thermal deformation due to temperature gradient will lead to fatigue failure. When the temperature cycle of components changes more than ±20 °C, their failure efficiency can be increased to more than 8 times that of the basic constant temperature. In the case of a constant temperature, no matter whether the rate of temperature change per unit time is positive or negative, the greater the relative change, the more drastic the temperature change is, the greater the impact on the failure probability.

In the fuzzy set with double input of temperature and temperature change rate per unit time, the grade subscripts corresponding to different fuzzy language values meet the following criteria. Then, Equation 7 can be used to generate the fuzzy rule table as shown in Figure 4.

$$Z(i,j) = \text{Round}[(a \times e^{bi}) + c \times j^2)] \tag{7}$$

		T								
		NB	NM	NS	NZ	ZE	PZ	PS	PM	PB
ΔT	NB	PS1	PS1	PS1	PS1	PS1	PS2	PS2	PM1	PB2
	NM	PZ2	PZ2	PZ2	PZ2	PZ2	PZ2	PS1	PS2	PB1
	NS	PZ1	PZ1	PZ1	PZ1	PZ1	PZ1	PZ2	PS1	PM2
	NZ	ZE	ZE	ZE	ZE	ZE	PZ1	PZ1	PZ2	PM1
	ZE	ZE	ZE	ZE	ZE	ZE	ZE	PZ1	PZ2	PM1
	PZ	ZE	ZE	ZE	ZE	ZE	PZ1	PZ1	PZ2	PM1
	PS	PZ1	PZ1	PZ1	PZ1	PZ1	PZ1	PZ2	PS1	PM2
	PM	PZ2	PZ2	PZ2	PZ2	PZ2	PZ2	PS1	PS2	PB1
	PB	PS1	PS1	PS1	PS1	PS1	PS2	PS2	PM1	PB2

Figure 4. Fuzzy rule I.

The most commonly used Arrhenius model to describe the reaction rate is Jacobus. The equation shows that the failure rate increases exponentially with the increase of temperature. However, very low temperatures also have a significant effect on failure probability. For example, low temperatures can cause some materials to break and become brittle. Therefore, the low temperature case, though not as big as the high temperature effect, should also be considered.

When considering the influence of low temperature on failure probability, in the fuzzy set with double-input temperature and temperature change rate per unit time, the grade subscripts corresponding to different fuzzy language values and their corresponding failure rates meet the following criteria. Then, equation 8 can be used to generate the fuzzy rule table as shown in Figure 5.

$$Z(i,j) = \text{Round}[a \times (\mathrm{i} + b)^2 + j^2)^{0.5}] \tag{8}$$

2.4. System Health Dynamic Evaluation Modeling

Detailed dynamic iterative calculation simulation flow chart is shown in the Figure 6. First, initialize the parameters required to set up the model. Specifically, it includes fuzzy segmentation in fuzzy reasoning and parameter setting of membership function, setting of conversion factor and design of fuzzy rules; and the simulation calculates the clock zero, generates the seed initialization settings for the random number. To characterize the randomness of a component's failure at each moment, we use a pseudo-random number to produce a uniformly distributed failure rate. The component is judged to be invalid by comparing the random failure rate of the component with the failure rate calculated by the fuzzy inference. Then, the health and failure factors of the subsystem or system are evaluated by counting the number of failed components at each moment.

		T								
		NB	NM	NS	NZ	ZE	PZ	PS	PM	PB
ΔT	NB	PB1	PM2	PM1	PM1	PM1	PM2	PB1	PB1	PB2
	NM	PM2	PM1	PS2	PS2	PS2	PM1	PM2	PB1	PB2
	NS	PM1	PS2	PS1	PS1	PS1	PS2	PM1	PM2	PB1
	NZ	PS2	PS1	PZ2	PZ1	PZ2	PS1	PS2	PM1	PB1
	ZE	PS2	PS1	PZ1	ZE	PZ1	PS1	PS2	PM1	PB1
	PZ	PS2	PS1	PZ2	PZ1	PZ2	PS1	PS2	PM1	PB1
	PS	PM1	PS2	PS1	PS1	PS1	PS2	PM1	PM2	PB1
	PM	PM2	PM1	PS2	PS2	PS2	PM1	PM2	PB1	PB2
	PB	PB1	PM2	PM1	PM1	PM1	PM2	PB1	PB1	PB2

Figure 5. Fuzzy rule II.

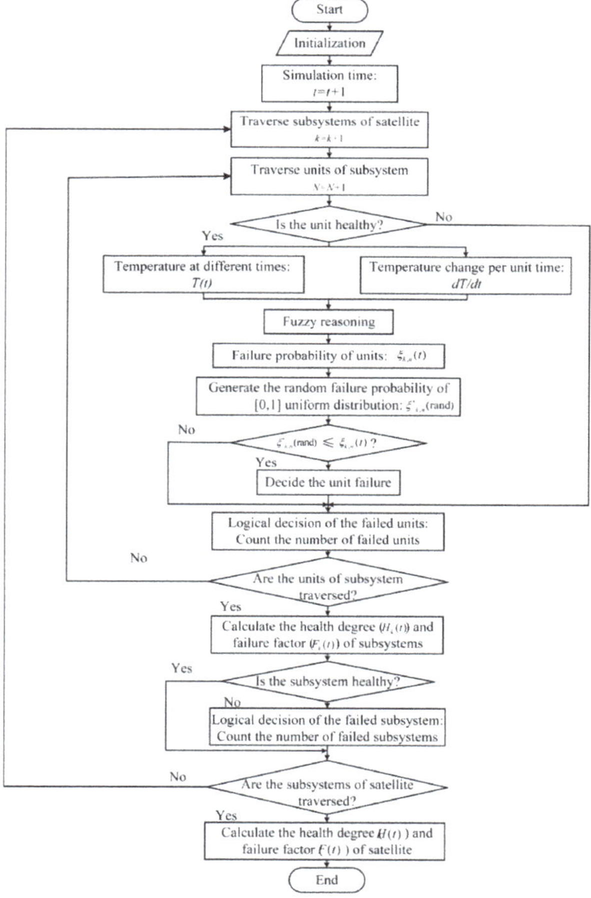

Figure 6. Algorithm process.

There are several key points:

(1) After the simulation starts, in each simulation time, it is traversed in order according to the topology of the whole star -> subsystem -> component unit;

(2) Each time the component unit of the subsystem is traversed, it is judged whether the unit is healthy, because once the unit is determined to be unhealthy at the last moment, it is not judged at the current moment;

(3) Each non-failed component unit correspondingly generates a random probability of random distribution of [0, 1], and compares with the random failure probability calculated by the component unit, thereby determining whether the component unit fails at the current moment;

(4) The logical determination method of the failed component is that if it is a tandem type component unit, as long as the failure occurs, its health deterioration effect on the subsystem composed of the unit is counted. However, the parallel type unit and the voting unit type reserve type unit do not malfunction when one system fails. Therefore, for the latter three types, the number of failed component units is counted only when the number of failures has an impact on the deterioration of the health state of the system.

3. Results and Discussion

3.1. Cases Design

This section will mainly describe the normal operation condition of the satellite in orbit and the damage conditions under different thermal loads. During the normal operation of the satellite in orbit, the satellite's thermal control system makes each component of the satellite work within its own normal temperature range, so the satellite's health will not change significantly in the short term. However, it should be noted that the time-varying and extremely high thermal loads coupled with the full sun/eclipse operating environment yield large temperature changes and large thermal gradients, which will seriously affect the satellite's health and life. So, we designed four groups of thermal load damage conditions that could reflect different time-varying characteristics to study the health changes of satellite typical subsystems.

3.1.1. Parameters Setting for Normal Orbital Operation

The external heat flux (solar radiation heat flux, earth albedo heat flux and earth infrared radiation heat flux) of the satellite in each direction (± X, ± Y, ± Z) varies periodically during the normal orbit cycle. The satellite orbit parameters used in the simulation analysis are shown in Table 2. The simulation time is one orbital period (1.63 h), in which, the initial moment of the entire simulation cycle is the time when the satellite-Z direction is subjected to the maximum solar radiation intensity. The satellite entered the earth's shadow at 1884.97 s and left the earth's shadow at 3968.28 s.

Table 2. Satellite orbit parameters.

Parameters	Value
Solar incident angle	17.23 °C
Orbit altitude	641.65 km
Average of solar radiation	1354 W/m^2
Albedo	0.35
Earth infrared radiation	221.484 W/m^2
Space temperature	4 K

The external heat flow is calculated using the software THERMAL DESKTOP based on the orbital parameters of the satellite. Then, by importing the calculated external heat flow results into THIEM, the dynamic temperature of each component node of the satellite can be solved, and the health index of the satellite in normal orbit can be obtained.

3.1.2. Design of Thermal Damage Conditions

In order to analyze the effects of thermal damage caused by time-varying and extreme thermal loads on satellite health during orbit operation, we designed four conditions as shown in Table 3. In these cases, case I serves as a reference case. By comparing the Cases II, III, IV and the reference case I, respectively, the effects of the magnitude, duty cycle and period of different thermal loads on satellite health can be analyzed. The pulse rectangular wave is used in the time-varying heat load. Where duty ratio refers to the proportion of thermal load time relative to the total time in an impulse cycle. For example, in Case I, the duty ratio is 1/6, indicating that the thermal load of 10 s is 2 W/cm^2 in an impulse cycle, and the remaining heat load of 50 s is 0 W/cm^2.

Table 3. Thermal damage conditions and value.

Cases	Amplitude	Duty Ratio	Cycle
I	2 W/cm^2	1/6 (Heat 10 s; Cool 50 s)	10
II	4 W/cm^2	1/6 (Heat 10 s; Cool 50 s)	10
III	2 W/cm^2	1/2 (Heat 10 s; Cool 10 s)	10
IV	2 W/cm^2	1/6 (Heat 10 s; Cool 50 s)	40

The target of thermal damage is the temperature controller in the thermal control subsystem. The temperature controller as a thermal control component guarantees the normal operation of other satellite core components, but it cannot withstand large thermal loads because it does not have good cooling. So, the final simulation selects 2 W/cm^2 and 4 W/cm^2. In the initial moment of simulation, the target temperature controller is loaded with extreme thermal loads and the satellite thermal health response is solved by using the THIEM.

3.2. Effects of Thermal Load Amplitude on Satellite Health

Figure 7 shows the effects of different thermal load amplitudes on the dynamic temperature of target damaged component. Under normal orbit operation, the temperature range of the temperature controller is 0 to 50. However, under the action of the pulse thermal load, the temperature of the target damaged component rose rapidly and changed periodically. After 10 cycles of thermal loading, the temperature of the damaged component increased cumulatively. This is because the component cooling time caused by the thermal load of 0 is short, resulting in insufficient cooling, which eventually leads to a cumulative increase in temperature. Comparing Case I and Case II, the temperature rise caused by different thermal load amplitude was variant in one operating period. This is because the temperature rise of the same component in the same time interval is positively correlated with the thermal load.

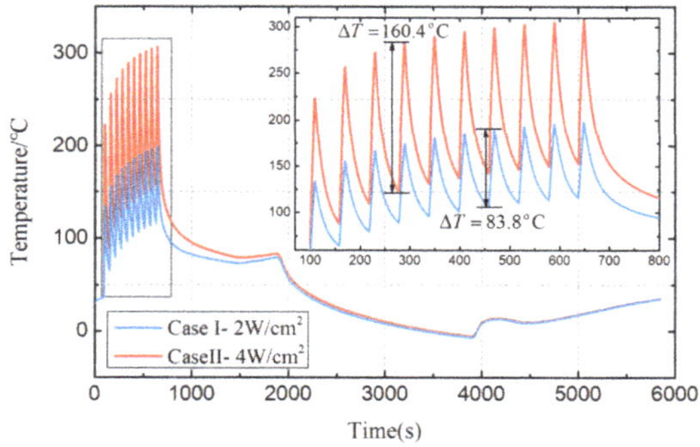

Figure 7. Dynamic temperature of damaged components under Case I and Case II.

When the heat flux density is 2 W/cm^2 and 4 W/cm^2 respectively, the health degree and failure factors of the satellite subsystems are shown in Figures 8 and 9. As shown in Figures 8a and 9a, the health degree of the satellite's thermal control system is exponentially decreasing, and the health degree of the satellite's payload is also slightly damaged. This is because the thermal damage of the temperature controller component will directly affect the health of the thermal control subsystem. As mentioned earlier, the temperature controller is designed to serve the payload, so when the controller is damaged, the payload health is also affected. As shown in Figures 8b and 9b, the failure factor of the satellite thermal control subsystem increased. According to the definition of failure factor, failure factor is not equal to 0, which means that a component will fail at that time. Therefore, the distribution of the failure factor can determine the speed of failure.

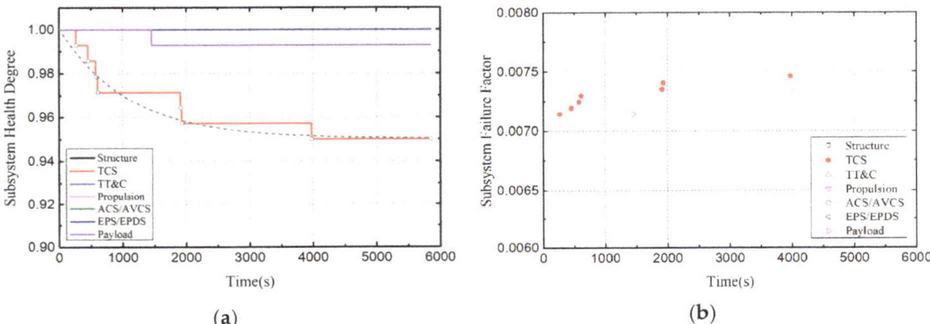

Figure 8. Health indicators of satellite subsystems (Case I): (**a**) Health degree. (**b**) Failure factor.

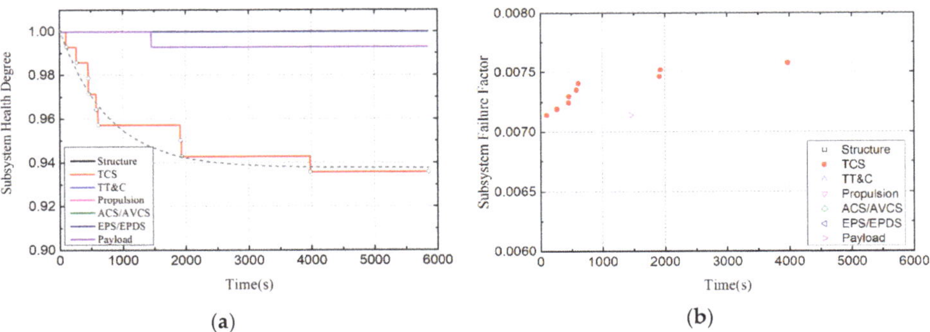

Figure 9. Health indicators of satellite subsystems (Case II): (**a**) Health degree. (**b**) Failure factor.

In order to visually compare and analyze the effects of different thermal load amplitudes on satellite systems, this section mainly discusses the health changes of thermal control subsystems, as shown in Figure 10. Compared with the normal orbital condition, the health of the thermal control system decreased with time, indicating that the health of the subsystem is deteriorated to some extent. Moreover, as the heat flux density increased, the health condition of the thermal control subsystem deteriorated more severely. In Case I, the health degree of the TCS eventually dropped to 0.95; in Case II, the health degree of the TCS eventually fell to 0.936. Moreover, the greater the thermal load, the earlier the system is damaged. This is because the higher the heat load, the higher the temperature of the damaged component, which makes the component more vulnerable to failure.

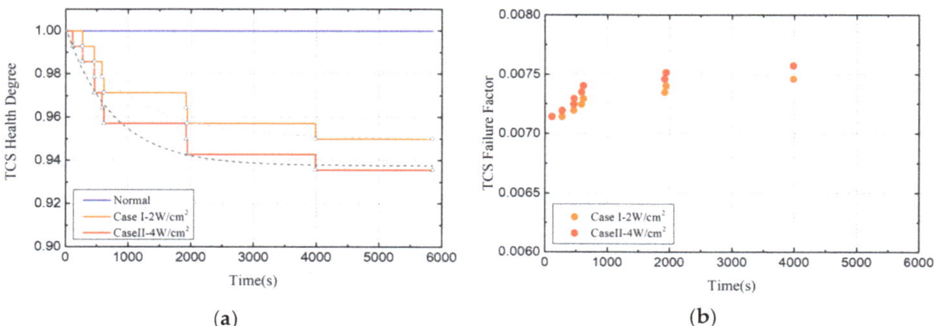

Figure 10. Health indicators of thermal control subsystem (TCS) (Compare Case I and Case II): (**a**) Health degree. (**b**) Failure factor.

3.3. Effects of Thermal Load Duty Ratio on Satellite Health

Figure 11 shows the effect of different heat load duty cycles on the dynamic temperature of the target damage component. In both cases, the thermal load amplitude and cycle number are the same, and the loading time of thermal load in one cycle is the same, which is 10 s. The duty ratio of pulsed thermal load directly affects the cooling time of a cycle, and finally leads to the difference of the total acting time of thermal load. The total acting time of pulse of case I is 600 s, while the total acting time of pulse of case III is 200 s. However, compared to case I, Case III causes the temperature of damaged components to rise more rapidly and eventually reach a higher temperature. This is because the pulse duty ratio is small, resulting in insufficient cooling of the damaged components and eventually continuous accumulation of temperature.

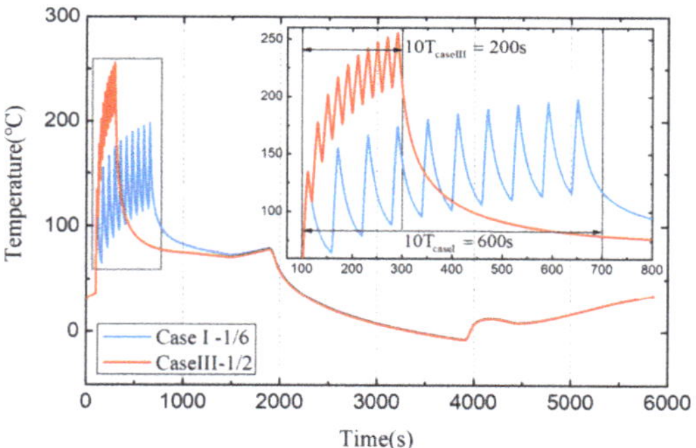

Figure 11. Dynamic temperature of damaged components under Case I and Case III.

Figure 12 shows the changes of satellite system health indicators when the duty cycle of thermal load is 1/2. The health degree of the satellite thermal control subsystem decreased exponentially to 0.97. In order to intuitively compare and analyze the impact of thermal load duty ratio on the health of satellite subsystem, this section shows the health impact of Case I and Case III on the thermal control subsystem in Figure 13. As shown in Figure 13a, when the duty ratio of thermal load is 1/2, the health degree of the thermal control subsystem decreased to 0.97, while when the duty ratio of thermal load is 1/6, and the health degree of the thermal control subsystem decreased to 0.95. Although the higher the thermal load duty ratio will lead to the higher temperature rise of the damaged components, the total loading time will be reduced. In general, components have the ability to respond to extreme environments in the short term, so short-term high temperatures do not necessarily lead to complete failure. As shown in Figure 13b, the higher the duty cycle of the thermal load causes the thermal control subsystem to fail earlier. In addition, it can also be found from the fit curve of health in Figure 13a that the slope of Case III is larger than that of Case I. This is because the heating time in a cycle is the same, the greater the thermal load duty ratio, the more drastic the temperature change of the damaged components.

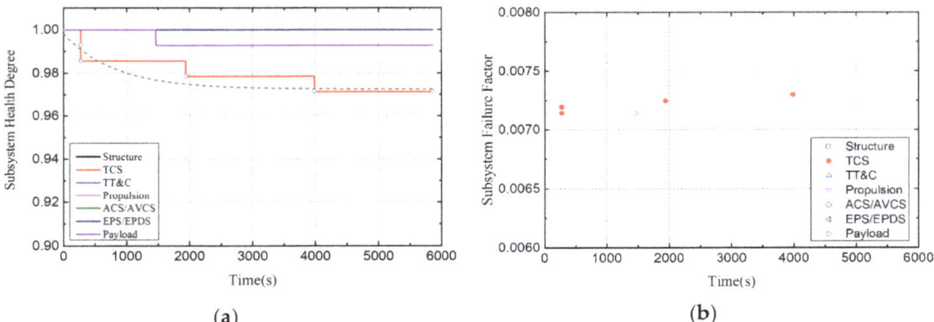

Figure 12. Health indicators of satellite subsystems (Case III). (**a**) Health degree. (**b**) Failure factor.

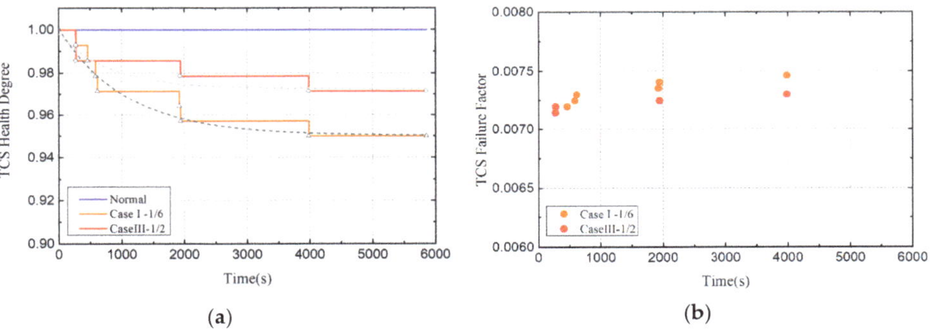

Figure 13. Health indicators of TCS (Compare Case I and Case III). (**a**)Health degree. (**b**)Failure factor.

3.4. *Effects of Thermal Load Cycle on Satellite Health*

Figure 14 shows the effect of different heat load cycles on the dynamic temperature of the target damage component. The number of thermal load cycles directly affects the loading time of thermal load. Under the action of thermal load cycle, the cumulative temperature of damaged parts rises to about 220°C. The pulsed thermal load causes an increase in the temperature of the satellite components, and the longer the loading time results in a higher temperature rise of the damaged component. In Case IV, the temperature of damaged component dropped after 1884 s. This is because the satellite entered earth's shadow in 1884 s and became colder.

Figure 15 shows the changes of satellite system health indicators when the cycle of thermal load is 40. The health degree of the satellite thermal control subsystem decreased exponentially to 0.87. In order to intuitively compare and analyze the impact of thermal load cycles on the health of satellite subsystem, this section shows the health impact of Case I and Case IV on the thermal control subsystem in Figure 16. As shown in Figure 16a, when the cycle of thermal load is 40, the health degree of the thermal control subsystem decreased to 0.87, while when the cycle of thermal load is 10, and the health degree of the thermal control subsystem decreased to 0.95. With the increase of thermal load cycle, the deterioration of thermal control subsystem becomes more serious.

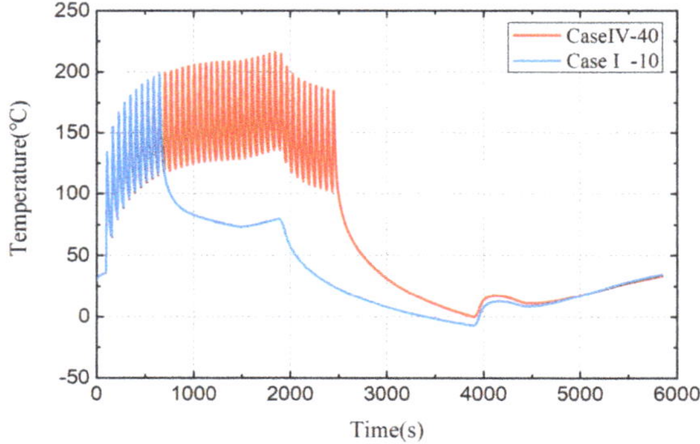

Figure 14. Dynamic temperature of damaged components under Case I and Case IV.

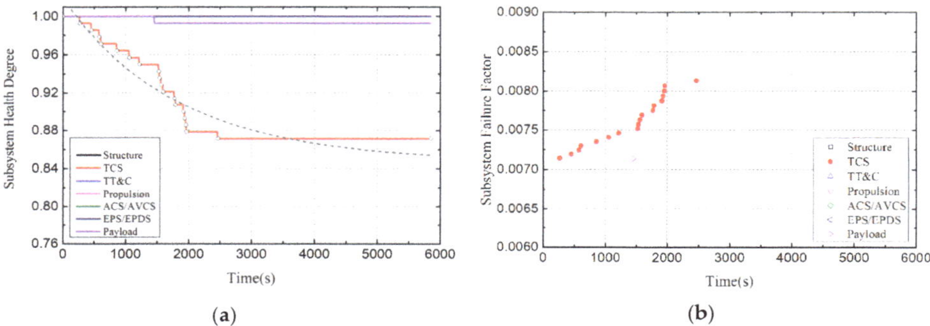

Figure 15. Health indicators of satellite subsystems (Case IV). (**a**) Health degree. (**b**) Failure factor.

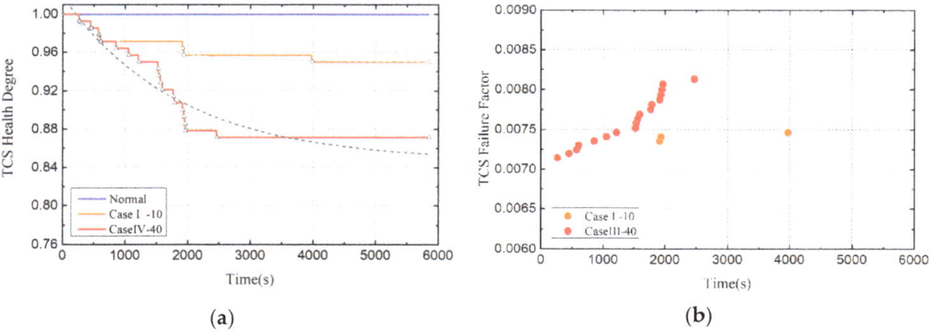

Figure 16. Health indicators of TCS (Compare Case I and Case IV). (**a**)Health degree. (**b**)Failure factor.

These results show that the health of the components in the satellite subsystem is affected by time-varying and extreme thermal loads. The analysis of cases accords with the qualitative law of theory.

4. Conclusions

This paper presents an intelligent evaluation model based on fuzzy logic for satellite thermal health analysis. A new evaluation index and topological system are proposed to evaluate satellite health status. Taking the temperature controller in the satellite thermal control subsystem as the target of thermal load, the health deterioration of the satellite under typical working conditions such as time-varying and maximum thermal load is analyzed. Based on the analytical investigations presented in this paper, the following conclusions may be made:

1. The fuzziness of the relationship between temperature and failure probability is considered, and the relationship between temperature and failure probability is quantitatively described by intelligent analysis method (fuzzy reasoning).
2. The model can quickly and accurately evaluate the effects of different thermal conditions on satellite health. The health deterioration of the system is characterized by the change of health degree and failure factor.
3. Multi-period, high heat flux density, and low duty ratio have great influence on satellite health.

The model solves the problem of collaborative solution of different models (satellite thermal analysis is continuous process simulation, satellite component failure is discrete probability event simulation), as well as the challenges of model absence and non-quantification in satellite thermal health analysis. The model can better reflect the dynamic deterioration process of the system health. In the future, it can be transported in many systems with strict life requirements, and only the topology system needs to be modified according to different systems.

Author Contributions: Conceptualization, Y.-Z.L.; methodology, E.-H.L.; software, E.-H.L. and T.-T.L.; validation, Z.-Z.Z. and T.L.; formal analysis, E.-H.L.; investigation, E.-H.L.; resources, Y.-Z.L.; data curation, T.-T.L. and Z.-Z.Z.; writing—original draft preparation, E.-H.L. and J.-X.L.; writing—review and editing, T.L.; visualization, E.-H.L.; supervision, Y.-Z.L.; project administration, Y.-Z.L.

Funding: The project was funded by the Open Research Fund of Key Laboratory of Space Utilization, Chinese Academy of Sciences through grant no.LSU-JCJS-2017-1.

Conflicts of Interest: The authors declare no conflict of interest.

References

1. Chudoba, B.; Coleman, G.; Gonzalez, L.; Haney, E.; Oza, A.; Ricketts, V. Orbital transfer vehicle (OTV) system sizing study for manned GEO satellite servicing. *Aeronaut. J.* **2016**, *120*, 573–599. [CrossRef]
2. Seo, H.S.; Rhee, J.; Han, E.S.; Kim, I.S. Thermal failure of the LM117 regulator under harsh space thermal environments. *Aerosp. Sci. Technol.* **2013**, *27*, 49–56. [CrossRef]
3. Barcena, J.; Maudes, J.; Vellvehi, M.; Jorda, X.; Obieta, I.; Guraya, C.; Bilbao, L.; Jimenez, C.; Merveille, C.; Coleto, J. Innovative packaging solution for power and thermal management of wide-bandgap semiconductor devices in space applications. *Acta Astronaut.* **2008**, *62*, 422–430. [CrossRef]
4. Wang, J.X.; Li, Y.Z.; Li, J.X.; Li, C.; Zhang, Y.; Ning, X.W. A gas-atomized spray cooling system integrated with an ejector loop: Ejector modeling and thermal performance analysis. *Energy Convers. Manag.* **2019**, *180*, 106–118. [CrossRef]
5. Swanson, T.D.; Birur, G.C. NASA thermal control technologies for robotic spacecraft. *Appl. Therm. Eng.* **2003**, *23*, 1055–1065. [CrossRef]
6. Yang, C.; Hou, X.; Wang, L. Thermal design, analysis and comparison on three concepts of space solar power satellite. *Acta Astronaut.* **2017**, *137*, 382–402. [CrossRef]
7. Palla, C.; Peroni, M.; Kingston, J. Failure analysis of satellite subsystems to define suitable de-orbit devices. *Acta Astronaut.* **2016**, *128*, 343–349. [CrossRef]
8. Gonzalo, J.; Domínguez, D.; Lopez, D. On the challenge of a century lifespan satellite. *Prog. Aerosp. Sci.* **2014**, *70*, 28–41. [CrossRef]
9. Kim, S.Y.; Castet, J.F.; Saleh, J.H. Spacecraft electrical power subsystem: Failure behavior, reliability, and multi-state failure analyses. *Reliab. Eng. Syst. Saf.* **2012**, *98*, 55–65. [CrossRef]
10. Tafazoli, M. A study of on-orbit spacecraft failures. *Acta Astronaut.* **2009**, *64*, 195–205. [CrossRef]
11. Pecht, M.; Kang, W.C. A critique of Mil-Hdbk-217E reliability prediction methods. *IEEE Trans. Reliab.* **1988**, *37*, 453–457. [CrossRef]

12. Bensoussan, A.; Suhir, E.; Henderson, P.; Zahir, M. A unified multiple stress reliability model for microelectronic devices-Application to 1.55 μm DFB laser diode module for space validation. *Microelectron. Reliab.* **2015**, *55*, 1729–1735. [CrossRef]
13. Castet, J.F.; Saleh, J.H. Satellite Reliability: Statistical Data Analysis and Modeling. *J. Spacecr. Rocket.* **2009**, *46*, 1065–1076. [CrossRef]
14. Lee, W.S. Fault tree analysis, methods, and applications: A review. *IEEE Trans. Reliab.* **1985**, *34*, 194–203. [CrossRef]
15. De Queiroz Souza, R.; Álvares, A.J. FMEA and FTA analysis for application of the reliability centered maintenance methodology: Case study on hydraulic turbines. *ABCM Symp. Ser. Mechatron.* **2008**, *3*, 803–812.
16. Wu, J.; Yan, S.; Xie, L. Reliability analysis method of a solar array by using fault tree analysis and fuzzy reasoning Petrinet. *Acta Astronaut.* **2011**, *69*, 960–968. [CrossRef]
17. Achutuni, R.; Menzel, P. Space systems considerations in the design of advanced geostationary operational environmental satellites. *Adv. Space Res.* **1999**, *23*, 1377–1384. [CrossRef]
18. Akita, T.; Takaki, R.; Shima, E. A new adaptive estimation method of spacecraft thermal mathematical model with an ensemble Kalman filter. *Acta Astronaut.* **2012**, *73*, 144–155. [CrossRef]
19. Pérez-Grande, I.; Sanz-Andrés, A.; Guerra, C.; Alonso, G. Analytical study of the thermal behaviour and stability of a small satellite. *Appl. Therm. Eng.* **2009**, *29*, 2567–2573. [CrossRef]
20. Li, Y.Z.; Lee, K.M. Thermohydraulic Dynamics and Fuzzy Coordination Control of a Microchannel Cooling Network for Space Electronics. *IEEE Trans. Ind. Electron.* **2011**, *58*, 700–708. [CrossRef]
21. Qian, X.; Zhang, Y.; Gendeel, M. State Rules Mining and Probabilistic Fault Analysis for 5 MW Offshore Wind Turbines. *Energies* **2019**, *12*, 2046. [CrossRef]

Article

Maximum Entropy Models for Fatigue Damage in Metals with Application to Low-Cycle Fatigue of Aluminum 2024-T351

Colin Young [1,2] and Ganesh Subbarayan [1,*]

[1] School of Mechanical Engineering, Purdue University, West Lafayette, IN 47907-2088, USA; cyoung8@ford.com
[2] Ford Corporation, Dearborn, MI 48124, USA
* Correspondence: ganeshs@purdue.edu

Received: 4 September 2019; Accepted: 30 September 2019; Published: 3 October 2019

Abstract: In the present work, we propose using the cumulative distribution functions derived from maximum entropy formalisms, utilizing thermodynamic entropy as a measure of damage to fit the low-cycle fatigue data of metals. The thermodynamic entropy is measured from hysteresis loops of cyclic tension–compression fatigue tests on aluminum 2024-T351. The plastic dissipation per cyclic reversal is estimated from Ramberg–Osgood constitutive model fits to the hysteresis loops and correlated to experimentally measured average damage per reversal. The developed damage models are shown to more accurately and consistently describe fatigue life than several alternative damage models, including the Weibull distribution function and the Coffin–Manson relation. The formalism is founded on treating the failure process as a consequence of the increase in the entropy of the material due to plastic deformation. This argument leads to using inelastic dissipation as the independent variable for predicting low-cycle fatigue damage, rather than the more commonly used plastic strain. The entropy of the microstructural state of the material is modeled by statistical cumulative distribution functions, following examples in recent literature. We demonstrate the utility of a broader class of maximum entropy statistical distributions, including the truncated exponential and the truncated normal distribution. Not only are these functions demonstrated to have the necessary qualitative features to model damage, but they are also shown to capture the random nature of damage processes with greater fidelity.

Keywords: MaxEnt distributions; fatigue damage; low-cycle fatigue; thermodynamic entropy

---

1. Introduction

The wrought aluminum alloy 2024-T351 is an important light structural metal commonly used in aerospace and other weight-critical applications [1]. A common approach to modeling the low-cycle fatigue (LCF) life of this material and many other metals is the Coffin–Manson relationship [1,2]:

$$\frac{\Delta\epsilon_p}{2} = \epsilon_f\prime\left(2N_f\right)^c \quad (1)$$

This equation is intended to cover the range of life from 1 to about 20,000 reversals, where macroscopic plastic strain is measurable. However, as has been pointed in the literature [2], Equation (1) is less successful in fitting data in the very low reversal count range of 1 to about 200. The inadequacy of Equation (1) for modeling a representative LCF data set for 2024-T351 is demonstrated below and motivates an alternative LCF modeling approach.

In Figure 1, the results from a sequence of low-cycle fatigue tests and two monotonic tension tests on tension specimens of 2024-T351 aluminum are shown. The data is also fitted to a Coffin–Manson model in the figure.

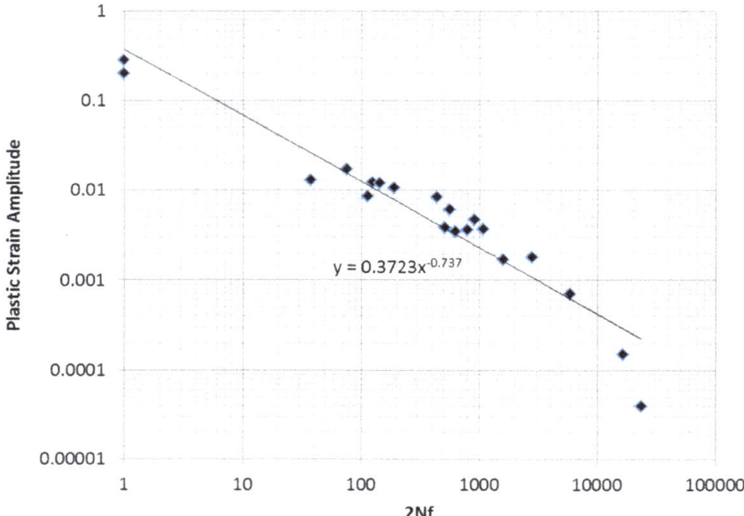

Figure 1. Coffin–Manson plot of data from eighteen low-cycle fatigue tests and two monotonic tests of aluminum 2024-T351 (R^2 = 0.92). The two data points to the single reversal are from monotonic tests.

It is clear that the data exhibits a curvature that is not captured by the straight line fit of the Coffin–Manson equation. An ideal model would be one based on a sound physical principle that assures the "best possible" fit to experimentally obtained fatigue test data, considering the statistical uncertainty inherent in the data. An ideal procedure would also provide systematic guidance on constructing the model form. Below, we argue that the maximum entropy concept may provide such a guiding principle.

The concept of entropy occurs in two different contexts in the literature reviewed below. The first case is represented by applications of a class of statistical methods based on information entropy (reviewed in detail in the following section), which may be applied to fatigue data or any other experimental data with inherent uncertainty. These applications may not refer to the physical entropy of the material. Alternatively, the physical entropy at a material point in a device or structure may be used to model the progress of damage at that point. In the latter instance, the process of damage and degradation in material behavior is a fundamental consequence of the second law of thermodynamics, resulting in the increase in entropy of isolated systems with time [3]. In contrast to the more commonly used parameters of stress and plastic strain, the argument is that specimen entropy has a deeper connection to the physics of the damage process.

One of earliest studies to use maximum entropy (or max entropy) probabilistic distributions to study fatigue fracture is [4]. Entropy as a purely statistical concept is used in [5] to model the variability of fatigue crack growth. A version of the maximum entropy method is shown to be a viable alternative to Bayesian updating for analyzing an evolving data population. However, the authors do not connect the concept of entropy to material damage. In [6], the maximum entropy method was used to build a statistical model of the strength distribution in brittle rocks. Since maximum entropy represents a general principle that can lead to many possible probabilistic distributions based on the choice of constraints, studies in the literature have included attempts at specifying constraints on either two or even four moments of the distribution [4,7] in an attempt to compute the parameters of the distribution.

In general, in [5–7], the thermodynamic entropic dissipation at a material point is not directly used to build a predictive fatigue life relationship.

Basaran and co-workers were among the first to make a connection, using the Boltzmann entropy formula, between physical entropy as measured by plastic dissipation and damage in ductile alloys [8,9]. Later, Khonsari and co-workers [10,11] demonstrated that the thermodynamic entropy generated during a low-cycle fatigue test can serve as a measure of degradation. They proposed that the thermodynamic entropy is a constant when the material reaches its fracture point, independent of geometry, load, and frequency. The hypothesis on critical thermodynamic entropy was tested in [10] on aluminum 6061-T6 through bending, torsion, and tension–compression fatigue tests. In our prior work [12], we used the maximum entropy statistical framework to derive a fatigue life model using material entropy as a predictive variable. This approach is inspired by the work of Jaynes [13], where the information theory concept of entropy was applied to the energy levels of a thermodynamic system, showing that known results from statistical mechanics could be obtained. Information theory entropy was, thus, proportional to thermodynamic entropy. While in some papers [8,9] the accumulated damage is empirically related to entropic dissipation, in [12], the damage $D(t)$ naturally results from the maximum entropy probability distribution as the corresponding cumulative distribution function (CDF). The fatigue life model in [12] is expressed as a damage function and is given in Equation (2) below. The authors describe this approach as a maximum entropy fracture model.

$$D(t) = 1 - \exp\left(-\frac{W_t}{\rho T k_\psi}\right) \tag{2}$$

In Equation (2), the damage parameter $D(t)$ is the non-decreasing CDF that ranges from zero (virgin state) to one (failed state). The independent variable is the inelastic dissipation in the material, which is proportional to the entropy of the material through the J2 plasticity theory and the Clausius–Duhem inequality. The single material parameter k_ψ in Equation (2) was obtained from isothermal mechanical cycling tests and then used to model fatigue crack propagation under thermal cycling conditions in an electronic assembly. Figure 2 shows a comparison of the estimated and actual number of cycles, as well as crack fronts, at an intermediate stage, with the same area of cracks from both the finite element simulation and thermal cycling fatigue test. To the best of the authors' knowledge, such a connection between physical entropy dissipation and fatigue crack propagation in ductile alloys has not been made in prior literature.

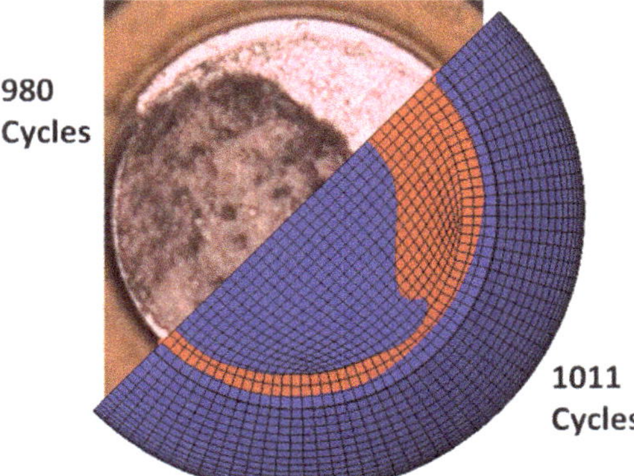

Figure 2. Comparison of crack fronts predicted by a single-parameter maximum entropy model against the experimentally observed creep–fatigue crack in a Sn3.8Ag0.7Cu solder joint under thermal fatigue cycling [12] (reproduced with permission). The single maximum entropy model parameter was extracted using isothermal mechanical tests.

In [12], it is demonstrated that it is possible to follow the physical process of fatigue crack propagation as a maximum entropy process. However, the arguments that led to the formation of Equation (2) assumed a constant dissipation rate, which in turn implies an exponential distribution for the form of the statistical distribution. More generally, while the use of maximum entropy principles provides the theoretical advantage of being maximally "non-committal" on the data that are unavailable from the experiments [13], the assumption of exponential distribution may be restrictive. Arguably, other distributions that conform to the max entropy principle may provide a better description of damage. However, systematic exploration of such maximum entropy functions, as well as thermodynamic entropy, in describing metal fatigue life data sets appears to be limited in the literature. Thus, in this paper, building on our prior work, we propose the development of a systematic procedure for development of maximum entropy models for describing metal fatigue based on measured thermodyanamic entropy. We demonstrate the approach using low-cycle fatigue experimental data for aluminum 2024-T351 material, and generalize the application of the maximum entropy principle using a broader class of statistical distributions, including the truncated exponential and the truncated normal distribution. We begin first with a brief review of the maximum entropy principle.

2. A Review of the Maximum Entropy Principle

The concept of entropy as applied to heat engines is due to Clausius, but the connection of entropy to the probability of the states of a thermodynamic system began with Boltzmann. Boltzmann demonstrated that the second law of thermodynamics for an ideal gas is a consequence of the mechanics of the collisions of the molecules [14]. He showed that a sufficiently large number of interrelated deterministic events will result in random states. He derived the following function, given in Equation (3), for a uniform distribution, and argued that this quantity had the same physical meaning as the entropy proposed by Clausius. This led to the Boltzmann H function:

$$H(p) = \sum_i p_i \ln p_i \;\; p_i = p = \frac{1}{n} \tag{3}$$

The above expression is closely related to Gibb's entropy formula:

$$S(p) = -k_b \sum_i p_i \ln p_i \tag{4}$$

Shannon's research in information theory led to a mathematical expression (discussed later in Equation (6)) that is strikingly similar to the thermodynamic entropy formulas of Boltzmann and Gibbs, described above. It is important to note that Shannon's argument was a purely statistical one and no physical significance was claimed. It was not until the work of Jaynes [13] that a connection between the information entropy of Shannon and thermodynamic entropy was established.

Here, we describe the abstract development of Shannon's formula based on a counting argument [15], considering the information content of a whole number, which can range in value from 0 to N. If we claim that each digit of the number is a unit of information, then it clearly takes $\log_b N$ digits to represent the number in a base b system. If the base of the logarithm is changed, the resulting information will change by a constant, but the ratios of information for different N will be preserved, provided the same base is used for all of them. Thus, $\log_b N$ is a reasonable measure of the information contained in a variable, which can range from 0 to N. If we consider a random experiment with N possible equally likely, mutually exclusive outcomes, then the information contained in a given outcome is still $\log_b N = -\log_b p$, with p being the probability of the event. We argue that the information in a given event is strictly determined by p, regardless of how the remaining $1-p$ probability is allocated to other events. Thus, even if the events do not have equal probabilities, the information for any given event is still $-\log_b p$ [15]. This function has the expected property that the information contained in the occurrence of two (or more) statistically independent events is the sum of the information in each of the events separately, as shown below in Equation (5). This property is fundamentally important (as pointed out in [13]) and further reinforces the argument for the $-\log_b p$ measure of information.

$$\begin{aligned} I(p) &= -\ln p_i \\ I(p_i p_j) &= I(p_i) + I(p_j) \quad : i \neq j \end{aligned} \tag{5}$$

If the events correspond to a discrete random variable, then they must be mutually exclusive, and the probability of the union of the sequence of the events is equal to one [16]. The entropy of the density function is taken as the expected value of the information in the events [17]. This leads to the Shannon information entropy formula:

$$H(p) = E[I(p)] = -\sum_i p_i \ln p_i \tag{6}$$

This function (and only this function) satisfies these three conditions:

1. Continuity: It is a continuous function of the p_i;
2. Monotonicity: It is an increasing function of n, if all the p_i are equal;
3. Composition: If an event can be decomposed into two or more lower level events, the function $H(p)$ will evaluate this identically, whether the lower or higher level events are used in the computation, provided that the appropriate conditional probabilities are used to relate the higher and lower level events.

Jaynes [13] noted that there is a symbolic similarity between the expressions for thermodynamic (Gibbs) entropy (Equation (3)) and Shannon's information entropy (Equation (6)), but commented that the similarity did not necessarily imply a deeper connection. Jaynes then proceeded to show that a connection did exist and that many results of statistical thermodynamics could be interpreted as applications of Shannon's information entropy concept to physical systems. The expression for the Gibbs entropy is the result of a development involving various physical assumptions—some based on experimental evidence, and some not. Conversely, Shannon's entropy is based on mathematical and logical reasoning, not physical evidence. Shannon's model was developed to model the abstract mathematical properties of digital communication, and prior to Jaynes, was not claimed to be applicable to the physical sciences. Shannon defined the entropy of a discrete probability distribution as Equation (6).

The maximum entropy method as set forth by Jaynes is as follows [13]: The probability mass function that maximizes Equation (6), subject to constraint from Equations (7) and (8), is the best choice if no other information is available to specify the probability distribution.

$$\sum_i p_i = 1 \tag{7}$$

$$E[g(x_i)] = \sum_i p_i g(x_i) \quad : x_i \in \{x_1, x_2, \ldots x_i \ldots x_m\} \tag{8}$$

where $E[g(x_i)]$ is the expected value of, $g(x_i)$. The following probability mass function (Equation (9)) can be shown to maximize Equation (6):

$$p_i = e^{-\lambda_0 - \lambda_1 g(x_i)} \tag{9}$$

The constants λ_0 and λ_1 are Lagrange multipliers associated with the constraints. Jaynes calls this approach the maximum entropy method and calls the derived probability functions maximum entropy distributions (MaxEnt method and MaxEnt distributions). Multiple expected value constraints may be applied (not simply moments, as is common in probability analysis), resulting in the following form of the MaxEnt distribution:

$$p_i = e^{-\lambda_0 - \lambda_1 g_1(x_i) - \ldots - \lambda_m g_m(x_i)} \tag{10}$$

The entropy of the resulting distribution is [13]:

$$S_{max} = \lambda_0 + \lambda_1 E[g_1(x)] + \ldots + \lambda_m E[g_m(x)] \tag{11}$$

Jaynes's argument was for the discrete case. The entropy of a continuous probability density function is also known and is defined as [16]:

$$H(f(x)) = -\int_{-\infty}^{\infty} f(x) \ln f(x) dx \tag{12}$$

The corresponding continuous version of Equation (10) is given below [16]:

$$f(x) = e^{-\lambda_0 - \lambda_1 g_1(x) - \ldots - \lambda_m g_m(x)} \tag{13}$$

One important point regarding Equation (13) is that it is only a probability density function for specific values of the parameters λ_k. This situation differs from the usual approach to representing probability density functions or distribution functions, where the functions are admissible for ranges of parameter values. Additionally, the method Jaynes sets forth assumes that the values used for moment function constraints are not estimates subject to sampling variation. They are taken as essentially exact values of the distribution moment functions. This assumption differs from traditional inferential statistics, where moments or quantiles are estimated from data and sampling errors are estimated.

Jaynes showed that if we choose the probability distribution for the system microstates based on maximizing Shannon entropy, known results from statistical mechanics can be obtained, without new physical assumptions, and in particular, the thermodynamic entropy of the system is found to be the Gibbs entropy of Equation (4). Shannon's entropy for the distribution is proportional to the physical entropy of the system, however, only if the probability distribution is applied to the thermodynamic states of the system. Jaynes [13] argues that this shows that thermodynamic entropy is an application of a more general principle. Further to this point, Jaynes argues that if a probability model is required for some application, where certain expected values are known but other details are not, the maximum entropy approach should be used to find the probability distribution. Jaynes uses the term "maximally non-committal" to describe probability distributions obtained by this process. What is known about the random variable in question is captured in mathematical constraints, while the principle of maximum entropy accounts for what is not known. While information entropy is only proportional to thermodynamic entropy in certain circumstances, Jaynes argues that choosing the probability density function that maximizes the Shannon entropy subject to various constraints is appropriate to any situation where a reference probability distribution is needed. The application could be physical or not, and need not necessarily have a relationship to thermodynamic states.

3. Maximum Entropy Distributions

We argue that if a given parametric family of distributions is selected for some reason (as is common practice), then within that family of distributions we should prefer the parameter values that maximize entropy (subject to any constraints) over those that do not. For example, if the Weibull distribution has already been chosen for some application, and the characteristic life is known, then the Weibull exponent should be chosen to maximize entropy. It is noteworthy that the exponential distribution and the normal distribution are the MaxEnt distributions corresponding to a prescribed mean value and to the prescribed mean and variance values, respectively [18]. Given the fundamental importance of these distributions in statistical theory, it is informative that they can be directly derived from the principles of maximum entropy. Just as Jaynes showed that statistical thermodynamic results derivable by other means could be obtained from maximum entropy methods, it has also been shown that the well-known and fundamental normal distribution, traditionally derived by other means, can also be based on a maximum entropy argument. Even the Weibull distribution can be derived from a maximum entropy approach if appropriate moment functions are chosen [18]. These MaxEnt distributions are listed in Table 1.

Table 1. Maximum entropy (MaxEnt) distributions corresponding to moment functions $g_r(x)$ [13,18–20].

Support	Type	$g_r(x)$	Distribution Function	Reference
[a, b]	Discrete	N.A.	Uniform	[13]
$[0, \infty)$	Discrete	x	Exponential	[13]
$[0, \infty)$	Continuous	x	Exponential	[18]
[0, a]	Continuous	x	Truncated Exponential	[19]
$[0, \infty)$	Continuous	x^2	Half Normal	[20]
$(-\infty, \infty)$	Continuous	x, x^2	Normal	[18]
$[0, \infty)$	Continuous	x, x^2	Left Truncated Normal	[20]
[0, a]	Continuous	x, x^2	Left and Right Truncated Normal	[20]
$[0, \infty)$	Continuous	$\ln(x), x^\beta$	Weibull	[18]

Note the references to truncated distributions in Table 1. A distribution is described as truncated if the value of its density or mass function is forced to zero (when otherwise it would be non-zero) outside of a specific range. Thus, the truncated normal distribution functions can be thought of as ordinary normal probability density functions (PDFs) that are clipped to zero probability outside of their non-zero range. As described later, they are multiplied by a normalizing constant to correct for the missing density. Truncation at $x = 0$ is necessary for applications to non-negative variables. The cumulative distribution function (CDF) of a truncated normal random variable has a finite slope at $x = 0$. If a second truncation at $x = a$ is specified, then the CDF is forced to be exactly equal to 1 for all $x \geq a$. We begin the discussion of MaxEnt distributions with the truncated exponential distribution.

3.1. MaxEnt Form of Truncated Exponential Distribution

The truncated exponential distribution can be constructed in an analogous fashion for positive values of λ (parent PDF is a decreasing function). An example is plotted in Figure 3. However, it is possible for a truncated exponential distribution to be an increasing function within its non-zero range (Figure 4). Clipping the positive exponent at some specified value enables its use as a PDF. This corresponds to a negative-valued lambda, which is not admissible in the non-truncated case. If the specified mean was to the right of the midpoint of the non-zero range, then the lambda would be negative.

It should also be noted that changing the location of a distribution function without changing its shape has no effect on the entropy value. Thus, a left endpoint other than zero could be used for any of the distributions that have zero value for negative x. Naturally, this shift would change the moment function values. Note that specifying a right truncation value changes the shape of the remaining distribution function and should be thought of as adding an extra parameter. Thus, a truncated exponential distribution is a two-parameter distribution.

Below is the truncated exponential distribution for PDF and CDF:

$$\begin{aligned} f_{trunc}(x,\lambda,a) &= \frac{\lambda\ Exp(-\lambda x)}{1-Exp(-\lambda a)} \quad for \quad 0 \leq x \leq a \\ F_{trunc}(x,\lambda,a) &= \frac{1-Exp(-\lambda x)}{1-Exp(-\lambda a)} \quad for \quad 0 \leq x \leq a \end{aligned} \tag{14}$$

Below is the expected value of a truncated exponential random variable:

$$E(x) = \frac{1}{\lambda}\left(\frac{1-(\lambda a+1)Exp(-\lambda a)}{1-Exp(-\lambda a)}\right) \tag{15}$$

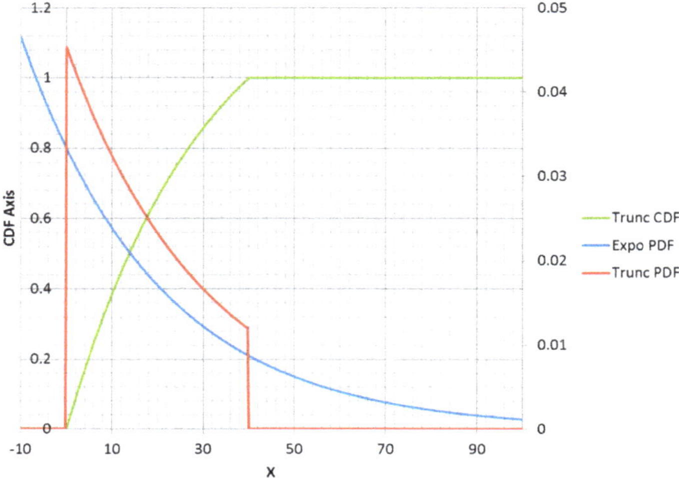

Figure 3. Truncated exponential distribution with $\lambda = 0.03$; $a = 40$. Note: CDF = cumulative distribution function; PDF = probability density function; trunc = truncated; expo = exponential.

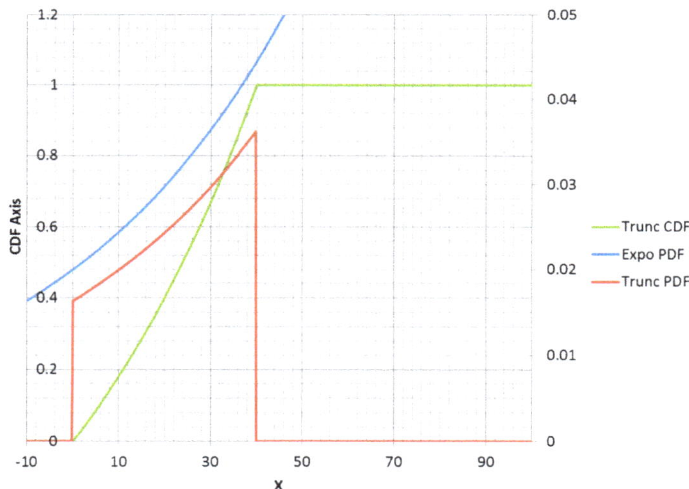

Figure 4. Rising truncated exponential distribution with $\lambda = -0.02$; $a = 40$.

Note that the uniform distribution is a limiting case of the truncated exponential distribution and corresponds to the lambda approaching zero. An example is shown in Figure 5.

$$\begin{gathered} \lim_{\lambda\to 0} f_{trunc}(x,\lambda,a) = \tfrac{1}{a} \quad for\ \ 0 \le x \le a \\ \lim_{\lambda\to 0} E(x) = \tfrac{a}{2} \end{gathered} \tag{16}$$

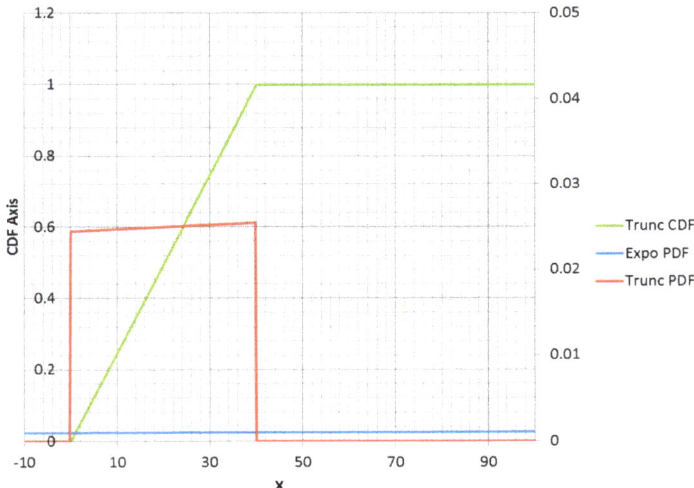

Figure 5. Rising truncated exponential distribution with $\lambda = -0.001$; $a = 40$.

3.2. MaxEnt Form of Truncated Normal Distribution

The truncated normal distribution can be explained in terms of the normal PDF. For $x \geq 0$, the PDF has the same shape as a non-truncated normal PDF, but scaled to make up the density lost for $x < 0$ (Figure 6). The truncation of the portion of the density less than zero changes the mean and standard deviation from the parameters that the truncated distribution inherits from the normal distribution. Adding a second truncation point at $x = a$ forces the function to be equal to 1 for all $x \geq a$ and adds a corner to the CDF at $x = a$ (Figure 7). Additionally, the correction factor must be larger to correct for missing density $x < 0$ and also $x \geq a$.

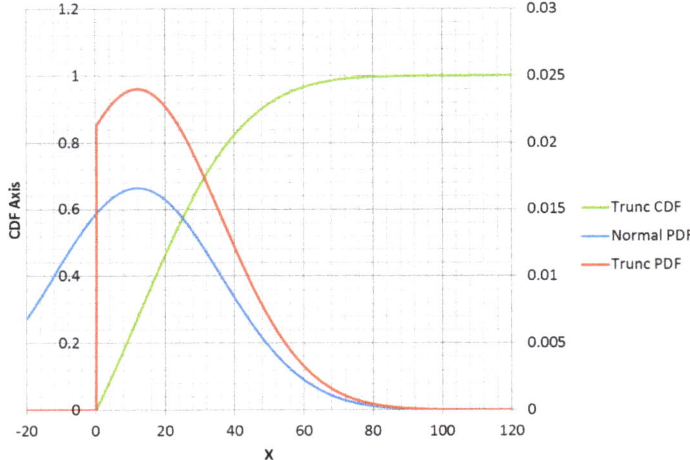

Figure 6. Truncated normal distribution plotted with the parent (non-truncated) normal distribution. Density correction for $x \geq 0$ is equal to 1.23.

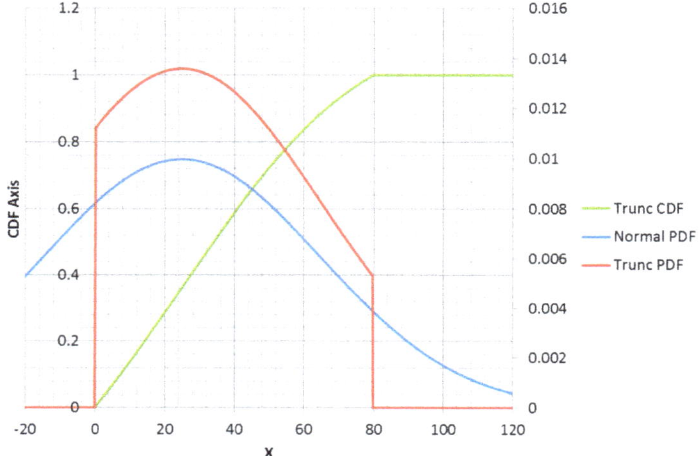

Figure 7. Left and right truncated normal distribution.

The PDF and the CDF for the left truncated normal distribution can be shown to be:

$$\begin{gathered}
\textit{Normal CDF in terms of standard normal CDF} \\
F_{norm}(x,\mu,\sigma) = \Phi\left(\frac{x-\mu}{\sigma}\right) \\
F_{trunc}(x,\mu,\sigma) = \frac{F_{norm}(x,\mu,\sigma) - F_{norm}(0,\mu,\sigma)}{1-F_{norm}(0,\mu,\sigma)} \quad for\ x \geq 0 \\
f_{trunc}(x,\mu,\sigma) = \left(\frac{1}{1-F_{norm}(0,\mu,\sigma)}\right)\frac{1}{\sigma\sqrt{2\pi}}Exp\left(-\frac{(x-\mu)^2}{2\sigma^2}\right) \quad for\ x \geq 0
\end{gathered} \tag{17}$$

The factor in the denominator of the CDF definition in Equation (17) is the area correction factor C.

$$C = \frac{1}{1 - F_{norm}(0,\mu,\sigma)} \tag{18}$$

Truncated normal distribution in two-parameter MaxEnt form is:

$$\begin{gathered}
f_{trunc}(x,\mu,\sigma) = Exp\left(-\lambda_0 - \lambda_1 x + \lambda_2 x^2\right) \\
\lambda_0 = -\frac{\mu^2}{2\sigma^2} - \ln\left(\frac{C}{\sigma\sqrt{2\pi}}\right) \quad \lambda_1 = -\frac{\mu}{\sigma^2} \quad \lambda_2 = \frac{1}{2\sigma^2}
\end{gathered} \tag{19}$$

Thus, just as the normal distribution is MaxEnt for moment functions x, x^2, where x ranges over $(-\infty,\ \infty)$, the truncated Normal distribution is MaxEnt for the same moment functions over the range $[0,\ \infty)$. Note that the μ and σ are the mean and standard deviation of the parent (un-truncated) normal distribution, not the truncated normal distribution.

3.3. MaxEnt Form of the Weibull Distribution

Since the Weibull distribution is widely used, it is useful to know what parameter value choices maximize the entropy of the function. It is often the case that only one of the two parameters is known and we seek a rational approach to assigning a value to the second parameter. In this case, we suggest that choosing the parameter value that maximizes the entropy of the distribution is the correct approach.

The entropy of the Weibull distribution is (Figure 8, derived from Equation (2.80c) in [21]):

$$\begin{gathered}
H = \gamma\left(1 - \frac{1}{\alpha}\right) + \ln\left(\frac{\beta}{\alpha}\right) + 1 \\
\gamma = 0.577216\ldots \quad \textit{Euler's constant}
\end{gathered} \tag{20}$$

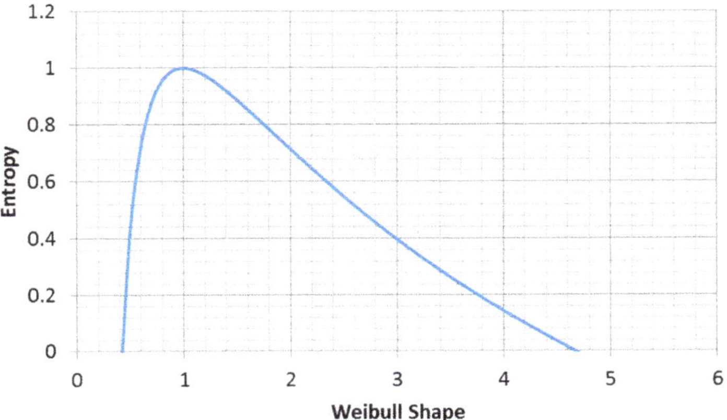

Figure 8. Entropy of a Weibull distribution with a fixed mean.

The mean of a Weibull distribution is [21]:

$$\mu = \beta\Gamma\left(1 + \frac{1}{\alpha}\right) \tag{21}$$

Thus, the entropy for a Weibull distribution with a fixed mean (moment constraint on x) is:

$$H_\mu = \gamma\left(1 - \frac{1}{\alpha}\right) + \ln(\mu) - \ln\left(\Gamma\left(1 + \frac{1}{\alpha}\right)\right) - \ln(\alpha) + 1 \tag{22}$$

Here, we maximize the entropy function:

$$\frac{dH_\mu}{d\alpha} = 0 \tag{23}$$

Then, we recall the properties of the digamma function [21]:

$$\begin{array}{c} \psi(x) = \frac{d}{dx}[\ln(\Gamma(x))] \\ \psi(1 + x) = \psi(x) + \frac{1}{x} \end{array} \tag{24}$$

Therefore:

$$\begin{array}{c} \frac{dH_\mu}{d\alpha} = \frac{\gamma}{\alpha^2} + \frac{\psi\left(1+\frac{1}{\alpha}\right)}{\alpha^2} - \frac{1}{\alpha} = 0 \\ \psi\left(\frac{1}{\alpha}\right) = -\gamma \end{array} \tag{25}$$

This is only true for $\alpha = 1$. Thus, within the Weibull family of distributions, for a given fixed mean, the exponential distribution has the highest entropy, in agreement with Jaynes's result.

The maximum entropy for fixed characteristic life is (Figure 9):

$$H = \gamma\left(1 - \frac{1}{\alpha}\right) + \ln\left(\frac{\beta}{\alpha}\right) + 1 \;\; for\; \beta = const. \tag{26}$$

Proceeding as above:

$$\begin{array}{c} \frac{dH}{d\alpha} = \frac{\gamma}{\alpha^2} - \frac{1}{\alpha} = 0 \\ \gamma = \alpha \end{array} \tag{27}$$

Thus, for the fixed characteristic life case, $\alpha = \gamma$ (the Euler's constant).

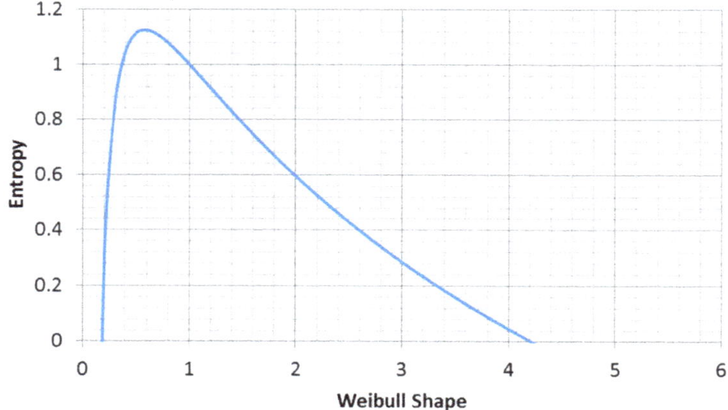

Figure 9. Plot of Equation (26) for $\beta = 1$.

4. Application of Maximum Entropy to Low-Cycle Fatigue of 2024-T351 Aluminum

When a specimen is subjected to axial load cycles of a magnitude sufficient to cause plastic deformation, the stress–strain history for the specimen can frequently be described as a loop, as shown in Figure 10. To determine the fatigue life of the specimen, the load cycles are applied until the specimen fails, or until its compliance exceeds some proportion of its initial compliance. The Coffin–Manson relationship (Equation (1)) is commonly used to model the relationship between plastic strain range and reversals to failure. The parameter $\epsilon_{f'}$ is determined by fitting the curve to fatigue data. It is frequently close in value to ϵ_f.

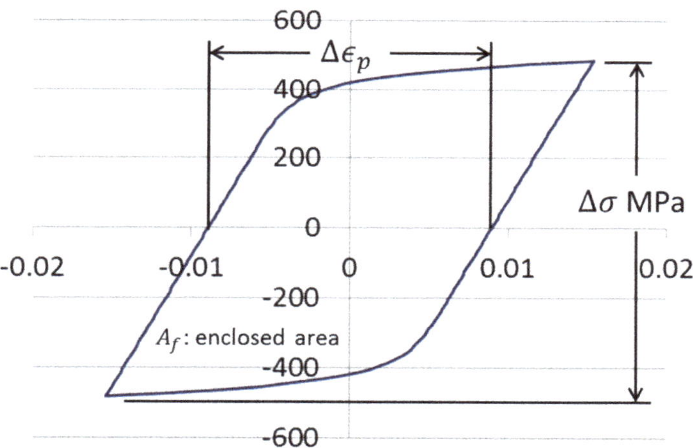

Figure 10. Stress–strain loop showing plastic strain. The variables are defined in Table 2.

As mentioned earlier, a sequence of low-cycle fatigue tests, along with two monotonic tension tests, was performed on tension specimens of 2024-T351 aluminum. Eighteen specimens were tested under constant-amplitude, fully reversed fatigue conditions. In five cases, representative stress–strain loops were collected at various cycle intervals. Two specimens were tested to failure monotonically. The data collected is summarized in Table 2. The data is fitted to a Coffin–Manson model, as shown in Figure 1.

As mentioned earlier, the data exhibits a curvature that is not captured by the straight line fit of the Coffin–Manson power law. An alternative approach to modeling data such as this, using concepts developed from maximum entropy, is developed below. The authors of [12] showed that material entropy is proportional to inelastic dissipation in experiments such as this, where the temperature of the specimens is essentially constant. Thus, inelastic dissipation is exploited as a surrogate for entropy in the development that follows.

Table 2. Low-cycle fatigue data summary [22].

k	$2N_f$	Stress Amplitude MPa	Plastic Strain Amplitude	Data
1	1	537.810	0.2	Values
2	1	558.495	0.28	Values
3	76	503.335	0.01725	Values
4	38	495.061	0.0129	S-20 Loop fitted
5	124	492.993	0.0123	Values
6	144	482.650	0.012	Values
7	190	475.755	0.01067	Values [22]
8	114	477.824	0.0085	S-17 loop fitted
9	440	466.792	0.0083	Values [22]
10	560	448.175	0.00606	Values [22]
11	920	437.143	0.00472	Values [22]
12	516	453.691	0.0038	S-12 loop fitted
13	1080	441.280	0.0037	Values
14	800	441.280	0.0036	Values
15	624	454.381	0.0035	S-11 loop fitted
16	2800	398.531	0.00178	Values [22]
17	1608	430.938	0.0017	S-18 loop fitted
18	5860	403.358	0.0007	Values
19	16336	351.645	0.00015	Values
20	23400	358.540	0.00004	Values

The variable D representing the ability of the material at a point to bear load is fundamental in the literature of damage mechanics [23]. The value of $D = 0$ (undamaged) represents virgin material, while $D = 1$ is taken to correspond to failed material. The variable D is a non-decreasing quantity, since damage is inherently irreversible. The Coffin–Manson equation can be rewritten in terms of damage, and doing so will be shown to provide a departure point for further development. We begin by rearranging Equation (1) into the following form:

$$\frac{1}{2N_f} = \left(\frac{\Delta\epsilon_p}{2\epsilon'_f}\right)^{-\frac{1}{c}} \tag{28}$$

Depending on the application, the damage variable D may be expressed as a function of various independent variables. In fatigue applications, it is common to use the following (applicable to constant damage per load cycle) Palmgren–Miner definition of damage. It is understood that N_f may depend on other variables, such as temperature or plastic strain amplitude.

$$D(N) = \frac{N}{N_f} \tag{29}$$

We can write the damage accumulation per reversal:

$$D_{rev} = \frac{1}{2N_f} \tag{30}$$

Finally, Equation (28) can be recast as a damage equation as follows:

$$D_{rev} = \left(\frac{\Delta\epsilon_p}{2\epsilon'_f}\right)^{-\frac{1}{c}} = f\left(\Delta\epsilon_p\right) \tag{31}$$

where $f(\cdot)$ denotes a functional relationship with the argument. Following [12], we propose developing a function of the form of Equation (31), in terms of energy per reversal rather than plastic strain range. This relationship will have the form:

$$D_{rev} = f\left(\frac{W_f}{2N_f}\right) \tag{32}$$

In the development that follows, a general approach to deriving functions of the form of the above equation will be proposed. In order to apply an equation of the above form to the data in Table 3, we first need to determine the inelastic dissipation per reversal corresponding to each of the test conditions of the form shown in Figure 10. The energy expended in inelastic dissipation for a cyclic test under constant conditions is given by the area enclosed by the loop. Note that in Table 3, actual loop data was only available for five of the 20 tests. In all cases, the plastic strain range and stress range (and reversals to failure) were collected. Fortunately, the shapes of the loops follow known trends, and thus it was possible to deduce the inelastic dissipation for the tests where loops were not available for measurement. The inelastic dissipation for the two monotonic tests was also deduced from the available loop data, although a different analytical approach was used.

Table 3. Inelastic dissipation and damage.

k	$2N_f$	Range Mpa	Range Ep	*1/n*	*rho*	$W_f/2N_f$	D_{rev}
1	1	538	2.00×10^{-1}	26.7	0.964	1.04×10^{2}	1.00
2	1	558	2.80×10^{-1}	26.7	0.964	1.51×10^{2}	1.00
3	76	1007	3.45×10^{-2}	26.7	0.928	1.61×10^{1}	1.32×10^{-2}
4	38	990	2.58×10^{-2}	26.7	0.928	1.19×10^{1}	2.63×10^{-2}
5	124	986	2.46×10^{-2}	26.5	0.927	1.12×10^{1}	8.06×10^{-3}
6	144	965	2.40×10^{-2}	26.4	0.927	1.07×10^{1}	6.94×10^{-3}
7	190	952	2.13×10^{-2}	25.9	0.926	9.40	5.26×10^{-3}
8	114	956	1.70×10^{-2}	25.1	0.923	7.50	8.77×10^{-3}
9	440	934	1.66×10^{-2}	24.7	0.922	7.15	2.27×10^{-3}
10	560	896	1.21×10^{-2}	21.2	0.910	4.94	1.79×10^{-3}
11	920	874	9.44×10^{-3}	19.1	0.900	3.72	1.09×10^{-3}
12	516	907	7.60×10^{-3}	17.6	0.893	3.08	1.94×10^{-3}
13	1080	883	7.40×10^{-3}	17.4	0.891	2.91	9.26×10^{-4}
14	800	883	7.20×10^{-3}	17.1	0.890	2.83	1.25×10^{-3}
15	624	909	7.00×10^{-3}	16.9	0.888	2.83	1.60×10^{-3}
16	2800	797	3.56×10^{-3}	13.6	0.863	1.22	3.57×10^{-4}
17	1608	862	3.40×10^{-3}	13.4	0.862	1.26	6.22×10^{-4}
18	5860	807	1.40×10^{-3}	13.4	0.862	4.87×10^{-1}	1.71×10^{-4}
19	16336	703	3.00×10^{-4}	13.4	0.862	9.09×10^{-2}	6.12×10^{-5}
20	23400	717	8.00×10^{-5}	13.4	0.862	2.47×10^{-2}	4.27×10^{-5}

Plotted loops for the five loop data samples are given below in Figures 11–15. In each case, several loops were provided.

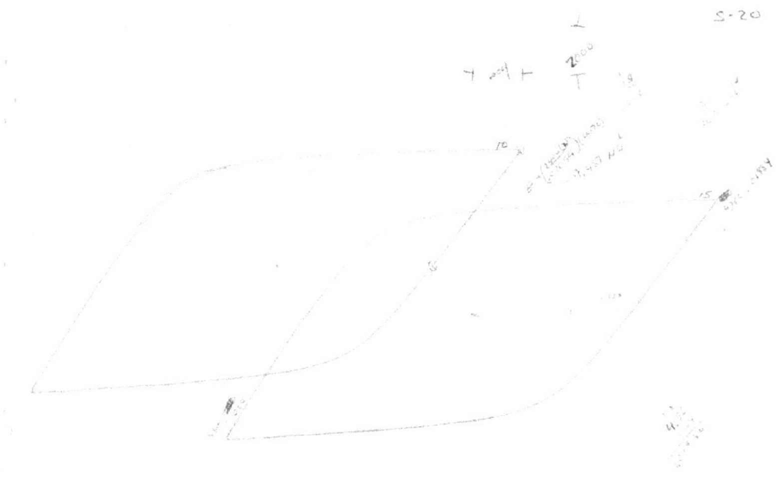

Figure 11. Raw data plotted on chart paper from Test 4, $2N_f = 38$.

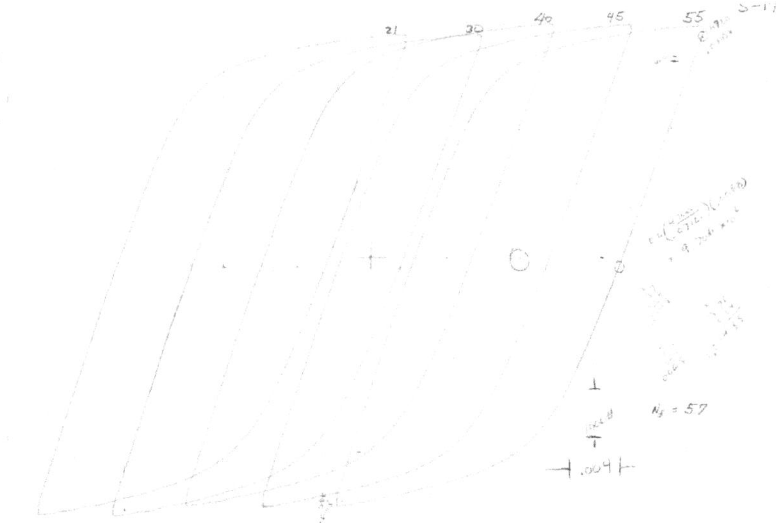

Figure 12. Raw data plotted on chart paper from Test 8, $2N_f = 114$.

Figure 13. Raw data plotted on chart paper from Test 12, $2N_f$ = 516.

Figure 14. Raw data plotted on chart paper from Test 15, $2N_f$ = 624.

The Ramberg–Osgood relationship (Equation (33)) is frequently successful for modeling data such as this. This model assumes that the plastic portion of the strain range is a power law of the stress range. There is no explicit yield point with this model. The total strain range is given by Equation (34) and is used to model the shapes of the loops. For the purposes of fitting Equation (34), the origin of the stress and strain range variables is placed at the lower left corner of the loop.

The Ramberg–Osgood plasticity model for stress–strain loops is [23]:

$$\Delta\epsilon_p = \left(\frac{\Delta\sigma}{K}\right)^{\frac{1}{n}} \tag{33}$$

$$\Delta\epsilon_{total} = \frac{\Delta\sigma}{E} + \left(\frac{\Delta\sigma}{K}\right)^{\frac{1}{n}} \tag{34}$$

Figure 15. Raw data plotted on chart paper from Test 17, $2N_f = 1608$.

The fits of Equation (34) to loop data were performed using the least squares approach and is shown in Figure 16. The fits to the data were of high accuracy, as demonstrated by the R^2 value of 0.997. This confirms that Equation (34) provides a reasonable model of the shape of the loops in Figures 11–15. The points are samples measured from the loops, while the line is the fit of Equation (34). A separate fit was performed for the parameters in Equation (34) for each of the five loops. A common value of Young's modulus was fit simultaneously to the five sets of data. Specific values of n and K were obtained for each loop.

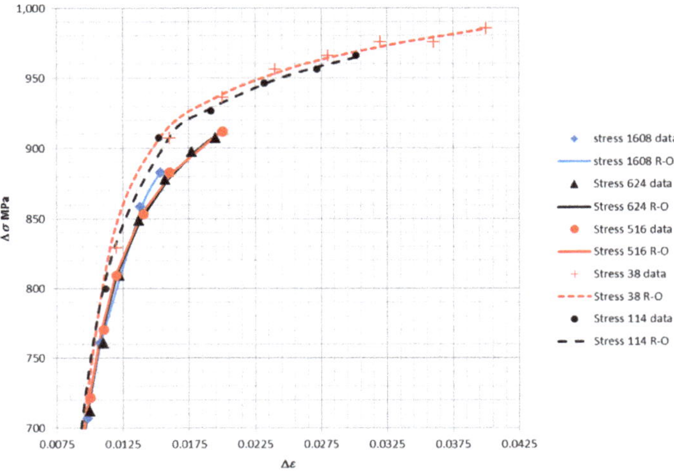

Figure 16. Fit of Equation (34) to 5 data sets (E = 73.800 GPa for all fits) (R^2 = 0.997).

The five sets of parameters obtained from the fitted loops were used to estimate the parameter $1/n$ for the remaining 15 tests. The fitted $1/n$ value was found to be a strictly increasing function of plastic strain range, and is plotted in Figure 17. The "interpolation" line markers show the values of $1/n$ used for the remaining 15 tests. The values were linearly interpolated between the maximum and minimum values. For plastic strain ranges outside the range of the measured data, the value of the nearest measured data value was used. As will be shown below, the predicted inelastic dissipation is mainly determined by the plastic strain range and the stress range, and is only weakly dependent on the value of $1/n$ used.

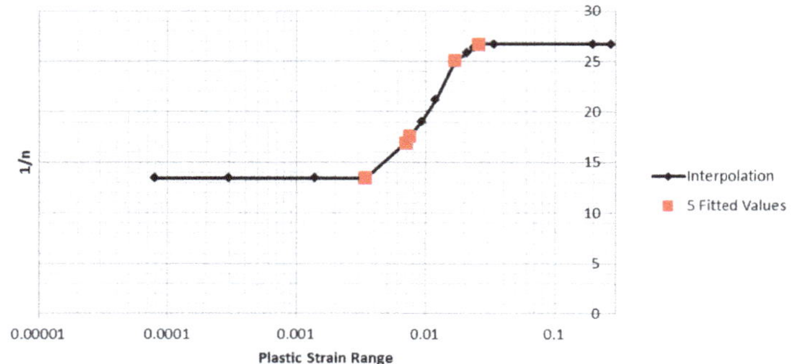

Figure 17. 5 Fitted values of $1/n$ with interpolation function.

The five loops (represented by Equation (34)) are plotted in Figure 18 below using the parameters fit to the corresponding loop data. The inelastic dissipation per cycle is the area enclosed by the loop.

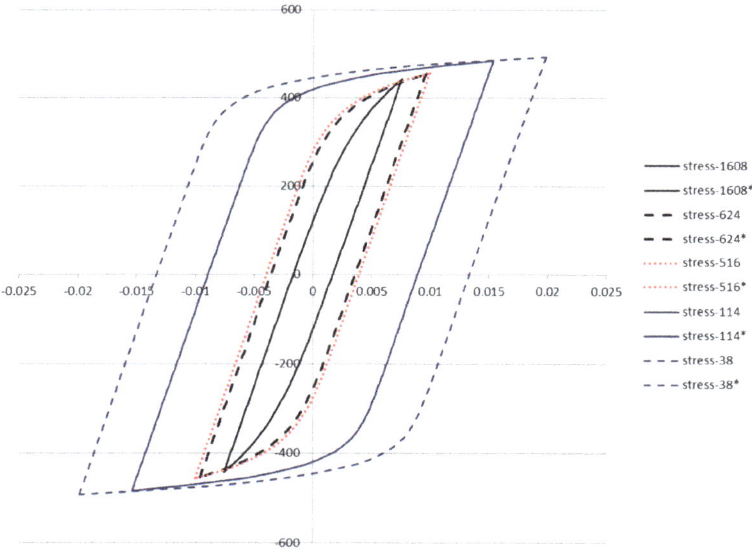

Figure 18. Ramberg–Osgood curves based on loop fits.

The area of the loop in terms of the parameters in Equation (34) and the loading parameters are given in Equation (35). The form of this equation has the advantage that it is relatively robust to errors in fitting the parameter n, since both of the actual measured values of the stress range and strain range are used.

The loop area (dissipation per cycle) in terms of n is [23]:

$$A_f = \frac{1-n}{1+n}\Delta\sigma\Delta\epsilon_p \tag{35}$$

In the present case, we wish to describe the evolution of damage in terms of reversals rather than cycles. It is apparent from Equation (36) that the inelastic dissipation per reversal is half the area of the loop given by Equation (35), and is given in Equation (37).

Total inelastic dissipation in terms of cycles and reversals:

$$W_f = N_f A_f = \left(2N_f\right)\left(\frac{1}{2}A_f\right) \tag{36}$$

Inelastic dissipation per reversal:

$$\frac{W_f}{2N_f} = \frac{1-n}{2(1+n)}\Delta\sigma\Delta\epsilon_p \tag{37}$$

For specimens subjected to a monotonic test, the inelastic dissipation is the area under the plastic portion of the stress–strain curve. If the plastic portion of the curve is modeled by an equation of the form of Equation (34), the area under the plastic portion is given by Equation (38). A monotonic test to fracture can be interpreted as a fatigue test, with failure occurring after a single reversal. Thus, the inelastic dissipation per reversal is given by Equation (39):

The monotonic area (dissipation per reversal) in terms of n:

$$A_f = \frac{1}{1+n}\sigma_f\epsilon_f \tag{38}$$

The inelastic dissipation for a monotonic test:

$$\frac{W_f}{2N_f} = A_f = \frac{1}{1+n}\sigma_f\epsilon_f \quad 2N_f = 1 \tag{39}$$

Note that in Equations (37) and (39), the area is computed from plastic strain range multiplied by stress range times a factor dependent on n. The functions are given in Equation (40) and the values of ρ are summarized in Table 4 and plotted in Figure 19.

$$\begin{aligned}\rho_{mono} &= \frac{1}{1+n} \\ \rho_{loop} &= \frac{1-n}{1+n}\end{aligned} \tag{40}$$

Note that the value of ρ does not change greatly as n is varied. This observation indicates that the computation of areas for the monotonic and cyclic tests is robust to errors in fitting the Ramberg–Osgood parameter n. Thus, the inference of inelastic dissipation for the 15 tests for which loop data was not available is justified.

Table 3 below includes values computed from Equations (37) and (39) for inelastic dissipation per reversal, as well as damage per reversal, according to Equation (30). These data are plotted in Figure 20. These points represent data corresponding to a relationship with the form of Equation (32). The lack of fit provided by the power law indicates that a different modeling equation is required for data of this type. In the development that follows, various expressions, including some based on MaxEnt principals, will be proposed to model the data plotted in Figure 20.

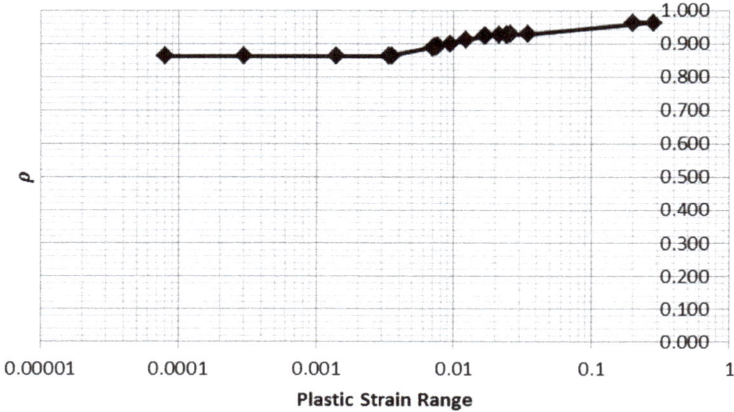

Figure 19. Here, ρ is shown as a function of plastic strain range.

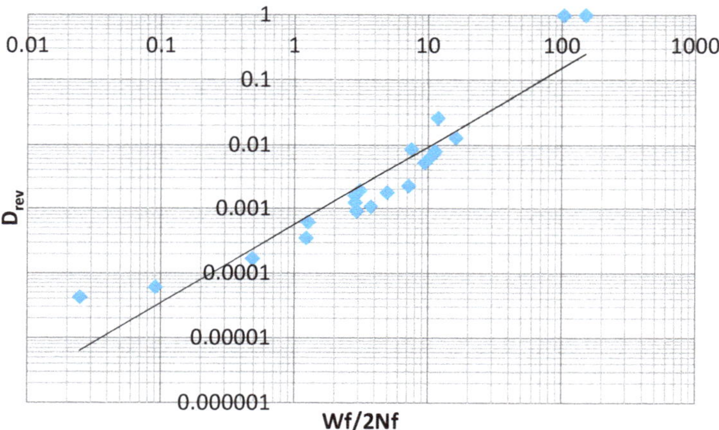

Figure 20. Damage per reversal as a function of inelastic dissipation per reversal with power law fit ($R^2 = 0.89$).

Table 4. Candidate function forms fit to data in Table 4.

Function	Form (for $0 \leq a$)	Sum of Sqr Error
Left Truncated Normal	$D_{rev} = \frac{F_{norm}(W_c,\mu,\sigma)-F_{norm}(0,\mu,\sigma)}{1-F_{norm}(0,\mu,\sigma)}$	5.17
Truncated Exponential	$D_{rev} = \frac{1-\exp(-\lambda W_c)}{1-\exp(-\lambda a)}$	5.45
Power law (Coffin–Manson form)	$D_{rev} = k(W_c)^{-\frac{1}{c}}\ for\ W_c \leq W_{c\ crit}$	14.8
Weibull	$D_{rev} = 1 - \exp(-kW_c{}^{\alpha})$	15.4
Smith–Ferrante form	$D_{rev} = 1 - (1 + kW_c)\exp(-kW_c)$	57.3

Discussion of Candidate Distribution Functions

Inelastic dissipation is a non-negative-valued function, so only distribution functions equal to zero for $x \geq 0$ are admissible candidates. Table 4 contains a summary of the fitted functions, as well as the sum of squares of error remaining after the fitting. The natural logs of the data were fitted to the natural logs of the predicted values. Plots of the fitted curves and the data are shown in Figure 21. Only the truncated forms of the normal distribution are considered. Distributions that are truncated on the right, such as the truncated exponential distribution, have the additional advantage that they are strictly equal one for $x \geq a$.

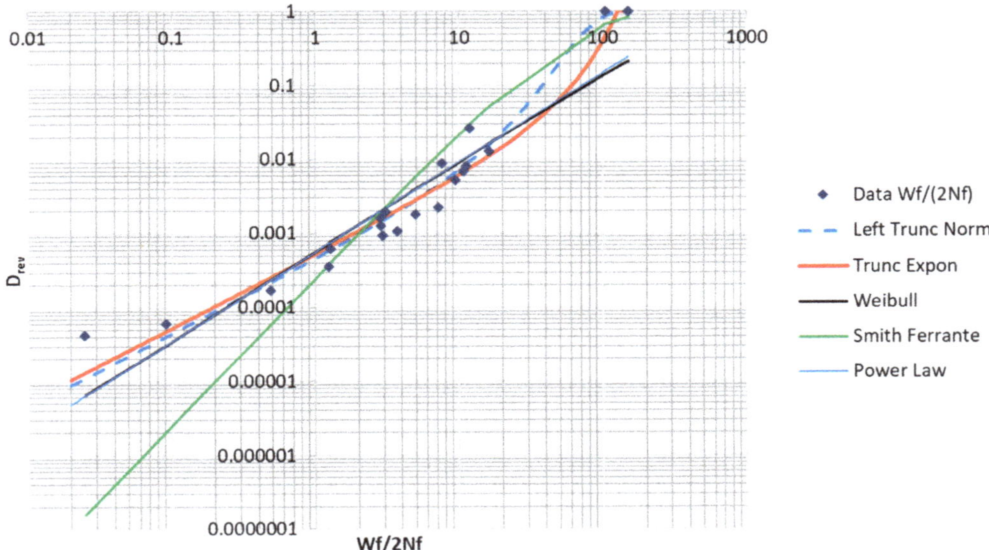

Figure 21. Plots of functions in Table 4.

The data set being fitted has some noteworthy features. Even though the data is of low-cycle fatigue, most of the samples still represent very small values of D_{rev}. Additionally, the data points show a concave upwards trend that limits the quality of the fit achievable by a power law relationship. The fit was notably better for the right truncated exponential distribution with a negative λ. The fitting procedure converged to a negative λ, which corresponds to a rising exponential curve that becomes constant at $D_{cycle} = 1$. The best fits were achieved by the truncated normal distribution and the truncated exponential distribution. The Smith–Ferrante function (popular in cohesive zone models of fracture) is typically used to represent the traction versus separation, and is founded on the relationship binding materials together at the microscopic scale [24]. Its integral is used here, which has the qualitative features of a damage function. The Weibull distribution function was also tried. Additionally, a power law expression having the form of the Coffin–Manson relation was tried. This function would be truncated at $D_{rev} = 1$.

Note that the Coffin–Manson expression typically relates plastic strain range to cycles to failure. In Table 4, it is shown in an inverted form and expressed in terms of W_c. It is clear from the sum of squared error column in Table 4 and from Figure 21 below that the truncated normal distribution provided the best fit to the data, followed by the truncated exponential distribution. The (inverted) Coffin–Manson expression and the Weibull distribution function provided the next best fits.

Parameters fit by numerical solver to the fatigue data for the truncated normal distribution (Equation (41)) and the truncated exponential distribution (Equation (42)) are given below:

$$D_{rev} = \frac{F_{norm}(x, 72.1,\ 27.3) - F_{norm}(0, 72.1,\ 27.3)}{1 - F_{norm}(0, 72.1,\ 27.3)} \quad for\ x \geq 0 \tag{41}$$

$$D_{rev} = \frac{1 - Exp(0.0325x)}{1 - Exp((-0.0325)(127.2))} \quad for \ 0 \leq x \leq 127.2 \tag{42}$$

Although the trunated normal distribution has the best fit, the truncated exponential distribution has some desireable properties. If monotonic tension data points are available, they can be used to directly constrain the point where the curve is strictly equal to 1.0. The parameter λ controls the shape of the curve between $x = 0$ and $x = a$. For λ close to zero, the curve is nearly a ramp function. For negative λ values, it has varying degrees of concave upwards curvature. Examples of a family of such curves are plotted in Figure 22. In the present case, $\lambda = -0.0319$, giving a strongly rising curve. A damage function of the mathematical form of Equation (42) exists in the literature [2]. The authors of [2] present Equation (43) as an improvement to the Coffin–Manson relationship (Equation (2)) for modeling LCF in the sub 100 cycle range (ϵ_{pa} is the plastic strain amplitude). The relationship is presented as an empirical improvement and is not derived from physical principles. The authors do not describe it as a truncated exponential distribution function. It is clear that Equation (43) can be rearranged to a form similar to Equation (42).

$$D_{cycle} = \frac{Exp\left(\frac{\lambda \epsilon_{pa}}{\epsilon_f}\right) - 1}{Exp(\lambda) - 1} \tag{43}$$

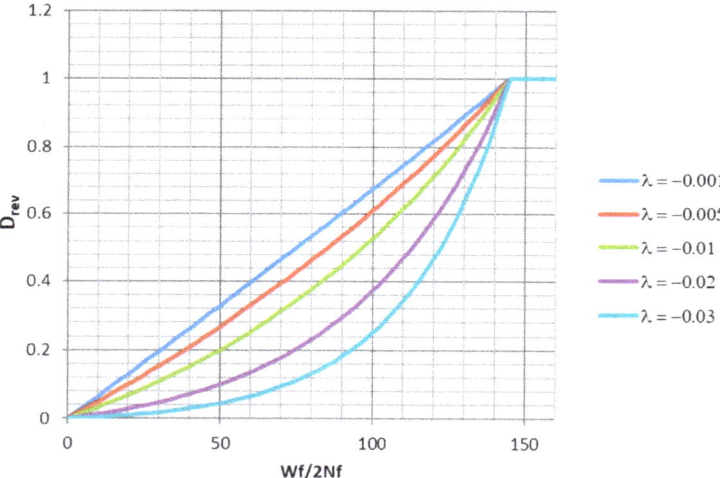

Figure 22. Plots of truncated exponential distribution with different shapes.

5. Concluding Remarks

In this study, the Maximum Entropy principle was shown to provide a systematic theoretical and philosophical basis for selecting a CDF to model damage. The method was demonstrated on an LCF data set for aluminum 2024-T351, but the proposed approach is equally applicable to ductile metals undergoing fatigue damage. In general, the relationship between the measured plastic dissipation per cyclic reversal and the damage per reversal is nonlinear, suggesting that the total work of fracture or the total entropy to cause fracture varies with the loading condition. We showed that several maximum entropy distributions, including the truncated exponential and the truncated normal distribution, are good choices for material damage modeling. Compared to the exponential distribution, the truncated exponential distribution has additional flexibility and can model concave upwards trending data. In the limit, it can approximate a uniform distribution. For the aluminum 2024-T351 alloy, the truncated normal distribution was shown to provide the best fit to the data, relative to the more common alternatives of Coffin–Manson equation or the Weibull distribution. Left truncation of the normal distribution extends its applicability to the many applications where data is non-negative. Finally, a Coffin–Manson function in terms of plastic strain (the standard form) was compared to the truncated normal distribution and shown to provide an inferior fit.

Author Contributions: Conceptualization C.Y. and G.S.; methodology C.Y. and G.S.; data analysis C.Y.; writing—original draft preparation, C.Y.; writing—review and editing, G.S.; funding acquisition, C.Y.

Funding: C.Y. was supported by Ford Motor Corporation during the course of this study.

Conflicts of Interest: The authors declare no conflict of interest.

Nomenclature

Variable	Definition
$\Delta\epsilon_p$	Plastic strain range
$\epsilon_f\prime$	Fatigue ductility coefficient
c	Fatigue ductility exponent
N_f	Total cycles (loops) to failure
$D(t)$	Material damage parameter as a function of time or pseudo-time
$\Delta\sigma$	Stress range—total height of stress–strain loop
$\Delta\epsilon_{total}$	Total elastic plus plastic strain range
σ_f	True fracture stress
ϵ_f	True fracture strain
W_f	Total inelastic dissipation (per unit volume) to failure
A_f	Inelastic dissipation (per unit volume) per stress–strain loop area of stabilized loop
$2N_f$	Total reversals to failure
H	The entropy of a probability distribution
S	Gibbs entropy
p_i	Probability mass function value of the ith random state
k_b	Boltzmann's constant
$I(p)$	The information associated with an event with probability p
$g_i(x)$	The ith moment function
λ_i	The Lagrange multiplier corresponding to the ith moment function
$f(x)$	The probability density function (PDF) of the random variable x
$F(x)$	The cumulative distribution function (CDF) of the random variable x
μ	Mean value of a random variable
σ	Standard deviation of a random variable
α	Weibull distribution shape parameter
β	Weibull distribution scale parameter
γ	Euler's constant
K	Ramberg–Osgood strength parameter
$\frac{1}{n}$	Ramberg–Osgood exponent

References

1. Dowling, N. *Mechanical Behavior of Materials*; Pearson Prentice Hall: Upper Saddle River, NJ, USA, 2007.
2. Xue, L. A unified expression for low cycle fatigue and extremely low cycle fatigue and its implication for monotonic loading. *Int. J. Fatigue* **2008**, *30*, 1691–1698. [CrossRef]
3. McPherson, J.W. *Reliability Physics and Engineering*; Springer International Publishing: Cham, Switzerland, 2013.
4. Gong, Y.; Norton, M. Materials Fatigue Life Distribution: A Maximum Entropy Approach. *ASTM J. Test. Eval.* **1998**, *26*, 53–63.
5. Guan, X.; Giffin, A.; Jha, R.; Liu, Y. Maximum relative entropy-based probabilistic inference in fatigue crack damage prognostics. *Probabilistic Eng. Mech.* **2012**, *29*, 157–166. [CrossRef]

6. Li, C.; Xie, L.; Ren, L.; Wang, J. Progressive failure constitutive model for softening behavior of rocks based on maximum entropy theory. *Env. Earth Sci.* **2015**, *73*, 5905–5915. [CrossRef]
7. Li, H.; Wen, D.; Lu, Z.; Wang, Y.; Deng, F. Identifying the Probability Distribution of Fatigue Life using the Maximum Entropy Principle. *Entropy* **2016**, *18*, 111. [CrossRef]
8. Basaran, C.; Yan, C. A Thermodynamic Framework for Damage Mechanics of Solder Joints. *ASME J. Electron. Packag.* **1998**, *120*, 379–384. [CrossRef]
9. Basaran, C.; Nie, S. An Irreversible Thermodynamic Theory for Damage Mechanics of Solids. *Int. J. Damage Mech.* **2004**, *13*, 205–224. [CrossRef]
10. Naderi, M.; Amiri, M.; Khonsari, M. On the thermodynamic entropy of fatigue fracture. *Proc. R. Soc. A* **2010**, *466*, 423–438. [CrossRef]
11. Naderi, M.; Khonsari, M. An experimental approach to low-cycle fatigue damage based on thermodynamic entropy. *Int. J. Solids Struct.* **2010**, *47*, 875–880. [CrossRef]
12. Chan, D.; Subbarayan, G.; Nguyen, L. Maximum-Entropy Principle for Modeling Damage and Fracture in Solder Joints. *J. Electron. Mater.* **2012**, *41*, 398–411. [CrossRef]
13. Jaynes, E. Information theory and statistical mechanics. *Phy. Rev.* **1957**, *106*, 620–630. [CrossRef]
14. Tolman, R. *The Principles of Statistical Mechanics*; Dover Publications: New York, NY, USA, 1979.
15. Mix, D.F. *Random Signal Processing*; Prentice Hall: Upper Saddle River, NJ, USA, 1995.
16. Papoulis, A.; Pillai, S. *Probability, Random Variables, and Stochastic Processes*; McGraw Hill: New York, NY, USA, 2002.
17. Shannon, C. Mathematical theory of communication. *Bell Syst. Tech. J.* **1948**, *27*, 379–423. [CrossRef]
18. Usta, I.; Kantar, Y. On the performance of the flexible maximum entropy distributions within partially adaptive estimation. *Comput. Stat. Data Anal.* **2011**, *55*, 2172–2182. [CrossRef]
19. Schroeder, A. *Accounting and Causal Effects: Econometric Challenges*; Springer: New York, NY, USA, 2010.
20. El-Affendi, M. Estimating Computer Performance Metrics when the Service and Interval Times are of the Truncated Normal Type. *Comput. Math. Appl.* **1992**, *23*, 35–40. [CrossRef]
21. Rinne, H. *The Weibull Distribution: A Handbook*; CRC Press: Boca Raton, FL, USA, 2009.
22. Leis, B.N. *Master's Thesis, Figure 17*; University of Waterloo: Waterloo, ON, Canada, 2001.
23. Lemaitre, J.; Chaboche, J. *Mechanics of Solid Materials*; Cambridge University Press: Cambridge, UK, 1990.
24. De-Andres, A.; Perez, J.; Ortiz, M. Elastoplastic finite element analysis of three-dimensional fatigue crack growth in aluminum shafts subjected to axial loading. *J. Solids Struct.* **1999**, *36*, 2231–2258. [CrossRef]

Article

Measures of Entropy to Characterize Fatigue Damage in Metallic Materials

Huisung Yun * and Mohammad Modarres

Center for Risk and Reliability, Department of Mechanical Engineering, University of Maryland, College Park, MD 20742, USA

* Correspondence: hsyun@terpmail.umd.edu; Tel.: +82-10-5087-6461

Received: 18 June 2019; Accepted: 14 August 2019; Published: 17 August 2019

Abstract: This paper presents the entropic damage indicators for metallic material fatigue processes obtained from three associated energy dissipation sources. Since its inception, reliability engineering has employed statistical and probabilistic models to assess the reliability and integrity of components and systems. To supplement the traditional techniques, an empirically-based approach, called physics of failure (PoF), has recently become popular. The prerequisite for a PoF analysis is an understanding of the mechanics of the failure process. Entropy, the measure of disorder and uncertainty, introduced from the second law of thermodynamics, has emerged as a fundamental and promising metric to characterize all mechanistic degradation phenomena and their interactions. Entropy has already been used as a fundamental and scale-independent metric to predict damage and failure. In this paper, three entropic-based metrics are examined and demonstrated for application to fatigue damage. We collected experimental data on energy dissipations associated with fatigue damage, in the forms of mechanical, thermal, and acoustic emission (AE) energies, and estimated and correlated the corresponding entropy generations with the observed fatigue damages in metallic materials. Three entropic theorems—thermodynamics, information, and statistical mechanics—support approaches used to estimate the entropic-based fatigue damage. Classical thermodynamic entropy provided a reasonably constant level of entropic endurance to fatigue failure. Jeffreys divergence in statistical mechanics and AE information entropy also correlated well with fatigue damage. Finally, an extension of the relationship between thermodynamic entropy and Jeffreys divergence from molecular-scale to macro-scale applications in fatigue failure resulted in an empirically-based pseudo-Boltzmann constant equivalent to the Boltzmann constant.

Keywords: physics of failure; prognosis and health management; entropy as damage; fatigue; entropy generation; acoustic emission; information entropy; thermodynamic entropy; Jeffreys divergence

---

1. Introduction

Prognostics and health management (PHM) is a promising method in reliability engineering to supplement traditional life assessments. The traditional damage measurements in fatigue, for example, crack growth and load-carrying capacity reduction, are detectable only in the later stages of life and are ineffective in characterizing damage during the earlier periods of life [1]. In contrast, PHM-based life estimation and prognosis incorporates related monitored damage variables into deterministic physics of failure (PoF) models [2–6]. In data-driven prognostics in PHM, observed damage precursors, such as initiation of very small cracks, are collected during system operation and are used to estimate the so-called remaining useful life (RUL) [5,7]. The approaches used to meet the requirements of early life prediction include the uses of entropy. Examples of entropic theories of damage for life prediction include the degradation-entropy generation (DEG) theorem [8] and the principle of maximum entropy (PME) [9–13]. The maximum entropy (MaxEnt) distribution, according to the PME, is the best choice to capture the state of knowledge and information about damage (e.g., measured PHM data). However, thermodynamic entropy, according to the DEG theorem, offers a direct representation of damage [8,14]. Both entropic representations offer powerful foundations for early life fatigue prediction.

Entropy, according to the DEG theorem, is based on irreversible thermodynamics and can be used to depict the endurance to failure, such as cycles to crack initiation or fracture [14,15]. Pioneering works in entropic approaches have verified successful applications to several failure mechanisms, such as fatigue, corrosion, and wear. These entropies are derived from sources of irreversible energy dissipation [14–17]. In the case of fatigue damage, irreversible energy dissipations include plastic mechanical work, heat, and acoustic emission [18]. A popular entropic approach in fatigue is to use plastic strain energy and surface temperature [19,20]. In this approach, the existence of a fixed entropic endurance, irrespective of the underlying conditions that lead to fatigue damage and failure, is experimentally verified. It has resulted in good agreement with the DEG theorem. Another approach has used acoustic energy dissipation during fatigue in the form of generated acoustic emission (AE) waveforms, where associated information entropy typically correlates well with the amount of the fatigue damage [21].

Strain energy dissipation during the cyclic fatigue loading and unloading also appears to apply to the relative entropy. Crooks et al. [22] have shown that the Kullback–Leibler divergence computed from loading/unloading distributions is equivalent to the thermodynamic entropy when distributions of loading/unloading processes are measurable. This concept was demonstrated by Collin et al. [23], who measured thermal dissipation in the unfolding/folding process of a ribonucleic acid (RNA) strand. Loading/unloading work distributions were also used by Douarche et al. [24] to measure a brass wire's cyclic torsional work and assess the Helmholtz free energy difference. In practical applications, relative entropy in cyclic mechanical work can be computed without the need for temperature information, which provides a potentially simpler entropic damage assessment than the classical thermodynamics.

This paper presents the entropic damage measurements from dogbone coupons that were fatigue tested using three energy dissipations: Plastic mechanical work, thermal energy, and AE. In these approaches, uses of the classical thermodynamics, information (Shannon), and relative entropy were evaluated and discussed in the context of PHM applications. In the proposed approach, relative entropy uses in fatigue damage is new; however, the paper also compares the relationship between these three entropic measures and discusses their applicability to fatigue failures.

In the remainder of this paper, Section 2 provides reviews of the three entropic theorems and discusses the relative entropy in the context of the fatigue damage process. Section 3 presents the experimental setup, including specimen design, cyclic loading conditions, sensor attachments, and data collection. Section 4 presents the results that support the DEG theorem demonstrated by each entropic approach and discusses the applications and results. Finally, the conclusion section summarizes the results.

2. Fatigue Damage Evaluation Using Three Entropy Measures

According to Lemaitre [1], measurements of fatigue damage include changes detected in crack length, elastic modulus, micro-hardness, ultrasonic wave, and electric resistance. These measurements, also called the markers of damage, are often only detectable when 10–20% of life remains, which is too late for effective prognostic and corrective actions [7]. Therefore, during the early period of life, the assessment of damage must rely on deterministic life models, which tend to be highly uncertain, variable, and conservative [16].

Amiri and Modarres [16] have summarized and delineated fatigue damage scales into nano-, micro-, meso-, and macro-scales. Thus, the damage measurement scale evolves from the very small to larger scales, and it is only in the macro-scale that damage can be detected. As such, the lack of detectable damage is highly scale-dependent. However, the damage measurement through the second law of thermodynamics suggests a universal methodology that applies to all the scales discussed above.

Entropic metrics of damage have been proposed and utilized in engineering applications. Basaran [17,25] and Bryant [14] introduced thermodynamic concepts to assess damage in specific failure modes. Based on irreversible thermodynamic processes, these studies considered the degradation-induced dissipated energy or entropy as a reflection of the cumulative damage process. Amiri and Modarres [16] reviewed entropy for various failure mechanisms, including fatigue, corrosion, and wear, and discussed the corresponding irreversible thermodynamic forces and fluxes used to calculate entropy generation. They reviewed in more detail the DEG theorem, including the concept of entropic endurance introduced by Imanian and Modarres [26]. Experimental results have supported this theorem for fatigue failures [15,19,20] by demonstrating that fatigue fracture occurs at a relatively fixed entropic endurance level regardless of the underlying loading profiles. Figure 1 presents an example confirming this entropic theorem.

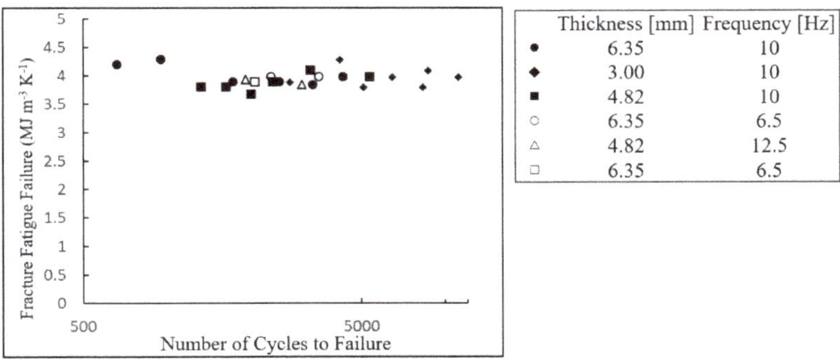

Figure 1. Cumulative entropy for various fatigue test loading conditions. The entropic data points show that the thermodynamic entropy has an endurance level to failure, irrespective of the path to failure. The cyclic bending loading was applied to each specimen with the amplitude of 25–50 mm. Entropic endurance raw data were from Figure 6 of Naderi et al. [19].

From the irreversible thermodynamics, the dissipative entropy generation may be expressed in the form of Equation (1) [15,16,26]:

$$\sigma = \sum_i X_i J_i \tag{1}$$

This equation is bilinear, where X_i is the thermodynamic force and J_i is the flux due to the dissipation mechanism i. Depending on the sources of energy dissipation, Amiri and Modarres [16] presented the entropy generation in its most general form, as shown in Equation (2) [15]:

$$\sigma = \frac{1}{T^2} J_q \cdot \nabla T - \sum_k J_k \left(\nabla \frac{\mu_k}{T} \right) + \frac{1}{T} \tau : \dot{\varepsilon}_p + \frac{1}{T} \sum_j \nu_j A_j + \frac{1}{T} \sum_m c_m J_m (-\nabla \psi) \tag{2}$$

where σ is the entropy generation rate, J_q is the thermodynamic flux due to heat conduction, J_k is the thermodynamic flux due to diffusion, μ_k is the chemical potential, τ is the mechanical stress, ε_p is the plastic strain, ν_j is the chemical reaction rate, A_j is the chemical affinity, c_m is the coupling constant, J_m is the thermodynamic flux due to the external field, and ψ is the potential of the external field.

In this equation, the five terms on the right-side are sources of thermodynamic entropy generation that include heat, diffusion, mechanical work, chemical reaction, and external field effect, respectively. In the fatigue damaging process, heat and mechanical work terms are involved. Naderi et al. [19] numerically calculated the dissipative entropy by using only the mechanical work term, assuming that plastic deformation is the dominant term and the heat conduction effect is negligible, as presented in Figure 1. This assumption was also empirically verified by Imanian et al. [15] and Ontiveros et al. [20], In addition, the concept of entropic endurance was further confirmed by Imanian et al. [15], who measured the interacting thermodynamic forces in a coupled failure mechanism, corrosion-fatigue. Summation of entropies from both mechanical work (fatigue) and chemical reactions (corrosion) contributed to the total entropic endurance at the point of fatigue failure.

In addition to the heat and mechanical work, AE has been considered another source of irreversible energy dissipation. Kahirdeh and Khonsari [18] regarded AE absolute energy as an AE waveform feature and a damage indicator. However, the entropic approach was not investigated as a part of their AE-based damage research. The recorded data from the AE sensor was digitized into the so-called waveforms. The AE information (Shannon) entropy may be characterized by the associated probability distribution in the form of a histogram representing each recorded waveform. Hughes [27] introduced information entropy from digitized waveform data collected from ultrasonic tests. Likewise, for specific features of the waveforms, such as the count rate, information entropy has been applied to AE waveforms to assess the entropy of the waveform signals and empirically establish any correlation between the increasing entropy and the ensuing progression of the fatigue damage observed. Digitized data is processed to a corresponding discrete histogram (expressed in $p(x_i)$), and the entropy is computed using Equation (3):

$$S = -\sum_i p(x_i) \log p(x_i) \tag{3}$$

Sauerbrunn et al. [21] used Equation (3) to calculate information entropy using collected AE waveforms from many fatigue tests. In their research, the AE waveform was shown to be a more appropriate damage indicator than the traditional AE features, such as count and energy.

In addition to the thermodynamic entropy and information (Shannon) entropy, the third approach to entropic damage explored as a new damage metric in this paper relies on the statistical mechanics definition of entropy, which provides relative entropy from energy dissipation modes during the fatigue damage process. Forward and reverse work distribution functions applied during the cyclic loading in fatigue can be related to the thermodynamic work and free energy. The so-called Crooks fluctuation theorem expressed in Equation (4) is one such relationship [28]:

$$\frac{\pi_f(+W)}{\pi_r(-W)} = \exp\left[\frac{W - \Delta F}{k_B T}\right] \tag{4}$$

where $\pi_f(+W)$ and $\pi_r(-W)$ in the content of the fatigue damage process may be interpreted as the forward and reverse work distributions over many load cycles, respectively, W is the net strain energy dissipated, ΔF is the Helmholtz free energy difference, k_B is the Boltzmann constant (1.381×10^{-23} J/K), and T is the temperature. Equation (4) has been applied to nano-scale systems, such as ribonucleic acid (RNA) strands, by introducing forward/reverse work to measure the Helmholtz free energy difference (ΔF) as the RNA system's inherent property [23]. This paper introduces an extension of this notion into a macro-scale system (i.e., fatigue) and examines its consistency with a fatigue damage assessment based on traditional thermodynamic entropy and information entropy.

By using the second law of thermodynamics and the Helmholtz free energy definition, Equation (4) can be converted to calculate the total entropy, as shown in Equation (5):

$$\Delta S_{tot} = k_B \ln\left(\frac{\pi_f(+W)}{\pi_r(-W)}\right) \tag{5}$$

According to the fluctuation theorem, the unloaded/fully-loaded points should be determined in thermodynamic equilibrium, whereas the loading/unloading in the fatigue process does not require the equilibrium condition. In addition, the source of the fluctuation is only thermal energy dissipation. However, these conditions may be invalid when applied to the macro-scale fatigue damage evaluation. Both the thermodynamic conditions and the mathematical implementation of Equation (5) may have limitations. Regardless of the unsettled extension of this theorem to the macro-scale, this research is inspired by the forward/reverse work convention and seeks to empirically investigate the application of this notion to assess fatigue damage. Crooks and Sivak [22] discuss measures of the trajectory ensemble. Consistent with Crooks and Sivak results, relative entropy and Jeffreys divergence (JD) effectively capture the symmetric hysteresis properties of the fatigue phenomenon. Furthermore, in the molecular-scale, JD is related to the classical thermodynamic entropy through the Boltzmann constant.

The relative (divergence) entropy in continuous distribution form is shown in Equation (6) [22]:

$$D(\pi_f \mid \pi_r) = \int \pi_f(+W) \ln\left(\frac{\pi_f(+W)}{\pi_r(-W)}\right) dW \tag{6}$$

Relative entropy may be interpreted in the classical thermodynamics as the total entropy difference [22]:

$$\begin{aligned} D(\pi_f \mid \pi_r) &= \tfrac{1}{k_B T}\left(\langle W_{diss}\rangle_f\right) = \tfrac{1}{k_B T}\left(\langle W\rangle_f - \Delta F\right) = \tfrac{1}{k_B T}\left(\langle W\rangle_f - \Delta\langle E\rangle_f + k_B T \Delta S_f^{sys}\right) \\ &= -\tfrac{1}{k_B T}\langle Q\rangle_f + \Delta S_f^{sys} = \Delta S_f^{env} + \Delta S_f^{sys} = \Delta S_f^{tot} \end{aligned} \tag{7}$$

where, in the nano-scale, k_B is the Boltzmann constant (1.381×10^{-23} J/K), T is the temperature, $\langle W\rangle_f$ is the mean work in the process f (forward work), $\langle W_{diss}\rangle_f$ is the mean dissipative work, $\Delta\langle E\rangle_f$ is the mean internal energy difference, $\langle Q\rangle_f$ is the mean heat dissipation, ΔS_f^{sys} is the entropy change within the system, ΔS_f^{env} is the entropy dissipated to the environment, and ΔS_f^{tot} is the total entropy during the process f. The relative entropy in the process f is interpreted as the product of the thermodynamic dissipative work $\left(\langle W\rangle_f - \Delta F\right)$ and the constant $\left(\frac{1}{k_B T}\right)$. Consistent with its definition, the Helmholtz free energy difference, ΔF, expands to the sum of internal energy $(\Delta\langle E\rangle_f)$ and the product of system entropy difference (ΔS_f^{sys}) and the constant $(-k_B T)$. Considering the first law of thermodynamics, the mean work and mean internal energy difference become the product of the mean heat dissipation $(\langle Q\rangle_f)$ and the constant $\left(-\frac{1}{k_B T}\right)$, which is expressed in terms of the entropy difference dissipated to the environment. Therefore, the relative entropy in the process f is expressed by the total entropy difference $\left(\Delta S_f^{tot}\right)$. The relative entropy of the reverse process is:

$$\begin{aligned} D(\pi_r \mid \pi_f) &= \tfrac{1}{k_B T}\left(\langle W_{diss}\rangle_r\right) = \tfrac{1}{k_B T}\left(\langle W\rangle_r + \Delta F\right) = \tfrac{1}{k_B T}\left(\langle W\rangle_r - \Delta\langle E\rangle_r + k_B T \Delta S_r^{sys}\right) \\ &= -\tfrac{1}{k_B T}\langle Q\rangle_r + \Delta S_r^{sys} = \Delta S_r^{env} + \Delta S_r^{sys} = \Delta S_r^{tot} \end{aligned} \tag{8}$$

For the reverse process r, it should be noted that, unlike the forward process, the Helmholtz free energy difference, ΔF, should be expressed with the positive sign.

Summing Equations (7) and (8) is defined as the JD and represents the dissipative thermodynamic entropy as related to the hysteresis associated with the cyclic loadings in fatigue [22]:

$$\begin{aligned} Jeffreys\left(\pi_f;\pi_r\right) &= D\left(\pi_f \mid \pi_r\right) + D\left(\pi_r \mid \pi_f\right) = \Delta S_f^{env} + \Delta S_f^{sys} + \Delta S_r^{env} + \Delta S_r^{sys} \\ &= \Delta S_f^{env} + \Delta S_r^{env} = \Delta S^{env}. \end{aligned} \tag{9}$$

In Equation (9), the terms ΔS_f^{sys} and ΔS_r^{sys} are canceled out, and the only term remaining is the dissipative entropy. Therefore, JD, from the statistical mechanics, corresponds to the thermodynamic entropy as described in the classical thermodynamics. Additionally, JD is only computed by strain energy distributions in fatigue.

3. Experimental Setup and Fatigue Damage Entropy Analyses

3.1. Specimen Preparation: Design, Evaluation, Manufacturing, and Surface Processing

In a series of uniaxial tensile fatigue experiments, stainless steel (SS) 304L was selected as the testing material. SS304L is a widely used structural material, especially in highly acidic environments. The properties of this alloy are shown in Table 1. The dogbone-shape specimen was selected and designed for fatigue testing under the American Society for Testing and Materials (ASTM) 406 guidelines [29]. To induce the crack formation at the center of the specimen, a V-shaped notch with $K_T = 4.04$ was designed. The stress concentration factor was calculated using a Peterson's plot, provided on the efatigue.com website [30]. The V-shape notch, which has a higher concentration-effect than a round-shaped notch, was selected in order not to have the crack around the loading hole. The V-shape notch was designed to minimize the AE noise by reducing the contact area from the noise source. After the design was selected, uniaxial stress distribution was investigated using the finite element method (FEM) with the ANSYS Workbench version R16.2 [31]. The maximum stress was detected at the notch center as expected, and no abnormal stress was found throughout the specimen geometry. Figure 2 shows the shape and dimensions of the specimen.

Table 1. Mechanical properties and chemical composition of specimen material (stainless steel (SS) 304L (SS304L)).

Mechanical Properties									
σ_U [MPa]		σ_Y [MPa]			Elongation [%]			Hardness [RB *]	
613.8		325.65			54.06			85.00	
Chemical Composition [w%]									
C	Cr	Cu	Mn	Mo	N	Ni	P	S	Si
0.0243	18.06	0.3655	1.772	0.2940	0.0713	8.081	0.0300	0.0010	0.1930

* RB: Rockwell hardness measured on the B scale.

The specimens were manufactured using electrical discharge machining (EDM). A total of 50 specimens were prepared for the series of tests, i.e., five loading conditions and 10 test repetitions. After cutting out the specimens, the specimen surface around the crack growth area was processed to clarify the surface image. First, the surface was sanded with increasingly larger grit numbers (grit # 400 → 800 → 2000), then the surface was polished with a polishing pad using one µm alumina solution. Finally, the etching process was employed using a Carpenters etchant.

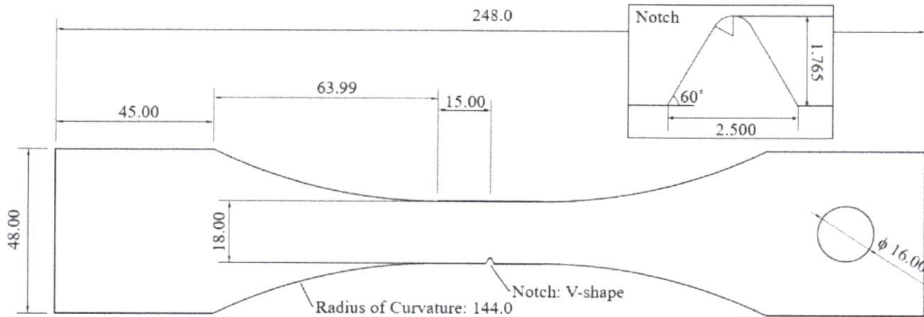

Figure 2. The geometry of the dogbone specimen. The specimen has a hole for loading with a 16 mm diameter pin and stress concentrated by a V-shape notch. Theoretical stress concentration factors (K_T) are 4.04 for the notch and 3.44 for the hole (pin in tension condition), respectively. The length unit is in millimeters.

3.2. Cyclic Loading Process

In this uniaxial loading test, a servo-hydraulic testing system was used. An Instron 8800 system was retrofitted on an MTS 311.11 frame. Each specimen was held and loaded by upper and lower wedge grips, and the actuator was connected to the lower wedge grip to apply cyclic tensile loading. The loading conditions were in the range of 16–24 kN maximum loads, 0.1 stress (or loading) ratio, and 5 Hz frequency. After every 1000 cycles, the cyclic loading was paused and clear microscope images were taken. Each test stopped at the pre-set limit of the actuator position (+1.5 mm). Table 2 summarizes the loading conditions.

Table 2. Five test conditions (test group) of uniaxial cyclic loadings. The test groups were categorized based on their maximum loads. Each group consists of 10 specimens tested successfully.

		Max. Load [kN]	Test Specimen IDs
Stress ratio	0.1	16	8VA43–8VA 52
Frequency	5 Hz	18	8VA33–8VA 42
# of cycle per block	1000	20	8VA23–8VA 32
Loading duration	200 s	22	8VA13–8VA 22
		24	8VA03–8VA 12

3.3. Measurement Setup

3.3.1. Stress and Strain

Load and extension data were collected by the Instron 8800 system [32,33]. A LEBOW 3116-103 load cell monitored loading applied in the specimen, and an Epsilon extensometer model 3542 measured the extension. The gauge length was 25 mm, and several rubber bands attached the extensometer to the specimen, centering it over the specimen's notch. The Instron 8800 system tabulated the load and extension data with 200 Hz frequency. The raw data of the load and extension were converted to stress and strain using the specimen geometry information (e.g., the cross-sectional area).

3.3.2. Acoustic Emission

Two Physical Acoustics Micro-30 s resonant sensors were symmetrically attached to the specimen surface 23 mm from the specimen center. The symmetric sensor placement made it possible to apply the delta T filtering technique [34] to filter out the AE signals generated, other than the area of interest. The electric signal from the piezoelectric AE sensors was amplified by the preamplifier in 40 dB gain mode. Overall control and recording of the AE signal were operated by AEWin SW [34].

3.3.3. Surface Temperature

A thermocouple (Omega 5TC-TT-K-40-36) [35] was attached to the surface of the specimen (close to the notch tip). The thermocouple was connected to a National Instrument 9211A module and controlled by NI Labview software [36]. The surface temperature was recorded every half second.

3.3.4. Crack Length Measurement

During the fatigue tests, an optical microscope system (Edmond 2.5–10X microscope body combined with OptixCam Pinnacle Series CCD digital camera) took images of the crack growth area. Images were taken every 5 s, controlled by OCView SW [37]. Every 1000 cycles, crack initiation and propagation were investigated. The crack length was monitored to collect data on the observable damage, and the material fatigue life was defined as specific crack lengths, e.g., 250 μm.

3.4. Data Analysis: Calculating Entropies

After the tests, entropies were calculated using the collected data. Acoustic emission waveform data were sorted after filtering, and the valid waveforms were converted to information entropy according to the equations discussed in Section 2. From the load and extension data, plastic strain energy was computed for each cycle. Classical thermodynamic entropy was calculated by combining surface temperature with the corresponding plastic strain energy. Jeffreys divergence was computed by relying on forward/reverse work distributions, of which the data were collected within the same test groups and the same proportions of life.

4. Results and Discussion

In this section, classical thermodynamic entropy is verified with its entropic endurance, then JD from the forward/reverse work distribution is computed, evaluated, and compared to the classical thermodynamic entropy results. Information (Shannon) entropy of the detected AE waveform is also computed, and its correlation with the classical thermodynamic entropy results is discussed.

4.1. Classical Thermodynamic Entropy (CTE)

4.1.1. Entropy Calculation Process

As described in Equation (2), thermodynamic entropy generation is computed by the bilinear equation of force and flux for each energy dissipation mode. In the fatigue damage process, mechanical work is the dominating term, as experimentally proved from previous studies [14,20,26]. Plastic strain energy is computed numerically using discrete stress-strain data. Figure 3 illustrates the process of plastic strain energy calculation for each cyclic loading. Summation of the forward and reverse work (strain energy) makes up the plastic strain energy. This forward/reverse work convention is further used in the JD calculation.

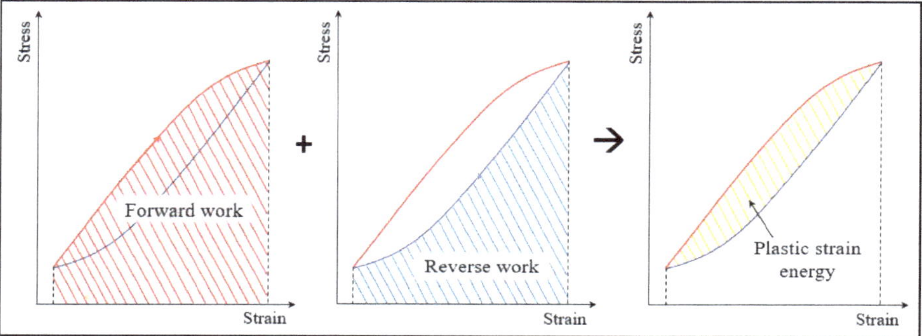

Figure 3. Strain energy calculation procedure. For each cyclic loading, the stress-strain path is divided into forward/reverse work processes, and strain energy is separately computed. The summation of two works is the plastic strain energy or hysteresis.

Temperature, measured by the thermocouple, was recorded every half second during each test. As an example, Figure 4 shows the temperature measurement of the test 8VA03. After acquiring both strain energy and temperature, classical thermodynamic entropy was calculated based on the third term of Equation (3) for each cycle.

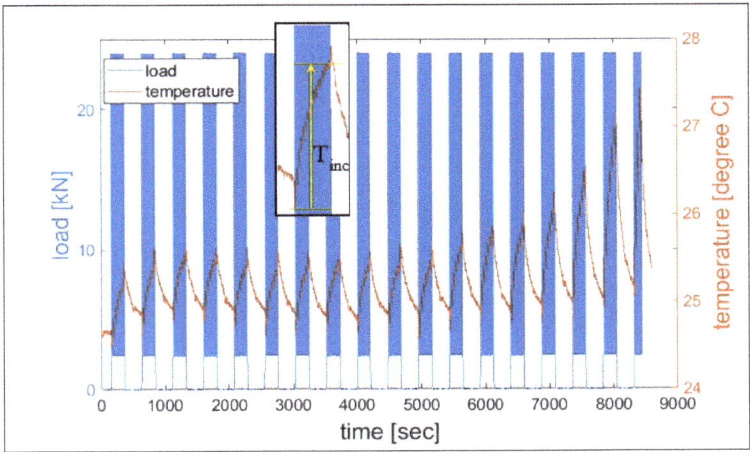

Figure 4. Temperature monitoring during the overall test (8VA03). The ninth damaging loading process is magnified to highlight the temperature rise during the loading process.

4.1.2. Results and Evaluation of Classical Thermodynamic Entropy

Figure 5a presents the cumulative classical thermodynamic entropy for a series of 10 tests with 22 kN maximum loading (i.e., tests 8VA13–22). For each cumulative entropy plot, the initial trend is nearly linear, then the slope rapidly increases. Using the calculated life data determined by the crack length, the cumulative entropy for each life was identified, as shown in Figure 5b.

Figure 6 presents the cumulative entropy at each defined life by the crack length, with respect to the fatigue loading conditions. Stress amplitude, according to the Smith–Watson–Topper (SWT) equation, was used as the representative fatigue loading condition [38,39]. The effect of the stress amplitude (slope in the regression line) diminishes as the crack length of the defining failure decreases.

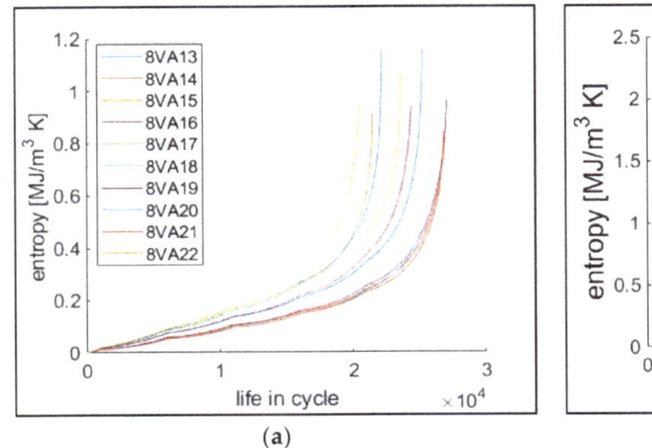

(a)

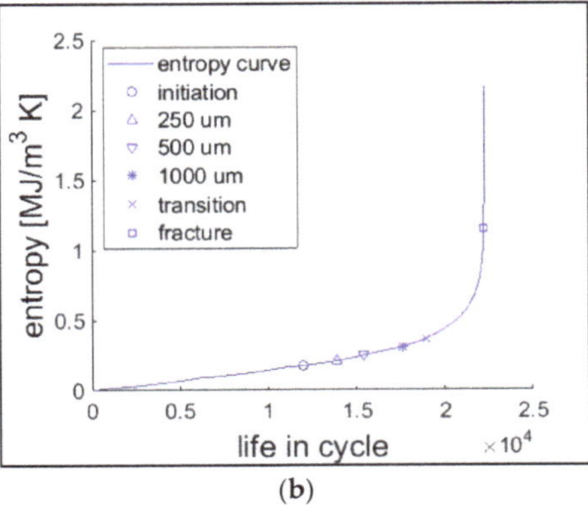

(b)

Figure 5. (**a**) Cumulative classical thermodynamic entropy for 10 tests with 22 kN maximum load. (**b**) The cumulative entropy measured by crack growth. After every 1000 cycles, the cyclic loading process was stopped to perform some measurements. This effect is seen as a slight discontinuity in the plotted curves.

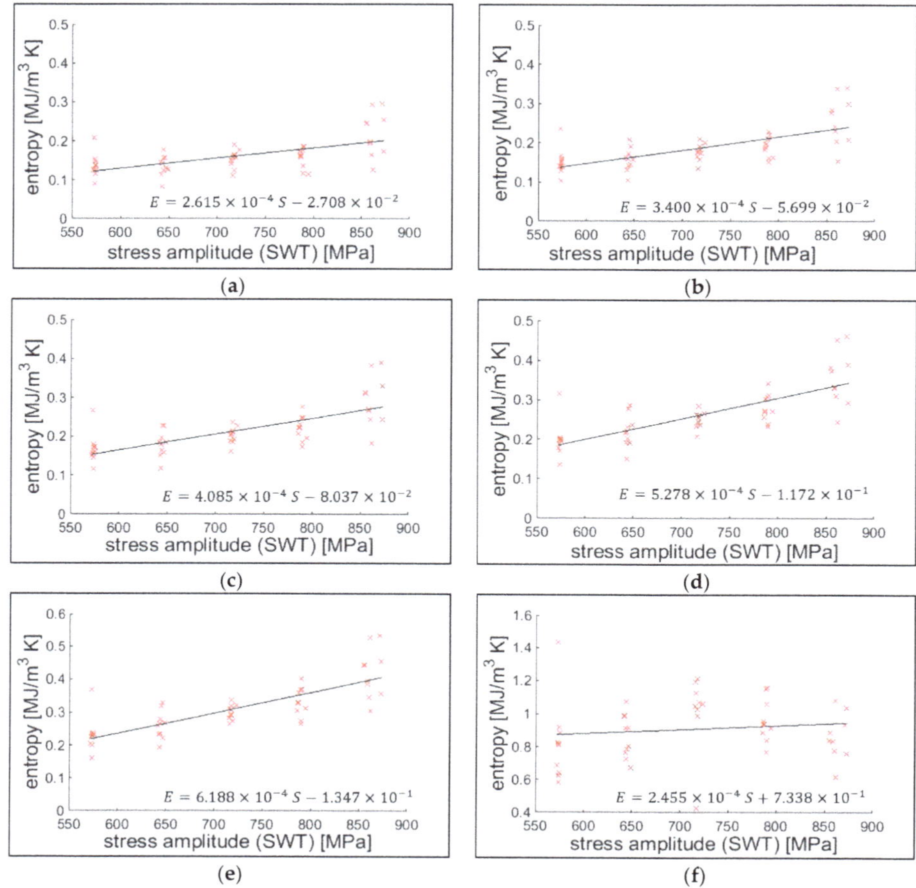

Figure 6. Classical thermodynamic entropy endurance for each defined life under crack growth. The life is determined at (**a**) crack initiation, (**b**) 250 µm crack, (**c**) 500 µm crack, (**d**) 1000 µm crack, (**e**) transition (from region II to III of linear elastic fracture mechanics), and (**f**) fracture, respectively.

The result indicates that entropic endurance has a small positive statistical correlation with the stress amplitude. The extensometer with 25 mm gauge length measured the strain (global strain), and the stress field is assumed to be proportional within the gauging area. This assumption is closer to reality before crack initiation. As the crack grows, the plastic zone area increases, and the stress distribution is more biased toward the plastic zone [38]. Nevertheless, endurances determined from the crack length criteria are also valid in the similar measurement setup applications. A similar entropic endurance behavior was also reported by Ontiveros et al. [20,40,41], who found that the cumulative strain energy or thermodynamic entropy at the crack initiation mildly increases with the stress amplitude.

4.2. *Jeffreys Divergence: The Entropy of Strain Energy Distributions*

4.2.1. Analysis and Results: Distribution of Forward/Reverse Work and JD Calculation

The first step to calculate JD using strain energy is to develop the forward and reverse work distributions. Forward/reverse work data within the same loading condition test group of fatigue tests, and strain energies with the same life ratio, were gathered. In this process, the life (cycles) was determined as a function of crack length, as described in Section 3.3.4. Ten strain energy data (i.e., from each test group of the same loading condition and the same life ratio) were fitted to the 3-parameter MaxEnt distribution [12,13]. Figure 7 shows an example of the estimated forward/reverse work mean and standard deviation with respect to the life ratio, and Figure 8 presents an example of forward/reverse work distributions at a given life ratio based on the estimated MaxEnt distribution parameters [12].

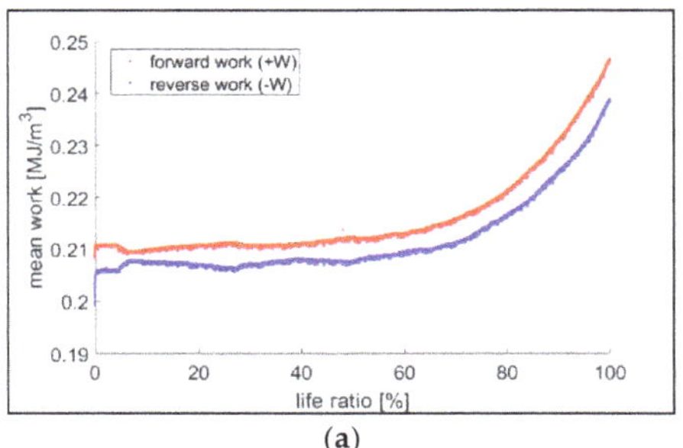

(a)

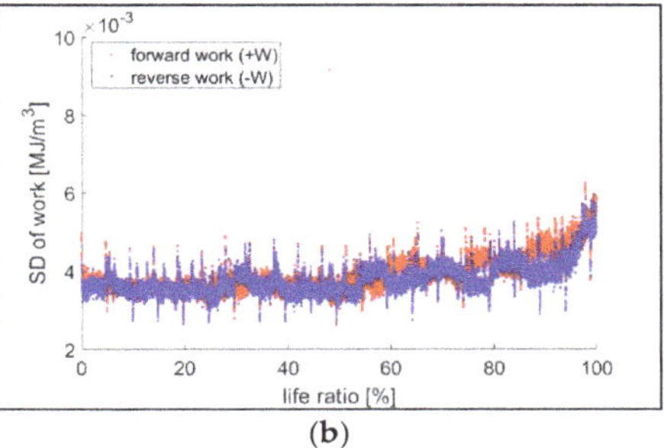

(b)

Figure 7. Mean and standard deviation of the collected forward/reverse work data. The data were collected from ten 22 kN maximum loading tests, and the failure (100% life ratio) was determined for an initial fatigue crack length of 1000 µm. (**a**) Shows mean (µ) and (**b**) shows the standard deviation (σ). As noted, standard deviations (SD) of work for forward/reverse normal distributions have a significant overlap.

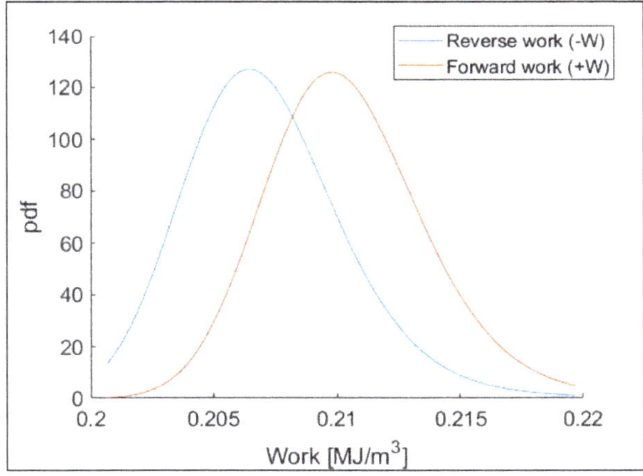

Figure 8. Forward/reverse work distributions of 22 kN maximum loading test group at 25% of life. The distributions were fitted in the maximum entropy (MaxEnt) distribution model.

After the parametric estimation for each strain energy data set, relative entropies (both $D(\pi_f \mid \pi_r)$ and $D(\pi_r \mid \pi_f)$) were computed using Equation (6). The cumulative JD was calculated and plotted, as shown in Figure 9, which presents the cumulative JD for the test group of 16 kN maximum load. Similar to the classical thermodynamic entropy, JD is initially linear, then the slope increases as the crack grows.

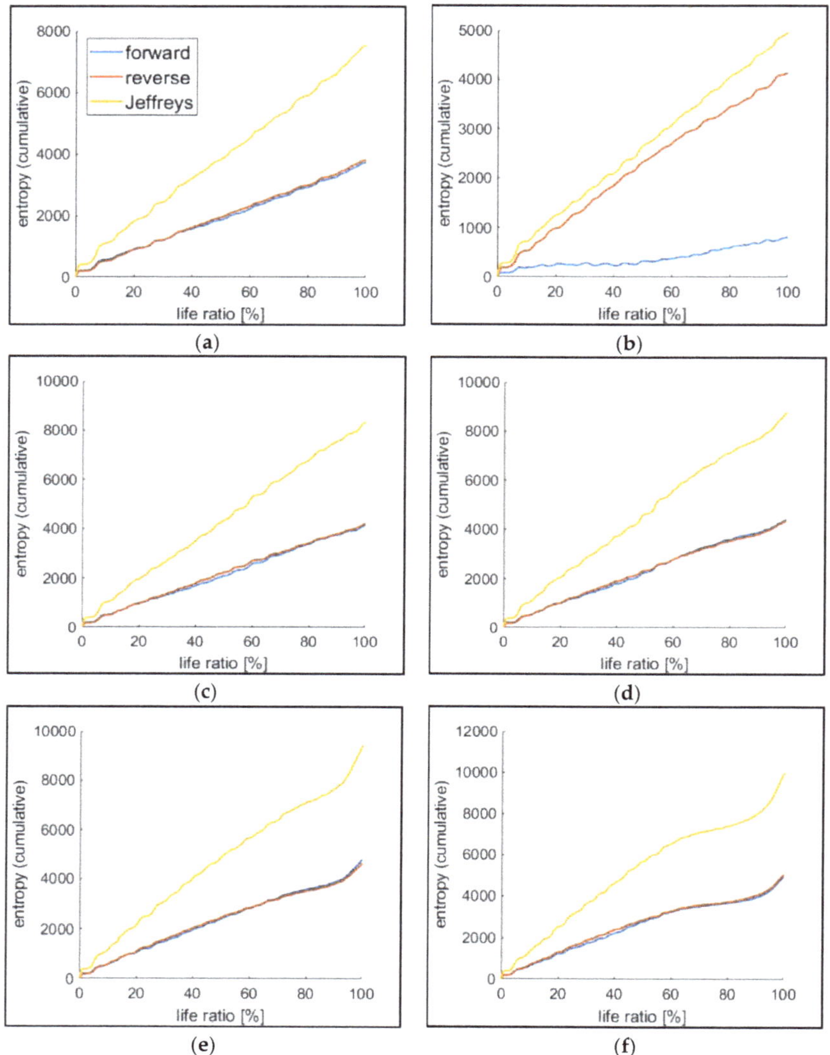

Figure 9. Cumulative relative entropy (Example: For the test group with 16 kN maximum load). Each plot represents the case of normalized life at various crack lengths: (**a**) Crack initiation, (**b**) 250 µm, (**c**) 500 µm, (**d**) 1000 µm, (**e**) transition, and (**f**) fracture.

4.2.2. Evaluation: Correlation to the Classical Thermodynamic Entropy

In the evaluation of possible fatigue damage measurements, the damage is normalized according to Equation (10) [15,21]:

$$D = \frac{M_i - M_0}{M_f - M_0} \tag{10}$$

where M_0 is the measured damage at time 0 or the pristine state of the specimen, M_f is the damage at the failure (e.g., fracture), and M_i is the damage at a given instance (loading cycle). Depending on which crack length is used to determine the failure, M_f was differently determined, meaning that, for example in the case of crack initiation, M_f corresponds to the measured damage at that point. The initial application of this damage measure was inspired by the Palmgren–Miner rule [42,43], in which the fatigue damage is measured in the proportion of the number of cycles. Not only the number of cycles, but also several measures, such as crack length, load-carrying capacity, and elastic modulus degradation have been utilized as measures of damage in the normalized damage [1]. Normalized entropic damage was first introduced by Imanian and Modarres [15] and used by Sauerbrunn et al. [21].

Figure 10 shows one of the five test groups (10 tests of 16 kN maximum loading) where normalized cumulative JD is linearly correlated to the normalized reference damage (classical thermodynamic entropy). The correlation between the JD and the classical thermodynamic entropy is consistent except at the point of fracture. All the loading groups present this inconsistency at the fracture failure. In case of large crack lengths, it is shown that the JD underestimates fatigue damage compared to the classical thermodynamic entropy. The cause of this inconsistency needs to be further investigated.

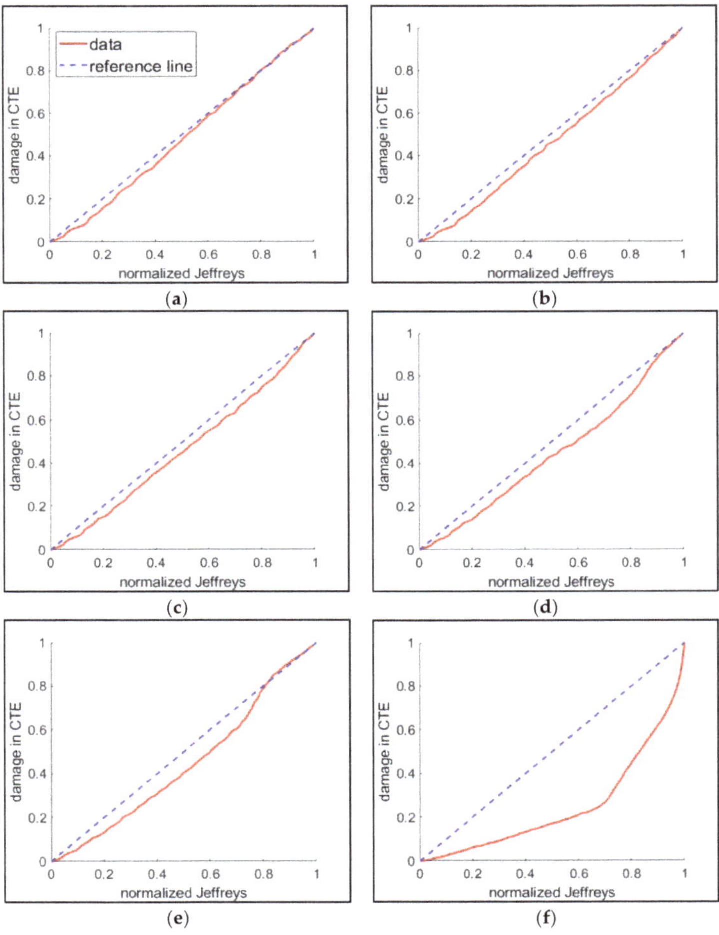

Figure 10. Evaluation of Jeffreys divergence (JD) by correlating to the reference damage (classical thermodynamic entropy (CTE)) as an example of the 16 kN maximum loading test group. Each correlation plot is drawn by the defined point of failure at (**a**) crack initiation, (**b**) 250 μm crack, (**c**) 500 μm, (**d**) 1000 μm, (**e**) transition, and (**f**) fracture.

Jeffreys divergence and thermodynamic entropy in molecular-scale are related through the Boltzmann constant (k_B). However, in the context of the macro-scale application in fatigue using Equations (5), (7), and (8), classical thermodynamic entropy (CTE) is empirically shown to be related to JD by the means of the pseudo-Boltzmann constant, k_{pB}, where, in Equation (11), k_B changes to k_{pB}.

$$CTE = k_{pB} \cdot JD \tag{11}$$

The pseudo-Boltzmann constant k_{pB}, which no longer has the same interpretation and unit as the Boltzmann constant in our macro-scale application, was computed from the slope of the fitted line relating the cumulative JD to the mean classical thermodynamic entropy, as shown in Figure 11, with the slope summarized in Figure 12.

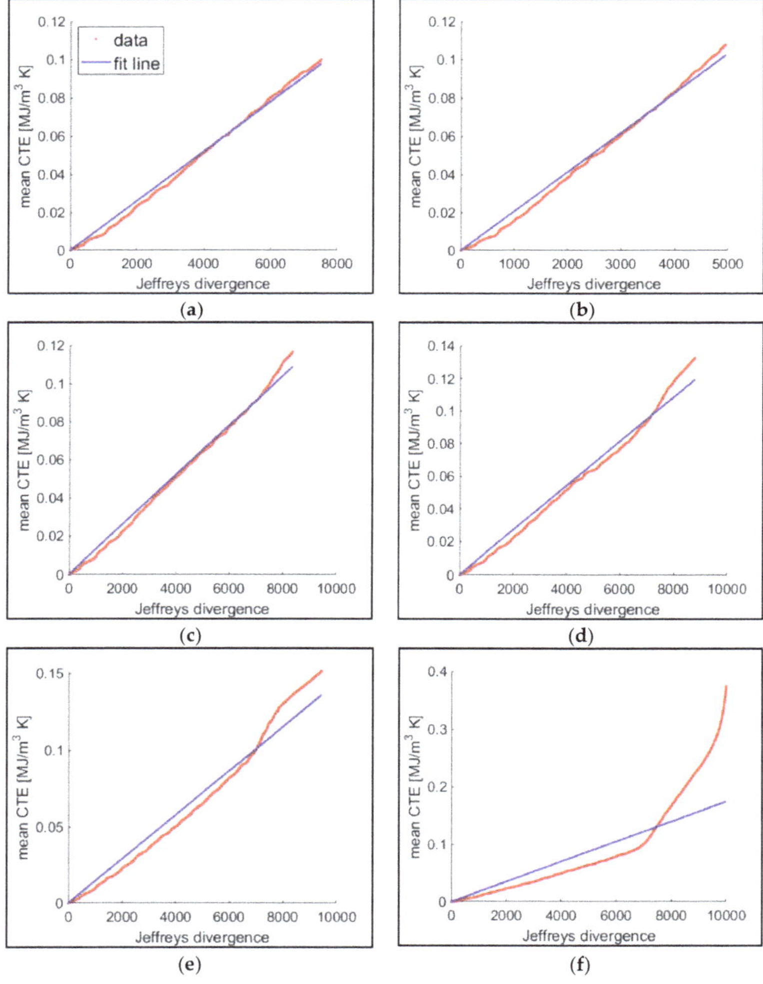

Figure 11. Linear correlation (with the 0 intercept) between mean CTE and JD (for the ten tests of the 16 kN maximum loading group). Using this correlation, the slope is estimated to correspond to k_{pB}. Failure is defined at (**a**) crack initiation, (**b**) 250 μm crack, (**c**) 500 μm, (**d**) 1000 μm, (**e**) transition, and (**f**) fracture.

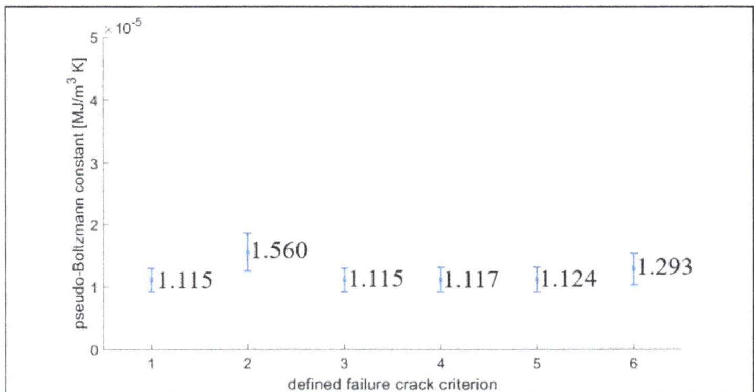

Figure 12. The slope (namely the k_{pB}) for each crack-length based failure. The bar of each data point shows one standard deviation above and below the mean shown. Failure is defined as (1) crack initiation, (2) 250 μm crack, (3) 500 μm crack, (4) 1000 μm crack, (5) transition, and (6) fracture.

The application of the fluctuation theorem to the macro-scale energy dissipation in the fatigue test has scale limitations. The comparison of the macro-scale applications in the fatigue tests to the reported RNA test is detailed in Table 3. In our experiments, the fluctuation source was extended from the molecular-scale to the macro-scale by changing the measurement mode from thermal to plastic strain energy in the macro-scale application. In this extension, the fluctuation was assumed to be caused by multi-scale dimensional variability. In our experimental investigations, the fluctuation was presented by the formation of forward/reverse strain energy distributions. Furthermore, the converting factor (namely, the pseudo-Boltzmann constant) shows statistical consistency that further supports our assumption that JD can be empirically applied as an alternative damage measurement. Further empirical surveys need to consider other conditions, such as material, geometry, damage mode, and stress conditions. The pseudo-Boltzmann constant, k_{pB}, can be generalized empirically.

Table 3. Comparison of Crooks fluctuation theorem application to RNA and metal fatigue test.

	RNA [23]	Metal Fatigue Test
Purpose	Finding Helmholtz free energy	Assessing the amount of damage
Source of fluctuation	Thermal energy Fluctuation in atomic distance	Plastic strain energy Multi-scale defects (e.g., point defect, dislocation, volumetric defect, inclusions, grain structure variability)
Test control	Controlled in displacement Thermal equilibrium at both end of displacement points	Controlled tensile load Thermal equilibrium not controlled
Test repetition	Hundreds of times. A specimen was repeated with unfolding/folding process without regarding the damage	10 fatigue tests repeated with a fixed loading condition, and strain energy data grouped in the corresponding damage
Correlating constant (JD to CTE)	Boltzmann constant (1.381×10^{-23} J/K)	Pseudo-Boltzmann constant estimated from tests $1.115 - 1.560 \times 10^{-5}$ J/m^{3}K (range of the mean values)

4.3. AE Information Entropy

AE sensors, attached on the specimen surface, collected acoustic energy dissipation in the form of elastic AE signals (waveform) represented by digitized voltage data. Each waveform file is transformed into its equivalent discrete probability distribution, represented by a histogram, and used to quantify the information entropy, as expressed by Equation (3).

4.3.1. Analysis of Information Entropy (IE)

To calculate information (Shannon) entropy from AE waveform data, we followed the approach reported by Sauerbrunn et al. [21] and Kahirdeh et al. [44], where information entropy is calculated from the discrete histogram of waveforms. Figure 13 presents the procedure for AE information entropy calculation. Variations in the bin size parameter of the histograms of the AE waveforms showed that the maximum entropy would be achieved by the selected bin size.

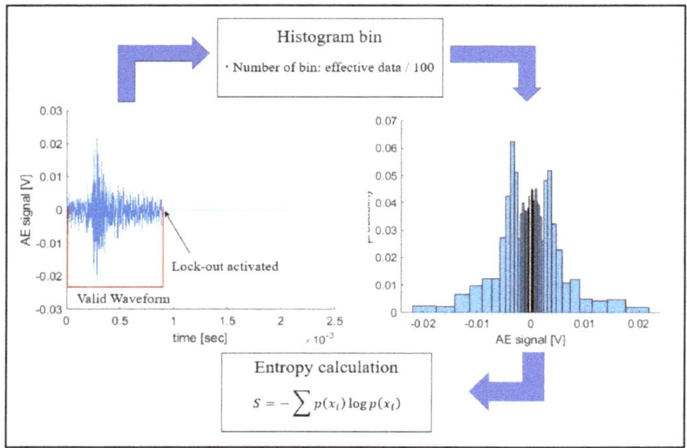

Figure 13. The procedure of AE information entropy calculation. By using the digitized waveform signal data, information entropy is calculated from the generated histogram.

Figure 14 presents an example of the individual and cumulative information entropies. On the cumulative entropy plot, the crack-length points were marked. It is observed that the cumulative entropy trend becomes far steeper around the point of crack initiation. This change in trend is useful information for PHM applications.

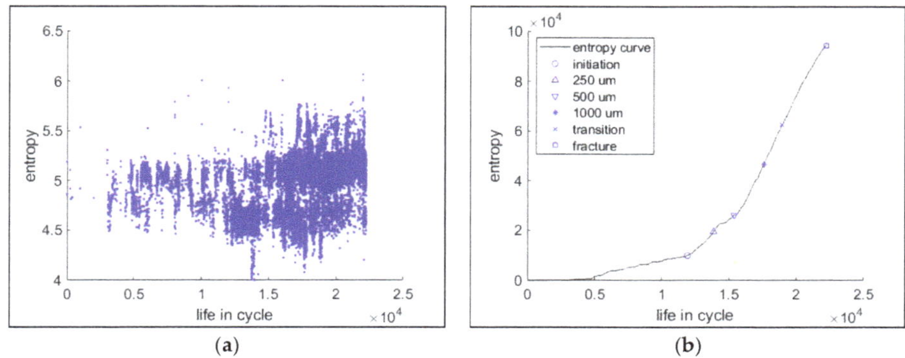

Figure 14. Acoustic emission (AE) information entropy (example: 8VA20). (**a**) Individual entropies for the collected waveforms. (**b**) Cumulative entropy through the life cycle.

4.3.2. Evaluation of AE Entropy and Correlation with Fatigue Damage

The AE count, absolute energy, and information entropy are compared to the classical thermodynamic entropy, as shown in Figure 15, where the failure is defined at the crack initiation. The overall at-a-glance observation shows that the AE information entropy is the closest to the CTE damage. The mean deviation (mean absolute distance from CTE damage to an AE feature) was computed for each test. The sign test was used to assess AE entropy performance using the mean deviation.

The sign test is a nonparametric statistical test that measure consistent differences between pairs of observations and calculates the tests statistic from the difference in the median of the two populations [45]. In this sign test, the left tail mode was utilized, and the entailed hypotheses are shown in Equation (12) (the sign test expressed in signtest(a,b)):

$$\mathrm{H}_0:\ a-b\geq 0;\ \mathrm{H}_1: a-b<0 \tag{12}$$

When the p-value from this statistic is less than a significance level (10% in this test), the null hypothesis is rejected, and the result concludes that the median of a is less than that of b.

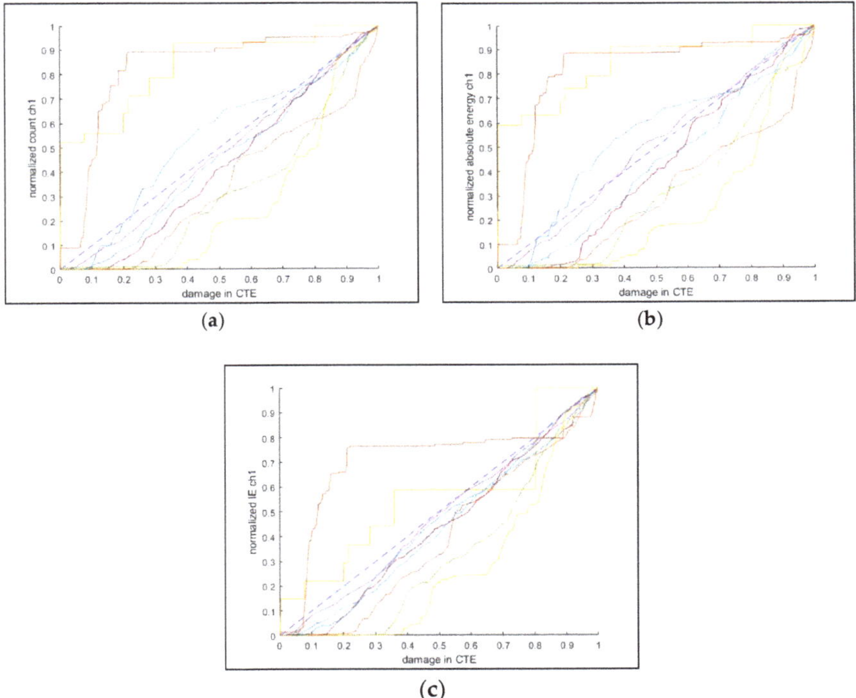

Figure 15. Correlation of AE features to the measured damage (classical thermodynamic entropy). The correlated features are (**a**) count, (**b**) absolute energy, and (**c**) information entropy. These correlation plots were drawn from the 24 kN maximum loading group and AE sensor channel 1 (the sensor more adjacent to the loading actuator).

Table 4 presents the sign test results for all the cases (failure defined by the crack length and AE sensor channel). From the results, one can conclude that the information entropy is better than the count and absolute energy, except for the case of fracture failure, and this result is also consistent with Sauerbrunn et al.'s [21] conclusions.

Table 4. Sign test results represented in p-value. The sign test rejects the null hypothesis (the former is not less than the latter) when the p-value is less than the significance level. In the 10% significance level, the cases of not rejecting the null hypothesis are underlined.

Failure Defined at	a:IE b: Absolute Energy		a:IE b: Count	
	ch1	ch2	ch1	ch2
Initiation	1.0×10^{-1}	3.3×10^{-2}	1.0×10^{-1}	6.6×10^{-2}
250 μm	7.7×10^{-3}	7.7×10^{-3}	6.0×10^{-2}	6.0×10^{-2}
500 μm	1.3×10^{-3}	1.3×10^{-3}	1.6×10^{-2}	7.7×10^{-3}
1000 μm	4.5×10^{-5}	1.5×10^{-4}	3.3×10^{-2}	6.0×10^{-2}
Transition	1.2×10^{-5}	1.5×10^{-4}	3.3×10^{-2}	6.0×10^{-2}
Fracture	$\underline{4.4 \times 10^{-1}}$	$\underline{2.4 \times 10^{-1}}$	6.0×10^{-2}	$\underline{1.6 \times 10^{-1}}$

4.4. Summary and Comparison

In Sections 4.1–4.3, three entropic approaches were reported for applications to fatigue damage assessment. Classical thermodynamic entropy was assessed in terms of the DEG theorem by presenting the existence of an entropic endurance indicating fatigue failure. The assessments of Jeffreys divergence and AE information entropy were followed using the CTE as the reference damage. From the assessment results, JD and AE information entropy exhibit reasonable correlations to the fatigue damage. Furthermore, JD quantitatively correlates with CTE through the pseudo-Boltzmann constant (k_{pB}). Correlation analyses show that JD has a better correlation to the reference damage than the AE information entropy. The analyzed entropic approaches are compared and summarized in Table 5. It is noted that the simulation of the entropic prediction model, for example, through a finite element approach, is more applicable to CTE and JD than the AE information entropy. For example, similar to Mozafari et al. [46], fatigue damage simulation modeling using mechanical plastic deformation can be equally applicable to CTE and JD.

Table 5. Comparison of entropic approaches and efficacy as the measure of fatigue damage.

	Classical Thermodynamic Entropy (CTE)	Jeffreys Divergence (JD)	AE Information Entropy (IE)
Analysis of source data	Plastic strain energy Surface temperature	Plastic strain energy	AE waveform
Calculation method	Bilinear irreversible thermodynamic entropy Equation (2)	Fluctuation theorem and relative entropy Equations (7)–(9)	Information theory Equation (3)
Evaluation	Consistent entropic endurance Used as the reference damage	Correlation to normalized measured damage Pseudo-Boltzmann constant (k_{pB})	Correlation to normalized measured damage
Effect	Endurance verified Linear relation to stress amplitude	Endurance verified Consistent k_{pB}	Better than AE count and absolute energy. Useful for early life in pre-crack initiation

5. Conclusions

In this paper, three entropic approaches for application to the metallic material fatigue damage process were explored and experimentally demonstrated. Three energy dissipations resulting from mechanistic degradation phenomena—plastic mechanical strain energy, heat (temperature), and acoustic emission—were monitored in multiple uniaxial cyclic fatigue tests. In these entropic approaches, the measured dissipations were quantified in terms of the classical thermodynamic entropy, Jeffreys divergence representing thermodynamic entropy, and information (Shannon) entropy of AE waveforms. Particularly, the application of Jeffreys divergence concept was extended to the macro-scale and applied to the fatigue damage assessment. Classical thermodynamic entropy showed a consistent entropic endurance to fatigue damage. Further, Jeffreys divergence and AE information entropy were adequately correlated to the fatigue damage. The contribution from this research was in limited extensions and applications of the three entropic methods, which resulted in the following findings:

- In classical thermodynamics, the entropic endurance showed a slight correlation with the cyclic stress amplitude. This entropy was shown to be an appropriate index of damage.
- Application of Jeffreys divergence in macro-scale was empirically explored and computed from the forward/reverse work distributions, which showed an excellent correlation to the normalized damage. The quantitative conversion factor (namely the pseudo-Boltzmann constant, k_{pB}) also showed consistency between the classical thermodynamic entropic damage and Jeffreys divergence-based entropic damage.
- Fatigue damage assessment using information (Shannon) entropy of the acoustic emission waveform data, compared well with the classical thermodynamic entropy. Similarly, using statistical tests, it was shown that the AE-based informational entropy of damage was more consistent than the two conventional AE features (i.e., count and absolute energy) used in the fatigue damage assessment.

Author Contributions: The theoretical concept and application were developed jointly by both authors. H.Y. planned, designed, executed, and analyzed results of the experiments. M.M. supervised the experiments and analysis of the data. Both authors have read and approved the final manuscript.

Funding: This work was partially sponsored by the U.S. Office Naval Research under grant number N00014140453.

Acknowledgments: Helpful discussions with Christopher Jarzynski are appreciated during the special application of Jeffreys divergence in this research. Christine Sauerbrunn is acknowledged for her support of the initial experimental design. Ali Kahirdeh is also recognized for his support in building some of the theoretical developments in this paper.

Conflicts of Interest: The authors declare no conflict of interest.

References

1. Lemaitre, L. *A Course on Damage Mechanics*, 2nd ed.; Springer: Berlin, Germany, 1990.
2. Tsui, K.L.; Chen, N.; Zhou, Q.; Hai, Y.; Wang, W. Prognostics and Health Management: A Review on Data Driven Approaches. *Math. Probl. Eng.* **2015**, *2015*, 1–17. [CrossRef]
3. Kim, N.; An, D.; Choi, J. *Prognostics and Health Management of Engineering Systems: An Introduction*; Springer: Cham, Switzerland, 2017.
4. Ekwaro-Osire, S.; Alemayehu, F.M.; Goncaleves, A.C. *Probabilistic Prognostics and Health Management of Energy Systems*; Springer: Cham, Switzerland, 2017.
5. Si, X.; Zhang, Z.; Hu, C. *Data-Driven Remaining Useful Life Prognosis Techniques*; Springer: Berlin, Germany, 2017.
6. Niu, G. *Data-Driven Technology for Engineering Systems Health Management*; Springer: Beijing, China, 2017.
7. Weiss, V.; Ghoshal, A. On the Search for Optimal Damage Precursors. *Struct. Health Monit.* **2014**, *13*, 601–608. [CrossRef]
8. Bryant, M.D. Modeling Degradation Using Thermodynamic Entropy. In Proceedings of the Annual Conference of the Prognostics and Health Management Society, Fort Worth, TX, USA, 29 September–2 October 2014.
9. Jaynes, E.T. Information Theory and Statistical Mechanics I. *Phys. Rev.* **1957**, *106*, 620–630. [CrossRef]
10. Jaynes, E.T. Information Theory and Statistical Mechanics II. *Phys. Rev.* **1957**, *108*, 171–190. [CrossRef]
11. Soize, C. *Uncertainty Quantification: An Accelerated Course with Advanced Applications in Computational Engineering*; Springer: Cham, Switzerland, 2017.
12. Mohammad-Djafari, A. A Matlab Program to Calculate the Maximum Entropy Distributions. In *Maximum Entropy and Bayesian Methods*; Springer: Seattle, WA, USA, 1991; pp. 221–233.

13. Li, H.; Wen, D.; Lu, Z.; Wang, Y.; Deng, F. Identifying the Probability Distribution of Fatigue Life Using the Maximum Entropy Principle. *Entropy* **2016**, *18*, 111. [CrossRef]
14. Bryant, M.; Khonsari, M.; Ling, F. On the Thermodynamics of Degradation. *Proc. R. Soc. A* **2008**, *464*, 2001–2014. [CrossRef]
15. Imanian, A.; Modarres, M. A Thermodynamic Entropy Approach to Reliability Assessment with Applications to Corrosion Fatigue. *Entropy* **2015**, *17*, 6995–7020. [CrossRef]
16. Amiri, M.; Modarres, M. An Entropy-Based Damage Characterization. *Entropy* **2014**, *16*, 6434–6463. [CrossRef]
17. Basaran, C.; Nie, S. An Irreversible Thermodynamic Theory for Damage Mechanics of Solids. *Int. J. Solids Struct.* **2004**, *13*, 205–223.
18. Kahirdeh, A.; Khonsari, M. Energy Dissipation in the Course of the Fatigue Degradation: Mathematical Derivation and Experimental Quantification. *Int. J. Solids Struct.* **2015**, *77*, 75–85. [CrossRef]
19. Naderi, M.; Amiri, M.; Khonsari, M. On the Thermodynamic Entropy of Fatigue Fracture. *Proc. R. Soc. A* **2009**, *466*, 1–16. [CrossRef]
20. Ontiveros, V.; Amiri, M.; Kahirdeh, A.; Modarres, M. Thermodynamic Entropy Generation in the Course of the Fatigue Crack Initiation. *Fatigue Fract. Eng. Mater. Struct.* **2016**, *40*, 423–434. [CrossRef]
21. Sauerbrunn, C.M.; Kahirdeh, A.; Yun, H.; Modarres, M. Damage Assessment Using Information Entropy of Individual Acoustic Emission Waveforms during Cyclic Fatigue Loading. *Appl. Sci.* **2017**, *7*, 562. [CrossRef]
22. Crooks, G.E.; Sivak, D.A. Measures of Trajectory Ensemble Disparity in Nonequilibrium Statistical Dynamics. *J. Stat. Mech. Theory Exp.* **2011**, *2011*, 1–10. [CrossRef]
23. Collin, D.; Ritort, F.; Jarzynski, C.; Smith, S.; Tinoco, I.J.; Bustamante, C. Verification of the Crooks Fluctuation Theorem and Recovery of RNA Folding Free Energies. *Nature* **2005**, *437*, 231–234. [CrossRef]
24. Douarche, F.; Ciliberto, S.; Petrosyan, A.; Rabbiosi, I. An Experimental Test of the Jarzynski Equality in a Mechanical Experiment. *Europhys. Lett.* **2005**, *70*, 593–599. [CrossRef]
25. Basaran, C.; Chandaroy, R. Mechanics of Ph40/Sn60 Near-eutectic Solder Alloys Subjected to Vibrations. *Appl. Math. Model.* **1998**, *22*, 601–627. [CrossRef]
26. Imanian, A.; Modarres, M. A Thermodynamic Entropy-based Damage Assessment with Applications to Prognostics and Health Management. *Struct. Health Monit.* **2018**, *17*, 240–254. [CrossRef]
27. Hughes, M. Analysis of Ultrasonic Waveforms Using Shannon Entropy. In Proceedings of the IEEE 1992 Ultrasonics Symposium Proceedings, Tucson, AZ, USA, 20–23 October 1992.
28. Crooks, G.E. Entropy Production Fluctuation Theorem and the Nonequilibrium Work Relation for Free Energy Difference. *Phys. Rev. E* **1999**, *60*, 2721–2726. [CrossRef]
29. ASTM E466. *Standard Practice for Conducting Force Controlled Constant Amplitude Axial Fatigue Tests of Metallic Materials*; American Society for Testing and Materials: West Conshohocken, PA, USA, 2015.
30. Socie, D. Efatigue, Altair. Available online: https://www.efatigue.com (accessed on 10 March 2019).
31. ANSYS. *ANSYS Workbench Release 16.2*; ANSYS: Canonsburg, PA, USA, 2016.
32. Illinois Tool Works Inc. *WaveMatrix V1.5*; Illinois Tool Works Inc.: Glenview, IL, USA, 2010.
33. Instron Inc. *8800 System Console Version 8.4*; Instron Inc.: Norwood, MA, USA, 2011.
34. Mistras Group. *AEWin for PCI2 Version E5.60*; Mistras Group: New Jersey, NJ, USA, 2014.
35. Omega Engineering. Ready-Made Insulated Thermocouples, Omega Engineering. Available online: https://www.omega.com/en-us/search/?text=5TC-TT-K-40-36 (accessed on 16 May 2019).
36. National Instrument. *Labview*; National Instrument: Austin, TX, USA, 2017.
37. *TShow Software, OCView Version 7.1.1.2*; OptixCAm: Roanoke, VA, USA, 2003.
38. Bannantine, J. *Fundamentals of Metal Fatigue Analysis*; Prentice-Hall: Upper Saddle River, NJ, USA, 1990.
39. Dowling, N.E. Mean Stress Effects in Stress-Life and Strain-Life Fatigue. In Proceedings of the Second SAE Brasil International Conference on Fatigue, Blacksburg, VA, USA, 22–27 January 2004.
40. Ontiveros, V.L. Strain Energy Density and Thermodynamic Entropy as Prognostic Measures of Crack Initiation in Aluminum. Ph.D. Thesis, University of Maryland, College Park, MD, USA, 2013.
41. Ontiveros, V.L.; Modarres, M.; Amiri, M. Estimation of reliability of structures subject to fatigue loading using plastic strain energy and thermodynamic entropy generation. *Proc. Inst. Mech. Eng. Part O J. Risk Reliab.* **2015**, *229*, 220–236. [CrossRef]
42. Palmgren, A.G. Life Length of Roller Bearings or Durability of Ball Bearings. *Z. Des Vreines Dtsch. Ing.* **1924**, *14*, 339–341.
43. Miner, M.A. Cumulative Damage in Fatigue. *J. Appl. Mech.* **1945**, *3*, 159–164.

44. Kahirdeh, A.; Sauerbrunn, C.; Yun, H.; Modarres, M. A Parametric Approach to Acoustic Entropy Estimation for Assessment of Fatigue Damage. *Int. J. Fatigue* **2017**, *100*, 229–237. [CrossRef]
45. Gibbons, J.D.; Chakraborti, S. *Nonparametric Statistical Inference*; Chapman & Hall/CRC Press, Taylor & Francis: Boca Raton, FL, USA, 2011.
46. Mozafari, F.; Thamburaja, P.; Srinivasa, A.R.; Moslemi, N. A rate independent inelasticity model with smooth transition for unifying low-cycle and high-cycle fatigue life prediction. *Int. J. Mech. Sci.* **2019**, *159*, 325–335. [CrossRef]

Article

A Copula Entropy Approach to Dependence Measurement for Multiple Degradation Processes

Fuqiang Sun [1], Wendi Zhang [1], Ning Wang [2] and Wei Zhang [1,*]

[1] Science and Technology on Reliability and Environmental Engineering Laboratory, School of Reliability and Systems Engineering, Beihang University, Beijing 100191, China
[2] Beijing Aeronautical Science & Technology Research Institute, Commercial Aircraft Corporation of China Ltd., Beijing 102211, China
* Correspondence: zhangwei.dse@buaa.edu.cn; Tel.: +86-10-82338651

Received: 25 June 2019; Accepted: 23 July 2019; Published: 25 July 2019

Abstract: Degradation analysis has been widely used in reliability modeling problems of complex systems. A system with complex structure and various functions may have multiple degradation features, and any of them may be a cause of product failure. Typically, these features are not independent of each other, and the dependence of multiple degradation processes in a system cannot be ignored. Therefore, the premise of multivariate degradation modeling is to capture and measure the dependence among multiple features. To address this problem, this paper adopts copula entropy, which is a combination of the copula function and information entropy theory, to measure the dependence among different degradation processes. The copula function was employed to identify the complex dependence structure of performance features, and information entropy theory was used to quantify the degree of dependence. An engineering case was utilized to illustrate the effectiveness of the proposed method. The results show that this method is valid for the dependence measurement of multiple degradation processes.

Keywords: copula entropy; measure; dependence; multiple degradation processes

1. Introduction

Degradation is seemingly fundamental to all things in nature [1]. Therefore, the failure mechanism of a highly reliable system usually can be traced to underlying degradation processes such as the fatigue and corrosion of metal materials, the wear of mechanical parts, the parametric drift of semiconductor devices, and other processes [2]. As a consequence, degradation modeling has become an efficient method to evaluate the reliability of long lifetime products, combining the probabilistic degradation process and the fixed failure threshold [3].

Modern engineering systems may have multiple degradation features because of their complex structures and functions [4], and any of them that exceed the defined threshold may cause product failure [5,6]. Since all degradation features of a product share several common factors (e.g., the same inside structures, user experience, environmental/operational conditions, and maintenance history), it is unavoidable that there is dependence among multiple degradation features. This dependence structure may be linear or nonlinear. When ignoring the existence of dependence among multiple degradation features, degradation modeling and lifetime estimation under the premise of the independence assumption may lead to less credible or erroneous results. Therefore, it is safer to describe and measure dependence accurately and reasonably among multiple degradation features prior to modeling.

The associated relationships of multiple random variables are usually described by correlation and dependence. The differences and relationships between dependence and correlation are elaborated as following firstly.

The correlation is usually utilized to describe linear relationships. It does not certainly mean that X and Y are independent of each other when there is no correlation between X and Y. The Pearson correlation coefficient, based on the multivariate normality assumption, is often employed to measure the linear relationships between two random variables in statistics [7]. Xu et al. [8] adopted the Pearson correlation coefficient to calculate the correlation between two degradation processes. However, the Pearson correlation coefficient can only be used to measure linear relationships. For example, the random variable X follows the standard normal distribution and the random variable $Y = X^2$. Obviously, there is a strong dependence between X and Y, and the value of Y can be completely determined by X. However, the correlation coefficient between them is 0. Therefore, the Pearson correlation coefficient has some shortcomings in measuring the associated relationships of random variables [9]. It will misestimate the dependence between two variables when the sample size is not large enough or the dependence relationship is nonlinear [10].

Dependence is the opposite of independence, which means that the random variables X and Y have no independence in probability characteristics. Dependence usually contains both linear and nonlinear relationships. Therefore, it is more appropriate to use dependence to describe the relationship between random variables [11]. Dependence measurement is how the dependence between variables or the dependence between distribution functions of variables is measured [12]. The traditional modeling method based on multidimensional joint distribution relies on the correlation coefficient. For two-dimensional normal random variables (X, Y), the correlation coefficient of X and Y is 0 means that X and Y are independent of each other, and it is not applicable to nonlinear relationships.

The rank correlation coefficient can be utilized to estimate the nonlinear dependence relationship between two variables, and it has no restriction regarding the distribution of variables. The rank correlation coefficient primarily includes the Kendall correlation coefficient and the Spearman correlation coefficient [13,14], and their original purpose was to measure and estimate dependence in the psychiatric symptom rating field. Nelsen [12] adopted the link function between the copulas and Kendall's τ (or Spearman's ρ) to assess the dependence of bivariate degradation data. Similarly, Wang and Pham [5], Sari et al. [15], and Sun et al. [16] also adopted the rank correlation coefficient and copulas to measure the dependence between two performance characters. One major disadvantage of the rank correlation coefficient is that there is a loss of information when the data are converted to ranks [17]. Furthermore, they cannot be used to detect dependence when more than two variables are involved [10]. In a multivariate context, in general, it is more important to study multivariate association than a bivariate association.

Therefore, it is difficult to use the existing methods to accurately measure the dynamic and nonlinear characteristics of dependence measurements of multiple degradation processes. Indeed, it is necessary to find a more suitable measurement method to calculate the dependence among multiple degradation features of a product. Schmid et al. [18] proposed a method of multivariate association measurement based on copula, which extended the commonly used bivariate measurement method to multivariate and applied copula to measure multivariate association. Ane et al. [19] applied copula to the financial area and proposed a measurement method of association between the financial risks based on copula. As a more useful alternative, the copula entropy, which combines information entropy and the copula theory, is proposed to measure dependence among multiple variables. Copula entropy can measure association information and dependence structure information simultaneously. Moreover, copula entropy does not impose constraints on the dimension of multiple variables. Due to these advantages, copula entropy has attracted much interest for its ability to measure multivariate dependence in many fields, and copula entropy has been gradually applied in hydrology, finance and other fields. Singh and Zhang [20] discussed the flexibility to model nonlinear dependence structure using parametric copulas (e.g., Archimedean, extreme value, meta-elliptical, etc.) with respect to multivariate modeling in water engineering. Zhao et al. [21] used the copula entropy model to measure the stock market correlations, compared with the linear correlation coefficient and mutual information methods, which have the advantages of dimensionless, and able to capture non-linear correlations.

Hao et al. [22] introduced the integration of entropy and copula theories to the hydrologic modeling and analysis area. Chen et al. [10,23] used the copula entropy to computed the dependence between the mainstream and its upper tributaries and also used the copula entropy coupled with an artificial neural network to calculate the correlation between each input and output of the neural network for rainfall-runoff simulation. Ma and Sun [24] proved the equivalence between copula entropy and mutual information, and mutual information is essentially an entropy. Xu et al. [25] proposed the copula theory to quantitatively describe the connection of bivariate variables or multivariate variables in the hydrometeorological field. Similarly, Huang et al. [26] applied copula entropy to measure dependencies between traffic noise and traffic flow. Salimi et al. [27] used copula entropy to capture the dependencies among the sub-components of the system in the modeling of complex service systems.

In this paper, a novel measurement method that uses copula entropy is proposed to measure the dependence among multiple degradation features. First, the copula function and information entropy theory were employed to build the copula entropy. The former was used to describe the dependence structure among variables, and the latter was utilized to quantify the dependence. Then, the copula entropy of multiple degradation processes was calculated. Parameter estimation of copula entropy was performed using the maximum likelihood estimate (MLE) method. The Akaike information criterion (AIC) was adopted to select the most suitable copulas. Finally, a case study with multivariable degradation data of a microwave electronic assembly was studied to validate the proposed method. The proposed copula entropy method could address two problems in the dependent measurement of multiple degradation processes: the first is how to measure and directly compare the dependence between every two pairs of the degradation processes, and the other is how to measure directly compare the dependence among multiple degradation processes at different phases.

The paper is organized as follows. Section 2 presents the theory of the copula function and information entropy. This section also combines these to build the copula entropy theory. Section 3 elaborates on the calculation methods of copula entropy, including the calculation of the cumulative distribution function, the method of parameter estimation, and the Monte Carlo simulation calculation. Section 4 provides the case study, and Section 5 concludes the paper.

2. Copula Entropy Theory

2.1. Multivariate Copula Function

It is difficult to identify the multivariate probability distribution because of the complexity and the high dimension of marginal distributions. The copulas separate the learning of the marginal distributions from the learning of the multivariate dependence structure to simplify this process [28].

The below theorem provides the necessary and sufficient conditions for copula theory. It explains the effect of copulas in expressing the relationship between the multivariate distribution and the relevant univariate marginal distributions [6].

Theorem 1. *(Sklar's theorem [29]): Let $X = (x_1, x_2, \ldots, x_n)$ be a random variable, and its marginal distributions are $F_1(x_1)$, $F_2(x_2)$, $\ldots$, $F_n(x_n)$, and H is their joint distribution function. Then, the copula function C is presented such that*

$$H(x_1, x_2, \ldots, x_n) = C(F_1(x_1), F_2(x_2), \cdots, F_n(x_n)) \tag{1}$$

The copula C is unique when $F_1(x_1)$, $F_2(x_2)$, $\ldots$, $F_n(x_n)$ are continuous. On the contrary, the function H, defined by Equation (1), will be the joint distribution function of the margins $F_1(x_1)$, $F_2(x_2)$, $\ldots$, $F_n(x_n)$ if $F_1(x_1)$, $F_2(x_2)$, $\ldots$, $F_n(x_n)$ are univariate distributions.

The multivariate copula function can be defined according to the theorem.

Definition 1. *(n-dimensional copula) [12]: An n-dimensional copula is a function C from $I^n = [0, 1]^n$ to I and it must have the following properties:*

(1) *If* $u = (u_1, \ldots, u_n) = 1$, *then* $C(u) = 1$;
(2) *For every* $u = (u_1, \ldots, u_n)$ *in* I^n, *if at least one coordinate of* u *is 0 then* $C(u) = 0$;
(3) *If all coordinates of* u *except* u_k *are* 1, *then*

$$C(u) = C(1, \ldots, 1; u_k; 1, \ldots, 1) = u_k \tag{2}$$

(4) *For each hyper rectangle* $B = \prod_{i=1}^{n} [u_i, v_i] \subseteq [0,1]^n$, *the C-volume of B is non-negative*

$$\int_B dC([u,v]) = \sum_{z \in \times_{i=1}^{n} \{u_i, v_i\}} (-1)^{n(z)} C(z) \geq 0 \tag{3}$$

where $n(z) = \#\{k{:}z_k = u_k\}$.

The density of a copula function C is denoted by c, which may be achieved by taking the partial derivatives as

$$c(u_1, u_2, \ldots, u_n) = \frac{\partial^n c(u_1, u_2, \ldots, u_n)}{\partial u_1 \partial u_2 \ldots \partial u_n} \forall u = (u_1, u_2, \ldots, u_n) \in I^n \tag{4}$$

Based on multivariate differentiation, the joint density function corresponding to the distribution function, $H(u_1, u_2, \ldots, u_n)$, can be calculated by

$$h(u_1, u_2, \ldots, u_n) = c(F_1(x_1), F_2(x_2), \cdots, F_n(x_n)) f_1(x_1) f_2(x_2) \cdots f_n(x_n) \tag{5}$$

where $u_1 = F_1(x_1)$, $u_2 = F_2(x_2)$, ..., $u_n = F_n(x_n)$ and $f_1(x_1), f_2(x_2), \ldots, f_n(x_n)$ are the probability density functions of marginal distribution function $F_1(x_1)$, $F_2(x_2)$, ..., $F_n(x_n)$, respectively.

Copula functions have many types, and different types can reflect different dependence structures. Table 1 shows a few typical copula functions.

Table 1. Some typical copulas.

Copulas	$C(u_1, \ldots, u_n)$	Parameter
Gaussian	$\Phi_\theta\left[\Phi^{-1}(u_1), \Phi^{-1}(u_2), \ldots, \Phi^{-1}(u_d)\right]$ [1]	$\theta \in (-1, 1)$
Clayton	$\left(\sum_{i=1}^{d} u_i^{-\theta} - d + 1\right)^{-1/\theta}$	$\theta \in (0, \infty)$
Frank	$-\frac{1}{\theta} \ln\left(1 + \frac{\prod_{i=1}^{d} [\exp(-\theta u_i) - 1]}{[\exp(-\theta) - 1]^{d-1}}\right)$	$\theta \in (-\infty, \infty) \backslash \{0\}$
Gumbel	$\exp\left\{-\left[\sum_{i=1}^{d} (-\ln u_i)^\theta\right]^{1/\theta}\right\}$	$\theta \in [1, \infty)$

[1] Φ is the standard normal distribution function; Φ_θ is the standard normal distribution function of d variables; u_i is the cumulative distribution function of each variable; θ is the parameter of the copula function.

2.2. Information Entropy

The entropy originated from the thermodynamics first and then gradually extended to the study of information theory. It is called information entropy in the information field and measures the uncertainty of information. Shannon [30] first proposed the concept of information entropy as follows:

(1) The function, S, is continuous and the probability is p_i;
(2) Under the condition of equivalence probability, S is a monotonically increasing function with the possible result quantity n;
(3) For two mutually independent events in S, the uncertainty between them is the sum of the uncertainties when considering them separately.

Then S can be named as the information entropy function.

Let X be a random variable with a probability p_i, the entropy of X is given by [31]:

$$S(x) = -k\sum_{i=1}^{n} p_i \log p_i \tag{6}$$

Information entropy has the following properties:

(1) $S_n(p_1, p_2, \ldots, p_n) \geq 0$;
(2) If $p_k = 1$, then $S_n(p_1, p_2, \ldots, p_n) = 0$, where $Sn(0, ..., 0, 1, 0, ..., 0) = 0$;
(3) $S_{n+1}(p_1, p_2, \ldots, p_n, p_{n+1} = 0) = S_n(p_1, p_2, \ldots, p_n)$;
(4) $S_n(p_1, p_2, \ldots, p_n) \leq S_n(1/n, 1/n, \ldots, 1/n) = \ln(n)$;
(5) $S_n(p_1, p_2, \ldots, p_n)$ is a symmetric concave function on all variables.

where $S(p_1, p_2, \ldots, p_n) = -\sum_{i=1}^{n} p_i \log p_i$.

Information entropy gives a quantitative measurement of the degree of uncertainty in the information. From its calculation formula, it can be seen that the probability distribution of p_i needs to be determined to carry out the calculation. However, the probability, p_i, in each case cannot be actually determined in practical calculations. Since the distribution of information cannot be directly obtained in many cases, only the average, variance, and other parameters of the distribution dependence information can be obtained through experiments.

In the case of an unknown distribution, the distribution needs to be determined according to the known distribution dependence information. Therefore, the final distribution must be a distribution that corresponds to the maximum entropy function under the premise of satisfying all known information. A maximum entropy method of estimation has been proposed by Behrouz [32] that is used to derive the minimum bias probability distribution for the given information based on constraints. It can be expressed as

$$\max S(x) \ s.t. \ E(g_j(x)) = c_i, j = 1, 2 \ldots m, \tag{7}$$

where $S(x)$ is given in Equation (6), $g_j(x)$ is a feature function, and c_j is the expected value of the j-th feature.

2.3. The Selection of Copulas

For the application of the multivariate copulas, an important question is how to select the most suitable copula from a set of given candidate copulas to describe the dependence structure.

One commonly used method is the Akaike information criterion (AIC). The Akaike information criterion [33] is a standard used to measure the goodness of statistical model fitting. It is based on the concept of entropy, and it can weigh the complexity of an estimated model and the goodness of the model-fit data. AIC is defined as follows:

$$\text{AIC} = -2\ln(L) + 2k, \tag{8}$$

where k is the number of parameters in the model; L is the likelihood function value. The smaller the value of AIC is, the fitter the dependence structure is.

Another commonly used criterion is the Bayesian Information Criterions (BIC) [34], which is defined as

$$\text{BIC} = -2\ln(L) + k \cdot \ln(n), \tag{9}$$

where n is the sample size. Similar to AIC, the smaller the BIC value is, and the better the fitting degree of the model is.

In addition, the likelihood function could be also used to select copulas. The essence is to compare the maximum value of the likelihood function under the constraint condition with the maximum value of the likelihood function without the constraint condition [35]. The larger the maximum value of the likelihood function is, the better the model fitting degree is.

2.4. Copula Entropy

2.4.1. Definition of Copula Entropy

James and Crutcheld [36] demonstrate that Shannon information measures can fail to accurately ascertain multivariate dependencies due to the conflation of different relationships among variables. Thus, we chose the copula entropy, which combines the information entropy and the copula function, to describe the dependence relationship of multivariate.

Copula entropy is a combination of copula theory and maximum entropy theory. The copula function is used to describe the dependence among variables, and information entropy theory is utilized to quantify the dependence. The entropy variables are mutually independent in the entropy model, which is a general assumption for the principle of maximum entropy [37]. However, the copula theory needs to be supported to describe the entropy variable with dependence. Based on the copula theory of Sklar [29], joint entropy can be expressed as the sum of n univariate entropy and copula entropy. From this, the functional form [21] of the copula entropy used in this paper is

$$Hc(u_1, u_2, \cdots, u_d) = -\int_0^1 \cdots \int_0^1 c(u_1, u_2, \cdots, u_d) \ln(c(u_1, u_2, \cdots, u_d)) du_1, \cdots, du_d, \quad (10)$$

where $c\,(u_1, u_2, \ldots, u_d)$ is the probability density function of the copula function; $u_i = F_i\,(x_i) = P\,(x_i \le X_i)$, $i = 1, 2, \ldots, d$, represents the marginal distribution function of random variables.

(1) The characteristics of copula entropy can be deduced based on the three properties of the entropy function [21,38], copula entropy should be continuous [39].
(2) If all the discrete probabilities of the copula are equal, then it should be a monotonically increasing function [40], and the measurement of uncertainty should be higher when there are more possible outcomes than when there are few.
(3) The monotonicity property of copula entropy can be deduced from that the copula function is monotonic [12].

Take Gumbel copula, for example, the mathematical expression for its copula entropy is given as below.

$$Hc(u_1, u_2, \cdots, u_d) = -\int_0^1 \cdots \int_0^1 \exp\left\{-\left[\sum_{i=1}^{d} (-\ln u_i)^{\theta}\right]^{1/\theta}\right\} \ln(\exp\left\{-\left[\sum_{i=1}^{d} (-\ln u_i)^{\theta}\right]^{1/\theta}\right\}) du_1, \cdots, du_d \quad (11)$$

As shown in Equation (11), the Gumbel function is monotonic and therefore its copula entropy is monotonic.

Copula entropy is dimensioned as entropy, and its unit of measurement is the nat [21]. The nat is the natural unit of information. Sometimes nit or nepit is also used as the unit of information or entropy and is based on natural logarithms and powers of e, rather than the powers of 2 and base 2 logarithms, which define the bit. It can be expressed as the following equation:

$$2^x = e^1 \Rightarrow x = \frac{1}{\ln 2}, \quad (12)$$

where x stands for one nat; and e is the base of the natural logarithm.

This unit is also known by its unit symbol, the nat. The nat is the coherent unit of information entropy. The International System of Units, by assigning the same units (joule per kelvin) both to heat capacity and to thermodynamic entropy, implicitly treats information entropy as a quantity of dimension one, with 1 nat = 1. Physical systems of natural units that normalize Boltzmann's constant to 1 effectively measure the thermodynamic entropy in nats. When the Shannon entropy is written using a natural logarithm, as in Equation (12), it is giving a value measured in nats.

According to information entropy theory, when the known information decreases, the corresponding entropy value becomes larger. In contrast, when the known information becomes larger, the corresponding entropy value becomes smaller. When the above properties are applied to copula entropies, the lower the dependence degree among the variables, the weaker the corresponding dependence information, and the larger the entropy value reflected in copula entropy. Similarly, the higher the dependence among variables, the stronger the corresponding dependence information, and the smaller the entropy of copula entropy. Copula entropy can be calculated using multidimensional integration, and its value range is a real number space.

2.4.2. Copula Entropy and the Pearson Correlation Coefficient

Copula entropy, as a newly developed measurement of dependence, has some advantages not found in other dependence measurements. As the copula function can describe nonlinear dependence, copula entropy can also measure the information of a nonlinear dependence structure. In addition, it is possible for copula entropy to obtain unitary results to achieve a direct comparison since entropy has a dimension. Therefore, copula entropy can measure the dependence of two or more variables.

Dependence among variables has been widely studied. The traditional dependence measurement method is based on the correlation coefficient. Although this method is currently widely used for dependence measurement, the correlation coefficient has some obvious limitations. In contrast, copula entropy theory is applied to nonlinear correlation modeling in this paper instead of relying on correlation coefficients [41]. In addition, a comparison of the copula entropy method and the correlation coefficient is given in Table 2.

Table 2. Comparison of copula entropy and the correlation coefficient.

Method	Application Scenarios	Concerns	The Number of Variables	Dimension
Correlation Coefficient	Linear	Degree of dependence	Bivariate	Dimensionless
Copula Entropy	Linear/nonlinear	Structure of dependence	Multivariable	Dimension

By comparing the information in the table, the following conclusions can be made. First, the correlation coefficient method only applies to the linear correlation. However, in practice, the relationship among the variables is not always an ideal linear relationship. However, nonlinear dependence is quite natural in many complex engineering applications; in this respect, copula entropy can be used to measure both linear and nonlinear correlation to solve dynamic nonlinear correlation measurement problems for multiple degradation processes.

Second, the correlation coefficient method often focuses on the degree of dependence. But another important aspect of this relationship is the structure of dependence, which is often omitted and ignored [42]. However, copula entropy focuses on not only the degree of dependence but also the structure of dependence. The copula entropy method can more accurately describe the relationship among variables.

Third, the correlation coefficient method is dimensionless, and it is difficult to compare in cases with more than two sets of variables. However, copula entropy has dimension and can be compared directly, and a comparison of the dependence among multiple variables can be obtained. For example, if the correlation coefficient between "A" and "B" is 0.6, and the correlation coefficient between "B" and "C" is 0.3, then one can conclude that "AB" is more correlated than "BC." However, if the correlation coefficient between "A" and "B" is 0.6, and the coefficient between "C" and "D" is 0.3, then the correlation between "AB" and "CD" cannot be compared. In contrast, copula entropy is comparable and is easily explained with entropy theory. In information theory, entropy has its own unit, the nat, which is used to measure the information obtained from variables. Therefore, if the value of the copula entropy between "A" and "B" is less than that between "C" and "D," this means that the dependence of "AB" is higher than that of "CD."

In summary, copula entropy can accomplish the following two issues that traditional dependence measurement methods cannot achieve. The nonlinear dependence will be measured using the copula method, and it can be used to analyze nonlinear dependence among variables, instead of just focusing on linear dependence among variables. The dependence between any two degradation processes can be directly compared without intermediate variables. Therefore, the dependence among three or more variables can be compared. In addition, it is possible to compare the dependencies of variables during different time phases and determine the time-varying law of dependence.

3. Dependence Measurement of Multivariate Degradation Processes

3.1. Problem Description

Some products with complex system structures, various performance features, and varying operational conditions tend to exhibit degradation of multiple performance features, and there are unavoidable dependencies among these degradation features. The dependencies among the degradation processes of multiple performance features often show dynamic and nonlinear statistical features. If these dependencies are ignored, product degradation modeling and lifetime estimation may result in less credible or even erroneous results.

Some products possess a simple degradation mechanism, and the reliability of the product can be directly derived by using the relationship among the degradation features amount and time (i.e., the product performance degradation trajectory). However, some products possess complex degradation mechanisms, and the quantitative relationship of the degradation model cannot be expressed directly. In this case, traditional dependence measurement methods cannot accurately obtain dependence information. This results in the inability to measure dependence or to obtain accurate results.

As mentioned above, simple linear correlation coefficients cover a wide range of values that can reveal a variety of dependency relationships. However, the information of the nonlinear relationships is ignored if they are not simply linearly correlated. The traditional dependence measurement method is only applicable to linear correlations and can only measure the dependence between two variables. The copula entropy proposed in this paper can be applied to correct this ignorance, and it can measure nonlinear dependence relationships among two or more variables. In addition, copula entropy has no dimensional constraints; hence, enough indicators can be chosen to measure dependence among the variables.

3.2. The Calculation Process

The essence of copula entropy is a multivariate integral that can be calculated using the integral method. However, when the dimension of the integrand is high and the form is complex, the calculation process will be very difficult. To this end, the Monte Carlo simulation method would be used to calculate the copula entropy. The method is divided into four steps, as shown in Figure 1.

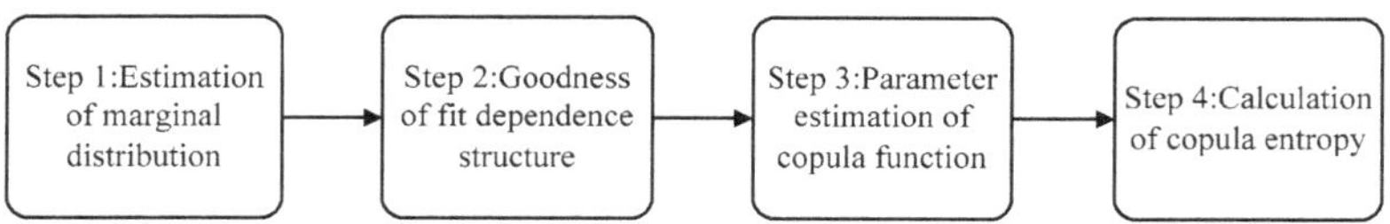

Figure 1. Dependence measurement of the copula entropy method flowchart.

The specific steps are described as follows:

- Step 1: The kernel density estimation method is used to estimate the marginal distribution and calculate the cumulative distribution function (CDF) of every performance feature degradation increasement.
- Step 2: A different type of copula function is adopted to combine the cumulative distribution functions of different performance feature degradation increasement separately. In addition, the Akaike information criterion (AIC) is performed to compare the goodness-of-fit of the copula function.
- Step 3: Parameter estimation of the copula function is performed using the maximum likelihood estimate (MLE) method. Also, the establishment of the copula entropy function is performed based on the chosen copula function and the determination of integrand function.
- Step 4: The Monte Carlo simulation method is utilized to calculate the copula entropy.

3.2.1. Estimation of Marginal Distribution

In the copula entropy method, the structure of the marginal distribution is an important issue. The first step is to estimate the marginal distribution. For degradation data of multi-performance features, the cumulative distribution function of the degradation increments for each performance feature needs to be calculated. In this step, the kernel density estimation method [43] is applied to estimate and calculate the cumulative distribution function of degradation increasement.

Kernel density estimation is used in probability theory to estimate unknown density functions. It is also one of the non-parametric test methods proposed by Rosenblatt [44] and Parzen [45]. Based on the univariate kernel density estimation, a risk value prediction model can be established. Different risk value prediction models can be established by weighting the variation coefficient of kernel density estimation. Since the kernel density estimation method does not use prior knowledge of the data distribution and does not attach any assumptions to the data distribution, it is a method for studying the data distribution features from the data sample itself. Therefore, it has received great attention in the field of statistical theory and application.

For a certain performance feature, the degradation data are $X_1, X_2, \ldots, X_m$, then the degradation increment is

$$\Delta X_i = X_i - X_{i-1}, \tag{13}$$

where i is the data coefficient, and $i = 1, 2, \ldots, m$, m represents the number of data observations.

The degradation increment of each performance feature is assumed to confirm the basic requirements of the statistical test used, such as independence and distribution. Then, the probability density function, $p_i(x)$, of the ith performance feature degradation increment is calculated as [46]

$$p_i(\Delta x) = \frac{1}{mh}\sum_{t=1}^{T} K\left(\frac{\Delta x - \Delta X_t}{h}\right) \tag{14}$$

$$K\left(\frac{\Delta x - \Delta X_t}{h}\right) = \frac{1}{\sqrt{2\pi}}\exp\left(-\frac{(\Delta x - \Delta X)^2}{2h^2}\right), . \tag{15}$$

where t is the time interval; T is the width of the time interval; h is the width of the form smooth parameter; and $K(\cdot)$ is a kernel function, which is a standard Gaussian distribution [47] with expectation 0 variance of 1, $i = 1, 2, \ldots, d$; and d is the number of performance features.

Therefore, the cumulative distribution function, u_i, of the i-th performance feature degradation increment is

$$u_i = \int p_i(\Delta x) d\Delta x. \tag{16}$$

3.2.2. Goodness of Fit Dependence Structure

As mentioned in Section 2.3, there are three commonly used methods for the selection of copulas, including AIC, BIC and the likelihood function method. Generally speaking, AIC is the most commonly used method in choosing copula. The difference between AIC and BIC mainly lies in the number of model parameters.

As the number of unknown parameters of the candidate copulas is the same [12], the more commonly used AIC method is chosen. Similarly, for the likelihood function method, the larger the likelihood function is, the smaller the AIC value is, and the fitter the dependence structure is. Therefore, ultimately the AIC method is chosen for the goodness of fit in the dependence structure.

3.2.3. Parameter Estimation of Copula Entropy

After the copula entropy is built, its internal parameters need to be estimated. The maximum likelihood estimation method (MLE) is the method used to solve this problem. Suppose the probability distribution function of variable x_i is expressed as $F_i(x_i; \varphi_i)$, $i = 1, 2, \ldots, n$, the probability density function is $f_i(x_i; \varphi_i)$, and φ_i is the unknown parameter in each function. Then the joint distribution function of $X = (x_1, x_2, \ldots, x_n)$ is

$$H(X, \varphi_1, \varphi_2, \cdots, \varphi_n, \theta) = C(F_1(x_1, \varphi_1), F_2(x_2, \varphi_2), \cdots, F_n(x_n, \varphi_n), \theta), \tag{17}$$

where C is the copula function.

The corresponding probability density function is

$$h(X, \varphi_1, \varphi_2, \cdots, \varphi_n, \theta) = c(F_1(x_1, \varphi_1), F_2(x_2, \varphi_2), \cdots, F_n(x_n, \varphi_n), \theta) \cdot \prod_{i=1}^{n} f_i(x_i; \varphi_i). \tag{18}$$

If the sample is known to be $\left\{(x_{1j}, x_{2j}, \ldots, x_{nj})\right\}_{j=1}^{k}$, then the unknown function's likelihood function is

$$L(\theta) = \prod_{j=1}^{k}\left\{c\left(F_1\left(x_{1j}, \varphi_1\right), F_2\left(x_{2j}, \varphi_2\right), \cdots, F_n\left(x_{nj}, \varphi_n\right), \theta\right) \cdot \prod_{i=1}^{n} f_i\left(x_{ij}, \varphi_i\right)\right\}. \tag{19}$$

The corresponding log-likelihood function is

$$\ln L(\theta) = \underbrace{\sum_{j=1}^{k} \ln c\left(F_1\left(x_{1j}, \varphi_1\right), F_2\left(x_{2j}, \varphi_2\right), \cdots, F_n\left(x_{nj}, \varphi_n\right), \theta\right)}_{L_c} + \underbrace{\sum_{j=1}^{k}\sum_{i=1}^{n} \ln f_i\left(x_{ij}, \varphi_i\right)}_{L_i}. \tag{20}$$

The above method uses the ordinary maximum likelihood estimation method. However, when there are multiple distribution parameters, the calculation becomes more complex. An improved method is to use a two-stage maximum likelihood estimation method. More precisely, the marginal distribution and copula function parameter estimation are used separately, so as to simplify the calculation. The specific algorithm is as follows.

First, the parameter estimation of each marginal distribution function is conducted, and the results are substituted into Equation (21)

$$\hat{\varphi}_i = ArgMaxl_i(\varphi_i) = ArgMax \prod_{j=1}^{k} f_i\left(x_{ij}, \varphi_i\right). \tag{21}$$

Then, the maximum likelihood estimation method is used to estimate the parameters of the copula function based on the following equation:

$$\hat{\theta} = ArgMaxl_c(\theta) = ArgMax\sum_{j=1}^{k} \ln c\Big(F_1\big(x_{1j}, \hat{\varphi}_1\big), F_2\big(x_{2j}, \hat{\varphi}_2\big), \cdots, F_n\big(x_{nj}, \hat{\varphi}_n\big), \theta\Big). \tag{22}$$

3.2.4. Calculation of the Copula Entropy Value

The copula entropy can be obtained by solving the multiple integrations of Equation (22). However, due to the complexity of the copula function, the form of the integrand function is also very complicated. When this method is used to calculate the copula entropy function, there is a situation where the calculated amount is too large to be calculated. Therefore, this paper uses the Monte Carlo simulation method for the calculation.

The key to the copula entropy calculation is to calculate the integral of the multivariate function using the idea of simulation sampling. Therefore, the main task of this step is to find the area that is completely surrounded by the coordinate surface with the range of V_0. The Monte Carlo method will be used in the surrounded area to sample N times ($N \geq 10{,}000$). The frequency of the sample points falling into the area of the integrand is then calculated. The percentage of the integral volume and the closed volume based on the frequency is then calculated. Finally, the copula entropy is calculated.

4. Case Study

In this section, the multiple degradation data of a microwave electronic assembly were used to verify the copula entropy measurement method. To evaluate the lifetime and reliability of the microwave electronic assembly, the degradation data of four performance features collected simultaneously were measured. These performance features were the power gains A, B, C, and D. The degradation data of each performance feature are shown in Figure 2, and the degradation increment data are shown in Figure 3.

Degradation data of the microwave electrical assembly under different operating conditions were used as the validation data in this section. During the same user experience and under certain environmental/operational conditions, the degradation data of the different performance features of the product showed nonlinear dependence relationships due to the complex internal structure of the product. If the dependence among the degradation data is ignored for degradation modeling and lifetime estimation, the result will have lower accuracy. Therefore, to model the degradation more accurately and estimate the service life and reliability of the product effectively, it was necessary to measure dependencies among the degradation features accurately.

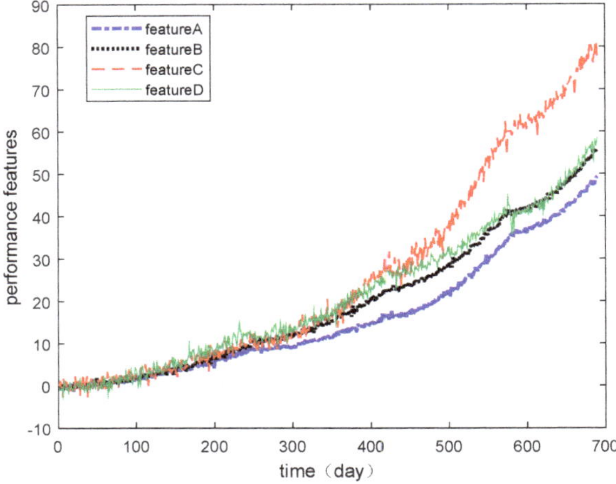

Figure 2. Multi-performance features of degradation data.

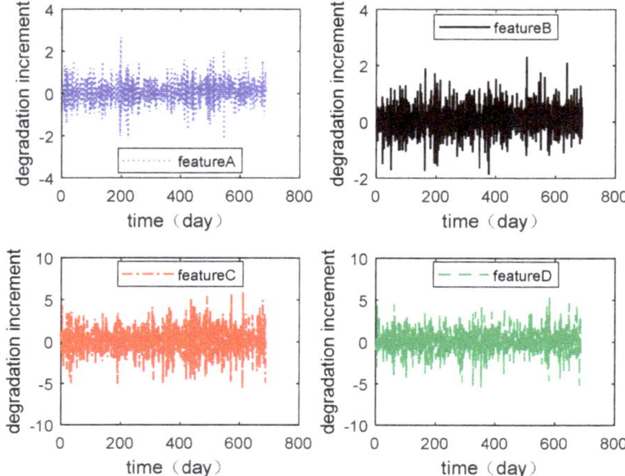

Figure 3. Degradation of various performance features.

4.1. Dependence Measurement of Bivariate Degradation Processes

One advantage of copula entropy is that it has dimension. Then the dependence among the variables can be directly compared. Therefore, this section will use the multivariate degradation data of a microwave electrical assembly to verify the copula entropy measurement of the dependence between binary variables.

The dependence of the degradation increment measured using the Pearson correlation coefficient method is shown in Figure 4. This is compared with the experimental results to verify the proposed method.

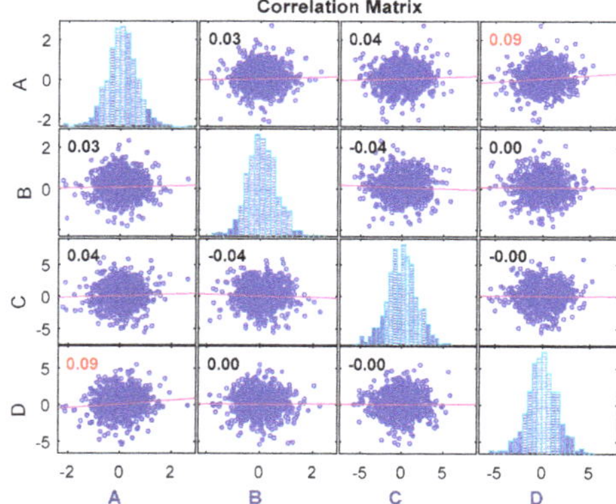

Figure 4. Pearson correlation coefficient measurement.

As can be seen in Figure 4, the correlation coefficient method can only measure the dependence among linear relationships, and the results cannot be compared to obtain different dependencies among the different variables. Therefore, the copula entropy method is proposed to measure the dependence between two sets of data and to compare the dependence between them.

First, the kernel density estimation method was used to calculate the cumulative distribution function (CDF) of each performance feature degradation data increment, as shown in Figure 5.

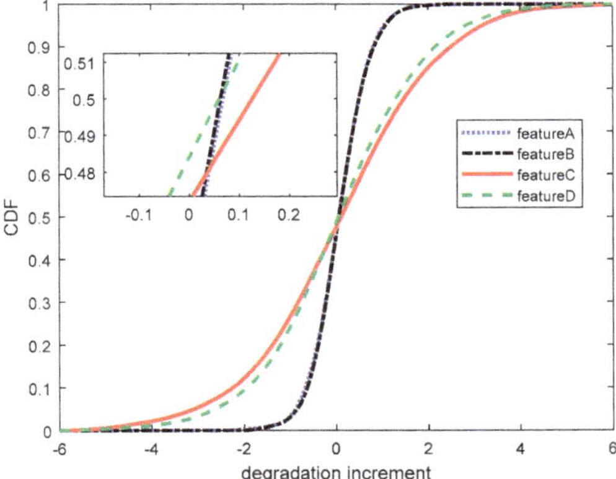

Figure 5. The CDF of degradation increments.

Second, the Gaussian copula, Frank copula, Clayton copula, and Gumbel copula are used to couple the degradation increase distributions of the different performance features. The AIC criterion is applied to select the most suitable copula function to calculate the copula entropy and quantify the dependence among them. The AIC results are shown in Table 3.

Table 3. AIC based on different copula functions.

Performance Features	Gaussian	Frank	Clayton	Gumbel
AB	−0.54	−10.27	−211.14	−232.55
AC	−0.47	−2.96	−5.10	−24.42
BC	−8.80	−2.90	−2.15	−24.59
AD	−48.55	−3.24	−12.48	−46.55
BD	−48.68	−3.17	−13.39	−46.76
CD	−11.52	−7.02	−20.83	−30.42

Third, the maximum likelihood estimation method is used to estimate the copula parameters for each of the two degradation increment CDFs. The results are shown in Table 4.

Table 4. Copula parameter estimation results.

Marginal Distribution Function	Parameter Estimation	Copula Function
AB	8.3911	Gumbel
AC	1.0412	Gumbel
BC	1.0347	Gumbel
AD	0.9538	Gaussian
BD	0.9532	Gaussian
CD	3.3042	Gumbel

Then, after the copula parameters were determined, the joint PDF of the multivariable was determined. The copula entropy was utilized to quantify the dependence among different features. The binary copula entropy can be calculated as follows:

$$Hc(u,v) = -\int_0^1 \int_0^1 c(u,v)\ln(c(u,v))du, dv. \tag{23}$$

Finally, since the form of $c(u,v)\mathrm{ln}c(u,v)$ is very complicated, the calculation of the integral will cause computational difficulties. Therefore, the Monte Carlo sampling method was utilized to calculate the copula entropy. Therefore, the integrand in Equation (23) needs to be calculated, namely $c(u, v)$ $\ln c(u, v)$. In addition, the Monte Carlo simulation method was used to calculate the copula entropy of each performance feature. According to Equation (23), the integrand functions based on different marginal distributions are shown in Figure 6.

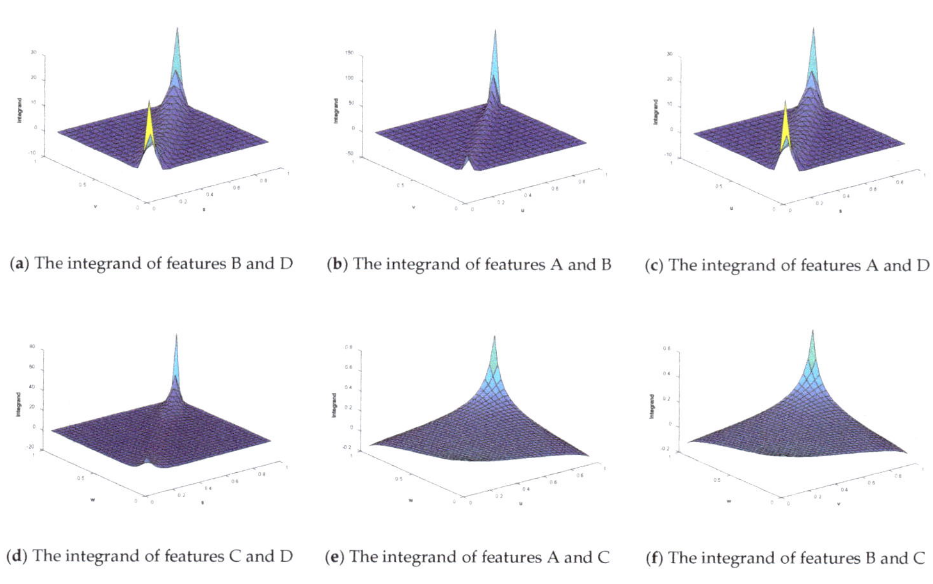

(**a**) The integrand of features B and D (**b**) The integrand of features A and B (**c**) The integrand of features A and D

(**d**) The integrand of features C and D (**e**) The integrand of features A and C (**f**) The integrand of features B and C

Figure 6. The integrand of different marginal distributions.

To make the results more obvious, the contours of the copula entropy are shown in Figure 7.

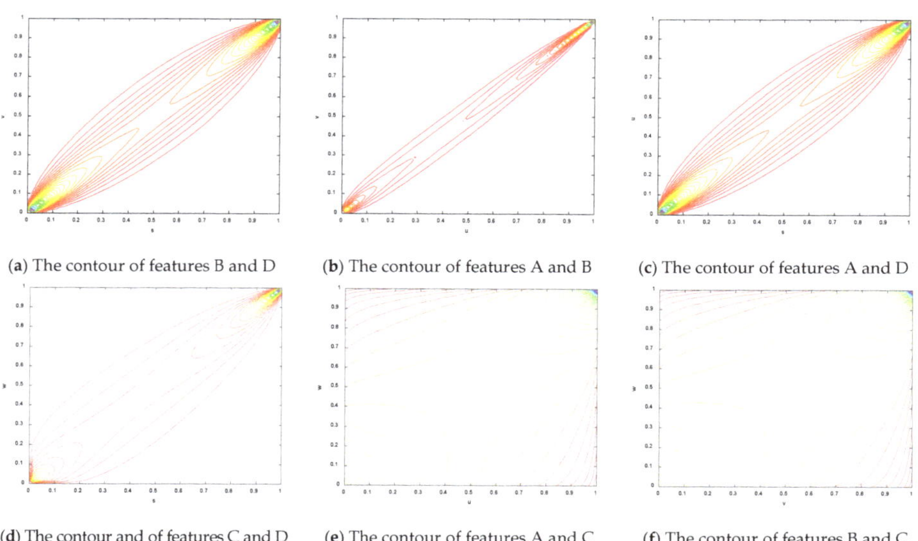

(**a**) The contour of features B and D (**b**) The contour of features A and B (**c**) The contour of features A and D

(**d**) The contour and of features C and D (**e**) The contour of features A and C (**f**) The contour of features B and C

Figure 7. The contours of different copula entropy.

The value of the volume is then calculated depending on the sampling method used. The calculated copula entropy results of each set of performance features are shown in Table 5.

Table 5. Copula entropy of binary performance feature.

Marginal Distribution Function	Copula Entropy (nat)	Copula Function
BD	−12.1314	Gaussian
AB	−9.3044	Gumbel
AD	−3.4229	Gaussian
CD	−2.6211	Gumbel
AC	−0.0717	Gumbel
BC	−0.0652	Gumbel

The principle of copula entropy shows that the value of entropy is negatively correlated with the degree of dependence. The dependence of any two variables can then be compared regardless if there exists an intermediate variable among them. In addition, the degree of dependence among variables can be sorted according to the value of the copula entropy among the variables. However, the existing dependence measurement methods fail to do this and cannot compare the degree of dependence without intermediate variables. Therefore, the data in the table show that the dependencies are arranged in descending order: BD, AB, AD, CD, AC, BC.

4.2. Dependence Measurement of Different Phases

One advantage of copula entropy is that it can be used to compare the dependencies of multiple feature degradation processes over time, and it can determine the time-varying law of the dependence. The dependence may be variational under different operational conditions. The influence of operational conditions on different degradation features is different. If the operational conditions transform, the degradation rules of different features will change in different ways, which will lead to variation in the dependence among the features.

In this case, the same data used in Section 4.1 were used. It can be seen in Figure 2 that the degradation trend of the variables is not invariable but presents different degradation rates. The dependence among these four sets of degradation data changes with time. However, if only one value that represents the dependence of the variables is obtained, then the degree of dependence among variables at different time phases cannot be compared. Therefore, subsection calculation was needed to compare the variation trend of the dependence of different subsections. Then the dependence among variables can be described more accurately.

In this case, the degradation data of performance features A, B, C, and D were divided into four phases in time with 247, 174, 155, and 114 sets of data, respectively. According to Equation (22), the copula entropies of the different stages were calculated to compare the regularity of dependence with time. Copula entropy was used to measure the dependencies for each group of data to analyze the data dependence during different phases. The cumulative distribution function of the four sets of degradation data increments is shown in Figure 8.

Degradation increments were coupled using the Gaussian copula, Frank copula, Clayton copula, and Gumbel copula at each stage, and the AIC for each set of data is shown in Table 6.

The estimation of the copula parameters and the copula entropy results of the total data and the four phrases for the four performance feature degradation increments are shown in Table 7.

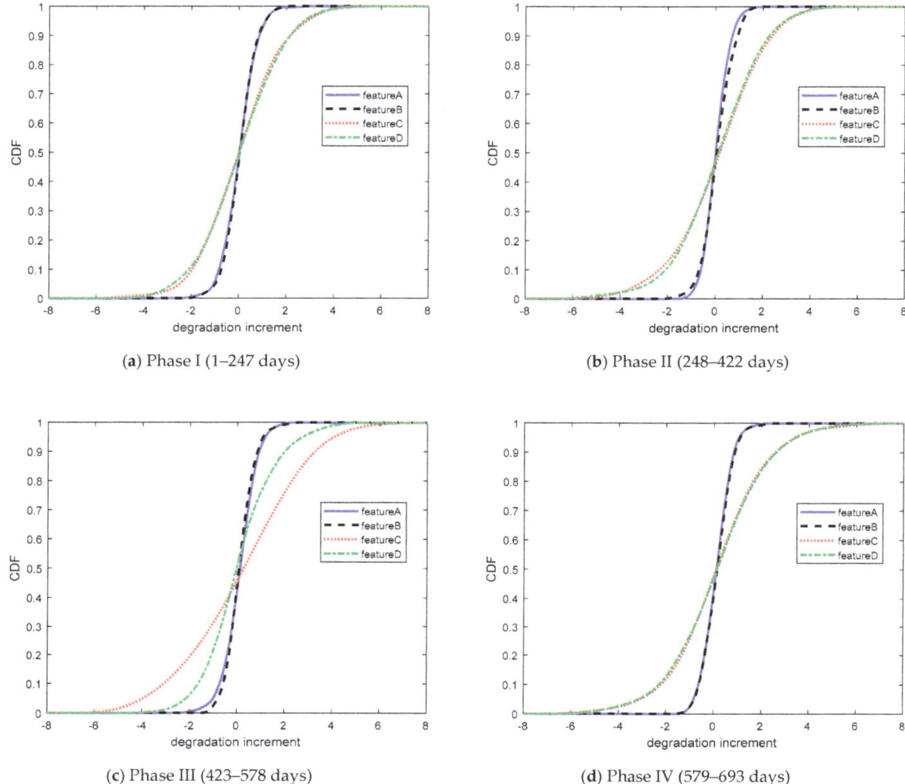

(**a**) Phase I (1–247 days)

(**b**) Phase II (248–422 days)

(**c**) Phase III (423–578 days)

(**d**) Phase IV (579–693 days)

Figure 8. The CDFs of the four sets of degradation data increments.

Table 6. AIC based on different copula families.

Phase	Gaussian	Frank	Clayton	Gumbel
I	−11.47	−194.59	−6.64	−491.19
II	−162.44	−231.15	−7.61	−425.70
III	−6.36	−330.28	−8.76	−8.40
IV	−14.65	−279.30	−9.11	−476.11

Table 7. Parameter estimations and the copula entropy calculation results.

Phase	**Total Data**	**I**	**II**	**III**	**IV**
Copula Function	**Clayton**	**Gumbel**	**Gumbel**	**Frank**	**Gumbel**
Parameter estimation	5.013325	3.359566	4.034455	25.51778	4.448991
Copula entropy	−2.7393	−88.0933	−81.7226	−0.0411	−123.3709

When Equation (22) was applied to calculate the value of the copula entropy for the total data in the case, only a value representing the dependence of four variables in this degradation process could be obtained. As shown in Table 7, the value of the copula entropy of the total data among the four variables was −2.7393, and it can be seen that there was dependence among the variables, but it was not significantly strong. In addition, the dependence did not seem to change over time. However as can be observed from the segmentation results in the table, the degree of dependence of the four performance features of the product continuously changed at different phases. The copula entropy of phase IV was the smallest, and the degree of dependence was the highest. Phase I was second, and the next one was phase II. Phase III was the least dependent. Thus, it can be seen that the dependencies of the four performance features changed dynamically with time during the entire degradation process.

4.3. Discussion

In the study of traditional theory, the dependence of four variables or multiple variables cannot be compared and analyzed. However, we can compare the dependence among four variables or multiple variables using the copula entropy dependence measurement method. As described in the definition of information theory, entropy is the measurement of disorder [21]. The higher the complexity of the system, the less knowledge we have about information, and the higher the value of entropy.

The conclusion obtained from copula entropy is a measurement of the dependence among the variables. This means that the less we understand the information of the dependence structure, the more disorder there is in the system and the higher the value of the copula entropy. The dependence information among variables becomes more obvious when the dependence becomes stronger and the value of the copula entropy decreases.

According to the data in Tables 5 and 7, the smaller the copula entropy value, the higher the dependence of variables. Therefore, the degree of the dependence can be directly determined according to the size of the copula entropy values among the variables. Furthermore, the dependence among variables can be analyzed. According to the data in Table 7, the copula entropy values among variables are different during the different phases. More succinctly, the dependence among the variables changes with time. The copula entropy method used to measure the dependence among variables can correctly describe the change in variable dependence during different stages.

An analysis of the above two cases shows that the proposed method can not only compensate for the deficiencies in the Pearson correlation coefficient method but also measure the linear and/or nonlinear dependence of multivariate degradation data. The proposed method can also be used to describe the dependence of data during different phases, suggesting that the dependence changes during different phases.

5. Conclusions

Degradation modeling has become an efficient method to evaluate the reliability of long lifespan products. Generally, a product may have multiple degradation features. It is unavoidable that there is dependence among multiple degradation features. The dependence structure may be linear or nonlinear. When ignoring the dependence among multiple degradation features, degradation modeling and lifespan estimation may lead to less credible or erroneous results. Therefore, it is safer to describe and measure dependence accurately and reasonably among multiple degradation features prior to modeling. The Pearson linear correlation coefficient and rank correlation coefficient are often used to measure the dependence between two variables. However, they will misestimate the dependence between two variables when the dependence relationship is nonlinear. Furthermore, they cannot be used to detect dependence when more than two variables are involved. There is no particularly suitable method to measure multiple degradation dependence in the present study. Therefore, we introduce copula entropy, which is used in statistics, to overcome the shortcomings of existing methods in measuring multiple degradation dependence.

In this paper, a measurement method for the dependence among multiple degradation processes based on copula entropy has been proposed for products with multiple performance features. The copula entropy was constructed using the copula function and the information entropy theory. Thus, the copula entropy has the advantages of both of them. It can be applied to measure not only the linear dependence but also the nonlinear dependence. Another advantage of the copula entropy method is that it is not confined to bivariate variables. It is valid to use it to compare dependence among two or more variables based on copula entropy. The copula entropy provides an effective way for us to solve the problem of multiple degradation dependence measurements and solves practical problems in engineering practice. A practical case was used to validate the proposed method and the results proved that the proposed method can effectively improve the accuracy of degradation modeling and life estimation in engineering applications.

Overall, the integration of the copula function and information entropy provides useful insights into dependence measurements in multiple degradation modeling. The effectiveness of the copula entropy method in other fields and a comparison with the traditional copula function fitting effect need to be conducted in the future. In addition, the copula method requires further investigation regarding the influence of different constraints on the fitting results during the process of solving the joint distribution equation.

Author Contributions: Conceptualization, F.S. and W.Z. (Wei Zhang); Data curation, W.Z. (Wendi Zhang) and N.W.; Funding acquisition, F.S.; Methodology, F.S., W.Z. (Wendi Zhang) and N.W.; Validation, W.Z. (Wendi Zhang); Writing—original draft, W.Z. (Wendi Zhang); Writing—review and editing, F.S. and W.Z. (Wei Zhang).

Funding: This research was funded by National Key R&D Program of China, grant number 2018YFB0104504 and National Natural Science Foundation of China, grant number 61603018.

Conflicts of Interest: The authors declare no conflict of interest.

References

1. McPherson, J.W. *Reliability Physics and Engineering Time to Failure Modeling*; Springer: New York, NY, USA, 2013.
2. Yang, G.B. Environmental-stress-screening using degradation measurements. *IEEE Trans. Reliab.* **2002**, *51*, 288–293. [CrossRef]
3. Ye, Z.S.; Xie, M. Stochastic modelling and analysis of degradation for highly reliable products. *Appl. Stoch. Models Bus. Ind.* **2015**, *31*, 16–32. [CrossRef]
4. Peng, W.W.; Li, Y.F.; Yang, Y.J.; Zhu, S.P.; Huang, H.Z. Bivariate Analysis of Incomplete Degradation Observations Based on Inverse Gaussian Processes and copulas. *IEEE Trans. Reliab.* **2016**, *65*, 624–639. [CrossRef]
5. Wang, Y.P.; Pham, H. Modeling the Dependent Competing Risks with Multiple Degradation Processes and Random Shock Using Time-Varying copulas. *IEEE Trans. Reliab.* **2012**, *61*, 13–22. [CrossRef]
6. Sun, F.Q.; Liu, L.; Li, X.Y.; Liao, H.T. Stochastic Modeling and Analysis of Multiple Nonlinear Accelerated Degradation Processes through Information Fusion. *Sensors* **2016**, *16*, 18. [CrossRef] [PubMed]
7. Sreelakshmi, N. An introduction to copula-based bivariate reliability Concepts. *Commun. Stat. Theory Methods* **2018**, *47*, 996–1012. [CrossRef]
8. Xu, D.; Wei, Q.D.; Elsayed, E.A.; Chen, Y.X.; Kang, R. Multivariate Degradation Modeling of Smart Electricity Meter with Multiple Performance Characteristics via Vine copulas. *Qual. Reliab. Eng. Int.* **2017**, *33*, 803–821. [CrossRef]
9. Kat, H.M. The Dangers of Using Correlation to Measure Dependence. *Inst. Investig. J.* **2003**, *6*, 54–58. [CrossRef]
10. Chen, L.; Singh, V.P.; Guo, S.L. Measure of Correlation between River Flows Using the copula-Entropy Method. *J. Hydrol. Eng.* **2013**, *18*, 1591–1606. [CrossRef]
11. Mari, D.D.; Kotz, S. *Correlation and Dependence*; Imperial College Press: London, UK, 2004.
12. Nelsen, R.B. *An Introduction to Copulas*; Springer Science & Business Media: New York, NY, USA, 2007.
13. Szmidt, E.; Kacprzyk, J. The Spearman and Kendall rank correlation coefficients between intuitionistic fuzzy sets. In Proceedings of the 7th Conference of the European Society for Fuzzy Logic and Technology, Aix-Les-Bains, France, 18–22 July 2011; Atlantis Press: Paris, France, 2011; pp. 521–528.
14. Arndt, S.; Turvey, C.; Andreasen, N.C. Correlating and predicting psychiatric symptom ratings: Spearman's r versus Kendall's tau correlation. *J. Psychiatr. Res.* **1999**, *33*, 97–104. [CrossRef]
15. Sari, J.K.; Newby, M.J.; Brombacher, A.C.; Tang, L.C. Bivariate Constant Stress Degradation Model: LED Lighting System Reliability Estimation with Two-stage Modelling. *Qual. Reliab. Eng. Int.* **2009**, *25*, 1067–1084. [CrossRef]

16. Sun, F.Q.; Wang, N.; Li, X.Y.; Cheng, Y.Y. A time-varying copula-based prognostics method for bivariate accelerated degradation testing. *J. Intell. Fuzzy Syst.* **2018**, *34*, 3707–3718. [CrossRef]
17. Gauthier, T.D. Detecting trends using Spearman's rank correlation coefficient. *Environ. Forensics* **2001**, *2*, 359–362. [CrossRef]
18. Schmid, F.; Schmidt, R.; Blumentritt, T.; Gaißer, S.; Ruppert, M. Copula-based measures of multivariate association. In *Copula Theory and Its Applications*; Springer: Berlin/Heidelberg, Germany, 2010; Volume 198, pp. 209–236.
19. Ane, T.; Kharoubi, C. Dependence Structure and Risk Measure. *J. Bus.* **2003**, *76*, 411–438. [CrossRef]
20. Singh, V.P.; Zhang, L. Copula–entropy theory for multivariate stochastic modeling in water engineering. *Geosci. Lett.* **2018**, *5*, 6. [CrossRef]
21. Zhao, N.; Lin, W.T. A copula entropy approach to correlation measurement at the country level. *Appl. Math. Comput.* **2011**, *218*, 628–642. [CrossRef]
22. Hao, Z.C.; Singh, V.P. Integrating Entropy and copula Theories for Hydrologic Modeling and Analysis. *Entropy* **2015**, *17*, 2253–2280. [CrossRef]
23. McCulloch, W.S.; Pitts, W. A logical calculus of the ideas immanent in nervous activity. *Bull. Math. Biol.* **1943**, *5*, 115–133. [CrossRef]
24. Ma, J.; Sun, Z. Mutual information is copula entropy. *Tsinghua Sci. Technol.* **2011**, *16*, 51–54. [CrossRef]
25. Xu, P.; Wang, D.; Singh, V.P.; Wang, Y.; Wu, J.; Wang, L.; Zou, X.; Chen, Y.; Chen, X.; Liu, J.; et al. A two-phase copula entropy-based multiobjective optimization approach to hydrometeorological gauge network design. *J. Hydrol.* **2017**, *555*, 228–241. [CrossRef]
26. Huang, K.; Dai, L.M.; Yao, M.; Fan, Y.R.; Kong, X.M. Modelling Dependence between Traffic Noise and Traffic Flow through An Entropy-copula Method. *J. Environ. Inform.* **2017**, *29*, 134–151. [CrossRef]
27. Salimi, E.; Abbas, A.E. A Simulation-Based Comparison of Maximum Entropy and copula Methods for Capturing Non-Linear Probability Dependence. In *2016 Winter Simulation Conference*; IEEE: New York, NY, USA, 2016; pp. 460–471.
28. Joe, H. *Multivariate Models and Multivariate Dependence Concepts*; CRC Press: Boca Raton, FL, USA, 1997.
29. Sklar, A. Fonctions de répartition à n dimensions et leurs marges. *Publ. Inst. Stat. Univ. Paris* **1958**, *8*, 229–231.
30. Shannon, C.E. A mathematical theory of communication. *Bell Syst. Tech. J.* **1948**, *27*, 3–15. [CrossRef]
31. Aczél, J.; Daróczy, Z. *On Measures of Information and Their Characterizations*; Academic Press: New York, NY, USA, 1975.
32. Behmardi, B.; Raich, R.; Hero, A.O. Entropy Estimation Using the Principle of Maximum Entropy. In Proceedings of the 2011 IEEE International Conference on Acoustics, Speech and Signal Processing (ICASSP), Prague Congress Center, Czech Republic, 22–27 May 2011.
33. Shono, H. Is model selection using Akaike's information criterion appropriate for catch per unit effort standardization in large samples? *Fish. Sci.* **2005**, *71*, 978–986. [CrossRef]
34. Chen, S.S.; Gopalakrishnan, P.S. Speaker, Environment and Channel Change Detection and Clustering Via The Bayesian Information Criterion. In Proceedings of the Broadcast News Transcription and Understanding Workshop (DARPA), Lansdowne, VA, USA, 8–11 February 1998; Volume 8, pp. 127–132.
35. Wilks, S.S. The large-sample distribution of the likelihood ratio for testing composite hypotheses. *Ann. Math. Stat.* **1938**, *9*, 60–62. [CrossRef]
36. James, R.G.; Crutchfiel, J.P. Multivariate Dependence beyond Shannon Information. *Entropy* **2017**, *19*, 531. [CrossRef]
37. Golan, A.; Maasoumi, E. Information theoretic and entropy methods: An overview. *Econ. Rev.* **2008**, *27*, 317–328. [CrossRef]
38. Wehrl, A. General properties of entropy. *Rev. Mod. Phys.* **1978**, *50*, 221–260. [CrossRef]
39. Durante, F.; Sempi, C. *Principles of copula Theory*; Chapman, Hall/CRC: Boca Raton, FL, USA, 2015.
40. Tenzer, Y.; Elidan, G. On the Monotonicity of the copula Entropy. *arXiv* **2016**, arXiv:1611.06714. Available online: https://arxiv.org/abs/1611.06714 (accessed on 23 July 2019).
41. Tibprasorn, P.; Autchariyapanitkul, K.; Chaniam, S.; Sriboonchitta, S. A copula-Based Stochastic Frontier Model for Financial Pricing. In *International Symposium on Integrated Uncertainty in Knowledge Modelling and Decision Making*; Huynh, V.N., Inuiguchi, M., Denoeux, T., Eds.; Springer: Berlin, Germany, 2015; Volume 9376, pp. 151–162.

42. Xu, Y. Applications of copula-based Models in Portfolio Optimization. Ph.D. Thesis, University of Miami, Miami-Dade County, FL, USA, 2005.
43. Tang, L.J.; Yang, H.; Zhang, H.Y. The Application of the Kernel Density Estimates in Predicting VaR. *Math. Pract. Underst.* **2005**, *35*, 31–37.
44. Rosenblatt, M. Remarks on Some Nonparametric Estimates of a Density Function. *Ann. Math. Stat. Its Interface* **1956**, *27*, 832–837. [CrossRef]
45. Parzen, E. On Estimation of a Probability Density Function and Mode. *Ann. Math. Stat. Its Interface* **1962**, *33*, 1065–1076. [CrossRef]
46. Yao, W.X.; Zhao, Z.B. Kernel Density-Based Linear Regression Estimate. *Commun. Stat. Theory Methods* **2013**, *42*, 4499–4512. [CrossRef]
47. Gasser, T.; Muller, H.-G.; Mammitzsch, V. Kernels for nonparametric curve estimation. *J. R. Stat. Soc.* **1985**, *47*, 238–252. [CrossRef]

Article

Thermodynamics of Fatigue: Degradation-Entropy Generation Methodology for System and Process Characterization and Failure Analysis

Jude A. Osara * and Michael D. Bryant

Mechanical Engineering Department, The University of Texas at Austin, Austin, TX 78712, USA
* Correspondence: osara@utexas.edu

Received: 14 May 2019; Accepted: 9 July 2019; Published: 12 July 2019

Abstract: Formulated is a new instantaneous fatigue model and predictor based on ab initio irreversible thermodynamics. The method combines the first and second laws of thermodynamics with the Helmholtz free energy, then applies the result to the degradation-entropy generation theorem to relate a desired fatigue measure—stress, strain, cycles or time to failure—to the loads, materials and environmental conditions (including temperature and heat) via the irreversible entropies generated by the dissipative processes that degrade the fatigued material. The formulations are then verified with fatigue data from the literature, for a steel shaft under bending and torsion. A near 100% agreement between the fatigue model and measurements is achieved. The model also introduces new material and design parameters to characterize fatigue.

Keywords: fatigue; system failure; degradation analysis; entropy generation; stress strain; plastic strain; thermodynamics; health monitoring

---

1. Introduction

All solids can yield or fail under continuous loading. For static loading, equilibrium and monotonic conditions facilitate evaluation of a component's strength. For dynamic loading, assessment of degradation leading to fatigue failure is complicated by various dynamic loads, material composition and load conditions. With metals under heavy structural loading, sudden failure can be catastrophic [1]. Cyclic loading causes about 90% of all metal failures [2–7]. Thermal cycle-induced stresses can fatigue electronic components.

Common fatigue analysis methods include stress-life (Wohler) curves for high-cycle fatigue (HCF) and strain-life curves for low-cycle fatigue (LCF). Vasudevan et al. [8] discussed deficiencies in structural fatigue life models involving crack growth da/dN and the challenges in implementing these models. Existing approaches sometimes give inconsistent results, and failure measures are usually component- or process-specific. Recent entropy-based fatigue studies [9–23] have shown high accuracy, establishing thermodynamic energies and entropies as measures of system damage, degradation and failure [7,24].

Thermodynamics-Based Fatigue Models

Lemaitre and Chaboche [7] coupled damage mechanics with irreversible thermodynamics to present a comprehensive breakdown of elastic, elastoplastic and elastoviscoplastic behavior of solids, and considered spatial rate-dependent and rate-independent response to loading. Chaboche [25,26] presented constitutive relations for isotropic and kinematic hardening (or softening) of metals, with experimental data obtained for stainless steel. Investigating size effects in low-cycle fatigue of solder joints, Gomez and Basaran [9,10] formulated thermodynamic models for isotropic and kinematic hardening, verified with experiments and finite elements. Via simulations and measurements, Basaran et al. [11–13] directly related entropy to damage evolution in solids. Combining Boltzmann's entropy $S = k \ln W$ as a measure of molecular disorder with Prigogine's entropy balance $dS = dS_e + dS\prime$, the authors defined a continuum damage mechanics damage variable

$$D = D_{cr} \frac{W - W_0}{W} = D_{cr} \left[1 - e^{-(m/R)(s - s_0)} \right] \tag{1}$$

similar to Einstein's oscillator energy of a nonmetallic crystalline solid [27]. Equation (1), where D_{cr} = critical disorder coefficient, W = disorder parameter, m = specific mass and R = gas constant, gives damage as a function of specific entropy change

$$s - s_0 = \int_{t_0}^{t} \frac{\sigma : \varepsilon_p}{T\rho} dt + \int_{t}^{t_0} \frac{k}{\rho} \frac{|grad\ T|^2}{T^2} dt + \int_{t_0}^{t} \frac{r}{T} dt. \tag{2}$$

Khonsari, Amiri and Naderi [14,23] related entropy to mechanical fatigue via extensive experiments and data, and proposed fatigue fracture entropy *FFE* as a consistent material property independent of load type, cycle frequency, amplitude or specimen size. Using thermodynamic formulations by Lemaitre and Chaboche [7], Khonsari et al. presented entropy generation rate

$$\dot{S'} = \frac{\dot{W}_p}{T} - \frac{A_k \dot{V}_k}{T} - J_q \frac{grad\ T}{T^2} \geq 0 \tag{3}$$

where the first right-hand side term is the plastic strain entropy from plastic strain energy W_p, the second term is the non-recoverable energy and the third term is heat conduction entropy. Assuming negligible non-recoverable energy and neglecting heat conduction within the specimen, the second and third right side terms were set to zero to give $\dot{S'} = \frac{\dot{W}_p}{T}$. By integrating up to the time of failure t_f, *FFE* was obtained as

$$S'_{TF} = \int_0^{t_f} \frac{W_p}{T} dt. \tag{4}$$

Data from bending and torsional fatigue measurements and Finite Element Analysis validated the constant process-independent, material-dependent *FFE*. Similar to Doelling et al. [28] for wear, the authors showed a linear interdependence between normalized entropy generation and normalized number of cycles as

$$\frac{s_i}{s_g} \approx \frac{N}{N_f} \tag{5}$$

where s_i and s_g are entropies at cycles N and failure N_f, respectively. Results came from over 300 specimens. Through Equation (5), damage accumulation parameter D [29] was related to entropy generation. Naderi and Khonsari [16] applied the approach in reference [15] to variable loading and proposed a universally consistent damage accumulation model. Amiri et al. [18] replaced entropy generation from plastic energy dissipation with entropy transfer out of the loaded specimen via heat. With thermal energy balance, heat transfer out of the specimen into the surroundings was evaluated from measurements of specimen and ambient temperatures during loading via

$$\left(\oint \sigma_{ij} d\varepsilon_{ij} \right) f = \dot{H}_{cd} + \dot{H}_{cv} + \dot{H}_{rd} + \rho c_p \frac{\partial T}{\partial t} + \dot{E}_p \tag{6}$$

where the first three right side terms represent heat transfer via conduction, convection and radiation. The authors described the last two right side terms as variation of internal energy, comprised of temperature-dependent change and a "cold" microstructural change assumed negligible at steady state, to simplify evaluation of entropy flow rate. They reported an uncertainty of 7.8% in their entropy values. Naderi and Khonsari [17] later developed a real-time fatigue monitoring system. With $FFE(\gamma_f)$ as failure parameter and a failure criterion $\gamma \leq 0.9\gamma_f$, failure was consistently predicted with about 10%

error, attributed to the difference between temperature measurement location on the sample and actual failure location. Naderi and Khonsari [19] demonstrated entropy-based fatigue analysis methods more consistent under varying load conditions than stress- and hysteresis energy-based models. Naderi and Khonsari's [20,21] entropy-based fatigue failure indicated stored energy in composite laminates comparable to dissipated heat, leading to the inclusion in their formulations of heat storage entropy and a crack-initiating damage entropy. Using hysteresis energy balance, entropy accumulation was

$$S\prime = \int_0^{tf} \frac{E_{th}}{T} + \int_0^{tf} \frac{E_{diss}}{T} + \int_0^{tf} \frac{E_d}{T} \tag{7}$$

where E_{th} is heat stored, E_{diss} is heat dissipated, and E_d is damage energy. Combining the first two terms of Equation (7) as mechanical entropy, experimental results compared each entropy component to the total entropy.

Russian works selected by Sosnovskiy and Sherbakov in reference [30] described the inadequacies of existing models in characterizing complex damage of tribo-fatigue systems due to simultaneously occurring degradation mechanisms, e.g., sliding friction, fretting, impact, corrosion, heating, etc. Using a cumulative general damage term ω' $(0 < \omega' < 1)$ including mechanical, thermal and electrochemical energy changes, they proposed a tribo-fatigue entropy

$$S'_{TF} = \omega\prime \frac{dW_D}{T} \tag{8}$$

where W_D is the absorbed damage energy at the failure site. Total entropy change summed thermodynamic entropy change and tribo-fatigue entropy, Equation (8), as

$$dS_T = dS + dS_{TF} = \frac{dU}{T} + \frac{\delta W}{T} - \frac{\mu dN'}{T} + \omega\prime \frac{dW_D}{T} \tag{9}$$

where the first right side term is internal energy change, the second term is boundary work, the third is chemical reaction and the fourth is damage. The authors related ω' to normalized time and predicted human death via stress/damage accumulation from birth, depicting an exponential relationship. They presented a human life version of the Wohler (S-N) curve showing a profile similar to metals. Naderi et al.'s Equation (7) and Sosnovskiy et al.'s Equation (9) are equivalent formulations of entropy evolution (with $dN' = 0$ in Equation (9)). Direct comparison shows damage energy $dE_D = \omega\prime dW_D$. Sosnovskiy et al. [31] further expanded and combined the above formulations with continuum damage mechanics to form mechanothermodynamics (MTD). Their data for isothermal fatigue of steel indicated an error of 15%.

Extensive data showed consistency of entropy measurements in estimating mechanical damage and failure in dynamically loaded components. Currently, most fatigue-entropy formulations apply to metal and composite laminate fatigue under mechanical loading only. Via thermodynamic principles and the DEG theorem, this article relates existing fatigue damage measures to instantaneous active process entropies to derive a fatigue model consistent with thermodynamics and natural laws. Data [15,18,32] will verify this DEG approach.

Subsequent sections are as follows:

- Section 2 introduces and reviews the DEG theorem and procedure.
- Section 3 reviews thermodynamics and introduces phenomenological entropy, consisting of a boundary work component and an internal fluctuation component.
- Section 4 couples fatigue analysis to thermodynamics.
- Section 5 uses published experimental data to validate and visualize the model.
- Section 6 discusses results and the models.
- Section 7 summarizes and concludes.

2. Degradation-Entropy Generation Theorem Review

In accordance with Rayleigh's dissipation function of mechanics [33], Onsager's reciprocity theorem in irreversible thermodynamics [34] and Prigogine's dissipative structures [35,36], a quantitative study of degradation of systems by dissipative processes [24] formulated the Degradation-Entropy Generation DEG theorem, establishing a direct relation between material/system degradation and the irreversible entropies produced by the dissipative processes that drive the degradation. Entropy measures disorganization in materials. Since degradation is advanced and permanent disorganization, entropy generation is fundamental to degradation.

2.1. Statement

Given an irreversible material transformation caused by $i = 1,2, \ldots , n$ underlying dissipative processes and characterized by an energy, work, or heat p_i. Assume effects of the mechanism can be described by an appropriately chosen variable

$$w = w(p_i) = w(p_1, p_2, \ldots , p_n),\ i = 1,2,\ldots,n \tag{10}$$

that measures the material transformation and is monotonic in the effects of each p_i. Then the rate of degradation

$$\dot{w} = \sum_i B_i \dot{S'}_i \tag{11}$$

is a linear combination of the rates of the irreversible entropies $\dot{S'}_i$ generated by the dissipative processes p_i, where the degradation/transformation process coefficients

$$B_i = \left.\frac{\partial w}{\partial S'_i}\right|_{p_i} \tag{12}$$

are slopes of degradation w with respect to the irreversible entropy generations $S'_i = S'_i(p_i)$, and the $|_{p_i}$ notation refers to the process p_i being active. The theorem's proof [24] is founded on the second law of thermodynamics. Integrating Equation (11) over time yields the total accumulated degradation

$$w = \sum_i B_i S'_i \tag{13}$$

which is also a linear combination of the accumulated entropies S'_i.

2.2. Generalized Degradation Analysis Procedure

Bryant et al.'s [24] structured DEG theorem-based degradation analysis methodology embeds the physics of the dissipative processes into the energies $p_i = p_i\left(\zeta_{ij}\right),\ j = 1,2,\ldots,m$. Here the p_i can be energy dissipated, work lost, heat transferred, change in thermodynamic energy (internal energy, enthalpy, Helmholtz or Gibbs free energy) or some other functional form of energy, and the ζ_{ij} are time-dependent phenomenological variables (loads, kinematic variables, material variables, etc.) associated with the dissipative processes p_i. The approach

(1) identifies the degradation measure w, dissipative process energies p_i and phenomenological variables ζ_{ij},
(2) finds entropy generation S' caused by the p_i,
(3) evaluates coefficients B_i by measuring increments/accumulation or rates of degradation versus increments/accumulation or rates of entropy generation, with process p_i active.

This approach can solve problems consisting of one or many variegated dissipative processes. Previous applications of the DEG theorem analyzed friction and wear [24,37,38] and metal fatigue [15,18,22,39] grease degradation [32] and battery aging [40].

3. Thermodynamic Formulations

This section reviews the first and second laws of thermodynamics applied to real systems [27,36,41–46].

3.1. First Law—Energy Conservation

The first law

$$dU = \delta Q - \delta W + \sum \mu_k dN_k \tag{14}$$

for a stationary thermodynamic system neglecting gravity, balances dU the change in internal energy, δQ the heat exchange across the system boundary, δW the energy transfer across the system boundary by work, and $\sum \mu_k dN_k$ the internal energy changes due to chemical reactions, mass transport and diffusion, where μ_k are chemical, flow and diffusion potentials, $N_k = N\prime_k + N^e{}_k + N^d{}_k$ are numbers of moles of species k with $N\prime_k$, $N^e{}_k$ and $N^d{}_k$ the reactive/diffusive and transferred species respectively. Inexact differential δ indicates path-dependent variables. For chemical reactions governed by stoichiometric equations, $\sum \mu_k dN_k = Ad\xi$ [36,43,47] where A is reaction affinity and $d\xi$ is reaction extent.

3.2. Second Law and Entropy Balance—Irreversible Entropy Generation

Known as the Clausius inequality, the second law of thermodynamics states: The change in closed system entropy

$$dS \geq \frac{\delta Q}{T}, \tag{15}$$

equal to or greater than the measured entropy transfer across the system boundary via heat. For open systems (having mass flow), the right side of Equation (15) would include a mass transfer term. For a reversible process

$$dS = dS_{rev} = \frac{\delta Q_{rev}}{T} \tag{16}$$

approximates a quasi-static (very slow) process in which total entropy change occurs via reversible heat transfer δQ_{rev}. The second law as the equality $dS = \delta S_e + \delta S'$ [12,34] equates the change in entropy dS to the measured entropy flow δS_e across the system boundaries from heat transfer and/or mass transfer (for open systems), plus any entropy $\delta S'$ produced within the system boundaries by dissipative processes. Entropy generation $\delta S'$ measures the permanent changes in the system when the process constraint is removed or reversed [27,43], allowing the system to evolve. For a closed system [11,33]

$$dS = dS_{irr} = \frac{\delta Q}{T} + \delta S' \tag{17}$$

where dS_{irr} is entropy change via an irreversible (real) path, $\delta Q/T$ is entropy flow by heat transfer which may be positive or negative, and T is the temperature of the boundary where the energy/entropy transfer takes place. The second law also asserts entropy generation $\delta S' \geq 0$.

3.3. Combining First and Second Laws with Helmholtz Potential

For a system undergoing quasi-static heat transfer and compression work, Equation (14) with $\delta Q = \delta Q_{rev} = TdS_{rev}$ from Equation (16) becomes [45]

$$dU = TdS_{rev} - P_{rev}d\mathcal{V} + \sum \mu_{k,rev} dN_k\,. \tag{18}$$

Here P is pressure and $\mathcal{V}$ is volume. Replacing entropy S with temperature T as the independent variable via a Legendre transform results in the Helmholtz free energy

$$A = U - TS, \tag{19}$$

an alternate form of the first law which can measure maximum work obtainable from a thermodynamic system. Differentiating Equation (19) and substituting Equation (18) for dU into the result give the Helmholtz fundamental relation

$$dA = dA_{rev} = -S_{rev}dT - P_{rev}d\mathcal{V} + \sum \mu_{k,rev}dN_k\,, \tag{20}$$

the quasi-static change in Helmholtz energy between two states, valid for all systems. Here $dA = dA_{rev}$ is the free energy change via the reversible (*rev*) path, maximum for energy transfer out of the system and minimum for energy transfer into the system.

Via the thermodynamic State Principle, the change in system energy/entropy due to boundary interactions and/or compositional transformation is path-independent. The change can be determined via reversible (linear) or irreversible (nonlinear) paths between system states. Equality of Equations (16) and (17) is based on this principle. Eliminating δQ from Equation (14) via Equation (17) gives, for compression work $Pd\mathcal{V}$, [36–38,42,43]

$$dU = dU_{irr} = TdS_e - T\delta S' - Pd\mathcal{V} + \sum \mu_k dN_k, \tag{21}$$

where reversible entropy change dS_{rev} was replaced by entropy flow dS_e and entropy generation $\delta S'$. Differentiating Equation (19) and substituting Equation (21) for dU into the result give the irreversible form of the Helmholtz fundamental relation

$$dA = dA_{irr} = -SdT - Pd\mathcal{V} + \sum \mu_k dN_k - T\delta S' \leq 0 \tag{22}$$

where $dA = dA_{irr}$ is the free energy change via irreversible (*irr*) path, maximum for energy transfer out of the system and minimum for energy transfer into the system. Equations (20) and (22) are equivalent representations of total change in Helmholtz free energy of all active systems, and show dA can be evaluated via an idealized change dA_{rev}, or a real spontaneous evolution dA_{irr}. From Equation (22), define phenomenological Helmholtz free energy change

$$dA_{phen} = -SdT - Pd\mathcal{V} + \sum \mu_k dN_k, \tag{23}$$

due only to changes in measurable intensive and extensive properties of a real system. With a known dA_{rev}, Equations (20) and (22) are combined to give

$$\delta S' = -\frac{SdT}{T} - \frac{Pd\mathcal{V}}{T} + \frac{\sum \mu_k dN_k}{T} - \frac{dA_{rev}}{T} \geq 0 \tag{24}$$

which satisfies the second law. During energy extraction or loading, $dT \geq 0, d\mathcal{V} \geq 0, dN_k \leq 0$ and $dA_{rev} \leq 0$, rendering $\delta S' \geq 0$. During energy addition or product forming process, $dT \leq 0, d\mathcal{V} \leq 0$, $dN_k \geq 0$ and $dA_{rev} \geq 0$, reversing the signs of the middle terms in Equation (24) to preserve $\delta S' \geq 0$ [43].

Equation (24) defines entropy generation or production as the difference between phenomenological $\delta S_{phen} = \frac{dA_{phen}}{T} = -\frac{SdT}{T} - \frac{Pd\mathcal{V}}{T} + \frac{\sum \mu_k dN_k}{T}$ and reversible $dS_{rev} = \frac{dA_{rev}}{T}$ entropies

$$\delta S' = \delta S_{phen} - dS_{rev} \geq 0 \tag{25}$$

where for energy extraction $dS_{rev} \leq \delta S_{phen} < 0$, and for energy addition $0 < dS_{rev} \leq \delta S_{phen}$.

Comparing Equations (16) and (17), (20) and (22), verifies that changes in entropy and energy between two states are path-independent, i.e.,

$$dS = dS_{rev} = dS_{irr} = \delta S_{phen} - \delta S';\ dA = dA_{rev} = dA_{irr} = dA_{phen} - T\delta S'. \tag{26}$$

In Equation (26), the change in Helmholtz energy $dA = dA_{rev}$ and entropy $dS = dS_{rev}$, evaluated for a reversible path requires only beginning and end state measurements of system variables. Contrast this for an irreversible path, wherein $dA = dA_{irr} = \delta A_{phen} - T\delta S'$ and $dS = dS_{irr} = \delta S_{phen} - \delta S'$ require instantaneous account of all active processes. Now dA and dS can be negative or positive, depending on energy flow TdS_e or entropy flow dS_e across system boundaries. Since neither dA nor dS measures the permanent changes in the system, this limits success of energy and entropy formulations in characterizing measurable permanent system changes. On the other hand, entropy generation, Equation (24) or (25), evolves monotonically per the second law. With $\delta S' = 0$ indicating an idealized system-process interaction, Equation (25) also indicates that a portion of any real system's energy is always unavailable for external work, $\delta S' > 0$. Equation (25) which gives the entropy generated by the system's internal irreversibilities alone, is in accordance with experience, similar to the Gouy-Stodola theorem of availability (exergy) analysis [44,46,48,49]. The foregoing equations are in accord with the IUPAC convention of positive energy into a system.

3.4. Entropy Content S and Internal Free Energy Dissipation "$-SdT$"

The Helmholtz fundamental relation, Equations (20) and (22), introduced "$-SdT$", free energy dissipated and accumulated internally by a loaded component, which can include effects of plastic work, chemical reaction heat generation and heat from an external source. Temperature change dT is driven by the system entropy content S. Equation (20) suggests Helmholtz-based entropy of a compressible system $S = S(T, \mathcal{V}, N)$ depends on temperature T, volume $\mathcal{V}$ and number of moles N. Via partial derivatives

$$dS = \left(\frac{\partial S}{\partial T}\right)_{\mathcal{V},N} dT + \left(\frac{\partial S}{\partial \mathcal{V}}\right)_{T,N} d\mathcal{V} + \left(\frac{\partial S}{\partial N}\right)_{T,\mathcal{V}} dN. \tag{27}$$

From Maxwell's thermodynamic manipulation of mixed partial second derivatives and Callen's derivatives reduction technique [27], Equation (27) can be re-stated using established and measurable system parameters [27,36]

$$\left(\frac{\partial S}{\partial T}\right)_{\mathcal{V},N} = \frac{C_{\mathcal{V}}}{T};\ \left(\frac{\partial S}{\partial \mathcal{V}}\right)_{T,N} = \left(\frac{\partial P}{\partial T}\right)_{\mathcal{V},N} = \frac{\alpha}{\kappa_T};\ \left(\frac{\partial S}{\partial N}\right)_{T,\mathcal{V}} = -\left(\frac{\partial \mu}{\partial T}\right)_{\mathcal{V},N} \tag{28}$$

where $C_{\mathcal{V}}$ is heat capacity (for solids, $C_P \approx C_{\mathcal{V}} = C$), $\alpha = \frac{1}{\mathcal{V}}\left(\frac{\partial \mathcal{V}}{\partial T}\right)_{P,N}$ is the volumetric coefficient of thermal expansion and $\kappa_T = -\frac{1}{\mathcal{V}}\left(\frac{\partial \mathcal{V}}{\partial P}\right)_{T,N}$ is isothermal compressibility. For a constant-composition system (no independent chemical transformations or phase changes), $\left(\frac{\partial \mu}{\partial T}\right)_{\mathcal{V},N} = 0$, to give

$$dS = \frac{C}{T} dT + \frac{\alpha}{\kappa_T} d\mathcal{V}. \tag{29}$$

Integrating with initial condition $S_0 = 0$ gives entropy content

$$S = C \ln T + \frac{\alpha}{\kappa_T} \mathcal{V} \tag{30}$$

and internal free energy dissipation

$$-SdT = -\left(C \ln T + \frac{\alpha}{\kappa_T} \mathcal{V}\right) dT. \tag{31}$$

4. Differential/Elemental Fatigue Analysis

The foregoing formulations will be applied to a component under cyclic mechanical, thermal and chemical loading [40].

4.1. Local Equilibrium

An extensively verified theorem by Prigogine [35,43,50] hypothesized that every macroscopic system is made up of elemental volumes wherein observable system properties can be instantaneously ascertained, and established equilibrium formulations valid for each elemental volume. If continuity or thermodynamic contact exists between measurement location and the region of interest, the evolution of locally defined state variables can adequately characterize the overall transformation of the component.

4.2. Helmholtz Energy Dissipation and Entropy Generation

Engineering Model: Thermodynamic boundary encompasses system only; loading occurs across system boundary; system is closed; heat transfers with surroundings (system is not isolated). Equation (22) gives the loss of Helmholtz energy in a compressible system. To represent all forms of dynamic loading, thermodynamic boundary work $\delta W = YdX$ replaces compression work $\delta W = Pd\mathcal{V}$. Here Y is generalized constraint/force/load potential, X is generalized response/displacement/loading, $\sum \mu_k dN_k$ (= μdN for a closed system with one reactive component) defines energy loss due to independent chemical processes such as corrosion or radioactive decay, where $dN = \frac{dm}{M_m}$, m is the component's mass and M_m is molecular mass. Equation (24) with generalized boundary loading and active chemical reaction

$$\delta S' = -\frac{SdT}{T} - \frac{YdX}{T} + \frac{\mu dm}{M_m T} - \frac{dA_{rev}}{T} \geq 0 \tag{32}$$

accumulates entropy generation of three simultaneous active processes. Note that derivations involving pressure-volume work in Equation (18) and subsequent Equations such as (27) and (29) originated from the general work term δW in the first law, Equation (14). Reformulating with generalized force-displacement work YdX instead of pressure-volume work $Pd\mathcal{V}$ allows replacement of pressure and volume terms in these formulations, without loss of generality.

Using generalized directional boundary work YX, Equation (30) gives entropy content

$$S = C \ln T + \frac{\alpha}{\kappa_T} X \tag{33}$$

which evolves monotonically in all systems. Note that the assumption of zero initial entropy content S_0 in Equation (33) is considered valid in a new component without defect, for analytical and characterization purposes. The first right side term is entropy from temperature changes (thermal energy storage). The second term emanates from internal changes in structure and configuration. Here generalized system/material properties $C = T\left(\frac{\partial S}{\partial T}\right)_Y > 0$, $\alpha = \frac{1}{X}\left(\frac{\partial X}{\partial T}\right)_Y$ and $\kappa_T = -\frac{1}{X}\left(\frac{\partial X}{\partial Y}\right)_T > 0$ are obtained as in Equation (28). While C and α measure system response to heat and temperature changes, generalized κ_T represents isothermal *loadability*, a measure of the material/component's "cold" response to boundary loading, which for a compressible system is compressibility.

4.3. Stress and Strain as Thermodynamic Variables

Most fatigue damage analyses involve evaluation of the impact of loading on a component. Energy-based formulations often define boundary work (e.g., thermal or mechanical cycling) as a volume integral of stress tensor σ times strain tensor ε with elastic and plastic components $\sigma = \sigma_e + \sigma_p$ and $\varepsilon = \varepsilon_e + \varepsilon_p$. For a non-reactive system undergoing boundary work $\sigma : d\varepsilon$ [7], Equation (31) becomes

$$-SdT = -\left(C \ln T + \frac{\alpha}{\kappa_T} \mathcal{V} \varepsilon\right) dT. \tag{34}$$

To clearly indicate the combined effect of thermal and structural changes due to loading, internal energy dissipation $-SdT$, expressed in terms of measured variables T, σ, ε in Equation (34), is named MicroStructuroThermal (MST) energy dissipation [32]. Here $\kappa_T = \frac{\partial \varepsilon_e}{\partial \sigma}$ is the isothermal *strainability* where ε_e is elastic strain and σ is stress. Similar to application in compression work, κ_T can be evaluated via the inverse of elastic or torsional modulus for normal or torsional loading. Torsional and frictional loads are described using shear stress τ and shear strain γ tensors. Similar terms as in Equation (34) were derived by Morris [51].

4.3.1. Cyclic Loading—High-and Low-Cycle Fatigue

Elastoplastic strain response to tensile stress is often modeled via the Ramberg-Osgood relation [52]: $\varepsilon = \frac{\sigma}{E} + K\left(\frac{\sigma}{E}\right)^n$. Fatigue failure results from dynamic loading. Fatigue measurements determine strain response to stress-controlled loading or stress response to strain-controlled loading. For stress-or strain-controlled cyclic loading, Morrow [53] experimentally showed that the corresponding strain or stress amplitude and strain energy are nearly constant throughout, except for the first few cycles, and last cycles before failure [7]. In systems subject to fatigue failure (high- and low-cycle fatigue HCF and LCF), the plastic component of the response to loading is significant (predominant in LCF), especially at critical locations on the system. To account for elastic and plastic loads, cyclic strain amplitude as a function of applied stress amplitude is [53] $\varepsilon_a = \frac{\sigma_a}{E} + \varepsilon\prime_f\left(\frac{\sigma_a}{\sigma\prime_f}\right)^{1/n\prime}$ where the first right side term is elastic strain and the second is plastic strain. Via the Coffin-Manson relation, this can be restated as [54–56]

$$\varepsilon_a = \frac{\sigma\prime_f}{E}\left(2N_f\right)^b + \varepsilon\prime_f\left(2N_f\right)^c \tag{35}$$

where N_f is the number of cycles to failure and $2N_f$ is the number of strain reversals. Here b and c are fatigue strength and ductility exponents. Cyclic elastic strain energy density $W_e = \sigma_N : \varepsilon_{eN}$ is often negligible in very low cycle failure [14–23,53]. Cyclic plastic strain energy density was given by Morrow [53] as

$$W_p = \sigma_N : \varepsilon_{pN}\left(\frac{1-n\prime}{1+n\prime}\right) \tag{36}$$

where n' is the cyclic strain hardening coefficient. With units J/m^3 equivalent to Pa, energy density is often described in mechanics as toughness [53]. Combining with cyclic elastic work gives the total cyclic boundary work or strain energy density

$$W = W_e + W_p = \sigma_N : \left[\varepsilon_{eN} + \varepsilon_{pN}\left(\frac{1-n\prime}{1+n\prime}\right)\right]. \tag{37}$$

For cyclic loading conditions, differential cyclic time or period [57]

$$dt_N = \frac{dt}{N_{dt}} = \frac{1}{h} \tag{38}$$

where h is the load cycle frequency and N_{dt} is the number of cycles in time increment dt. Fatigue loads are often defined per cycle as sinusoids with stress/strain amplitude or range per cycle. Here dt is replaced by $N_{dt}dt_N$ in integrals, such as upcoming Equation (47), for convenience and compatibility with differential thermodynamic formulations such as Equation (32), as done by Meneghetti [57] and Morris [51]. The measurement time step dt is often greater than dt_N when measuring phenomenological variables or parameters such as temperature, loads, etc. Entropy accumulates over cyclic loads. Via Equations (37) and (38), cyclic stress range $\sigma_N = \int_{t_N}^{t_{N+1}} \dot{\sigma} dt_N$ or $\dot{\sigma} = d\sigma_N/dt_N$ and cyclic strain range $\varepsilon_N = \int_{t_N}^{t_{N+1}} \dot{\varepsilon}_e dt_N + \int_{t_N}^{t_{N+1}} \dot{\varepsilon}_p dt_N$ together give the differential work density

$$\delta W_N = \sigma_N : \left[d\varepsilon_{eN} + \left(\frac{1-n'}{1+n'}\right)d\varepsilon_{pN}\right]. \tag{39}$$

Using Equation (38), boundary work done during time increment dt is

$$\delta W = N_{dt}\delta W_N = N_{dt}\sigma_N : \left[d\varepsilon_{eN} + \left(\frac{1-n\prime}{1+n\prime}\right)d\varepsilon_{pN}\right]. \tag{40}$$

Total strain accumulation over dt is

$$\varepsilon = \int_0^t (d\varepsilon_N / dt_N)dt. \tag{41}$$

Dividing Equation (34) by volume $\mathcal{V}$ and combining with Equation (40) gives the change in Helmholtz energy density or toughness under high- or low-cycle fatigue loading. For stress-controlled loading, i.e., constant σ_N, and constant N_{dt}, Helmholtz energy dissipation density

$$dA = -\Big(\rho c \ln T + \frac{\alpha}{\kappa_T}\varepsilon\Big)dT - N_{dt}\sigma_N : \Big[d\varepsilon_{eN} + \Big(\frac{1-n\prime}{1+n\prime}\Big)d\varepsilon_{pN}\Big] \tag{42}$$

and Helmholtz entropy generation density

$$\delta S' = -\Big(\rho c \ln T + \frac{\alpha}{\kappa_T}\varepsilon\Big)\frac{dT}{T} - N_{dt}\frac{\sigma_N}{T} : \Big[d\varepsilon_{eN} + \Big(\frac{1-n'}{1+n'}\Big)d\varepsilon_{pN}\Big] + \frac{\sigma'_f : d(\sigma'_f/E)}{T}. \tag{43}$$

For strain-controlled loading, σ and ε are interchanged. When available, measurements of stress/strain response to loading should be used in place of Equations (35) and (36), which assume constant cyclic strain and strain energy. In Equation (43), the first term is the elemental microstructurothermal MST entropy density $\delta S'_{\mu T}$ characterizing internal material-dependent dissipation, the second is the boundary loading term $\delta S'_W$ characterizing energy dissipation across the system boundary via useful work output and environmental conditions, and the third is the reversible entropy S'_{rev} defined using the component's fatigue strength coefficient $\sigma\prime_f$. From Equation (42), MST energy density change $\delta A_{\mu T} = -\big(\rho c \ln T + \frac{\alpha}{\kappa_T}\varepsilon\big)dT$ and boundary work density $\delta A_W = -N_{dt}\sigma_N : \big[d\varepsilon_{eN} + \big(\frac{1-n\prime}{1+n\prime}\big)d\varepsilon_{pN}\big]$.

In renewable energy systems, the maximum work obtainable from a system, its Helmholtz free energy change dA_{rev} or Gibbs free energy change dG_{rev} may be defined cyclically. In all other systems $\int_{t_0}^t dA_{rev}dt = \Delta A_{rev}$ is constant and defined globally at manufacture as the maximum energy in the system or component from its newly manufactured state to full degradation, or locally just before onset of loading as the maximum energy change in the system/component before and after loading. This term is relatively inactive in the characteristic path-dependent evolution of entropy generation [58]. Neglecting the constant (between 2 states) reversible term in Equation (43) as in Prigogine et al.'s irreversible entropy generation formulations for active process/work interactions [42,43], phenomenological entropy generation or production in a mechanically loaded system is given as

$$\delta S'_{phen} = -\Big(\rho c \ln T + \frac{\alpha}{\kappa_T}\varepsilon\Big)\frac{dT}{T} - N_{dt}\frac{\sigma_N}{T} : \Big[d\varepsilon_{eN} + \Big(\frac{1-n'}{1+n'}\Big)d\varepsilon_{pN}\Big]. \tag{44}$$

The above considers a loading rate h different from sampling rate $1/dt$. If cyclic loading and data sampling rates are the same, $N_{dt} = 1$. Similar expressions can be obtained for shear stress τ and shear strain γ, for torsion.

4.3.2. Infinite Life Design

In infinite life design, loading and material behavior are predominantly in the elastic region, hence elastic formulations are reliable [4–6]. The Wohler (S-N) curve and the Goodman diagram show the region below the fatigue limit in which certain materials may be loaded indefinitely without failure. Others such as the Soderbeg criteria are based on the component's elastic response. For bending, normal strain $\varepsilon_e = \frac{\sigma}{E}$. For torsion, shear strain $\gamma_e = \frac{\tau}{G}$. For simultaneous loads such as combined bending and torsion, von Mises formulations can be used. Predominant elastic interactions are nearly isothermal, so the Helmholtz energy density change from Equation (42) with $d\varepsilon_{pN} = 0$ becomes

$$dA = N_{dt}(\sigma_N : d\varepsilon_{eN}), \tag{45}$$

and phenomenological Helmholtz entropy generation density from Equation (44)

$$\delta S'_{phen} = -\frac{1}{T}N_{dt}(\sigma_N : d\varepsilon_{eN}). \tag{46}$$

Equation (46) is the minimum entropy generation in a dynamically loaded system (in terms of stress and strain) defined by Prigogine's stationary non-equilibrium theorem [43]. At the reversibility limit or for a fully reversible (elastic) system—which would imply a "true" infinite life design—boundary temperature T is constant, giving uniform $\delta S'_{phen}$. Metals such as steel exhibit nearly reversible characteristics (infinite life) when loaded below fatigue limits [2–7]. Equation (46) also applies to isothermal loading conditions.

4.4. Degradation-Entropy Generation (DEG) Analysis

Rewriting Equations (23) and (24) in rate form without the compositional change term, and integrating over time gives the total change in Helmholtz energy from t_0 to t as $\Delta A = -\int_{t_0}^{t} S\dot{T}dt - \int_{t_0}^{t} Y\dot{X}dt$, and phenomenological entropy generation as

$$S'_{phen} = -\int_{t_0}^{t} \frac{S\dot{T}}{T}dt - \int_{t_0}^{t} \frac{Y\dot{X}}{T}dt. \tag{47}$$

Via the DEG formulations in Section 2, system degradation measured by fatigue parameter w is directly related to phenomenological entropy generation as

$$w = B_{\mu T}\int_{t_0}^{t} -\frac{S\dot{T}}{T}dt + B_W\int_{t_0}^{t} -\frac{Y\dot{X}}{T}dt = B_{\mu T}S'_{\mu T} + B_W S'_W. \tag{48}$$

Via Equation (12), DEG coefficients

$$B_{\mu T} = \frac{\partial w}{\partial S'_{\mu T}};\; B_W = \frac{\partial w}{\partial S'_W} \tag{49}$$

which pertain to MST entropy $S'_{\mu T} = \int -\frac{S\dot{T}}{T}dt$ and boundary work entropy $S'_W = \int -\frac{Y\dot{X}}{T}dt$, respectively, can be evaluated from measurements of slopes of w versus entropy production components S'_i.

4.4.1. Applying the Degradation-Entropy Generation Theorem to Cumulative Strain (or Stress)

Assuming the cyclic effects of measured strain are cumulative (to account for all simultaneous variable and complex loading) and vary with strain intensity, a strain measure may be defined for the DEG theorem (using Equation (43) for S'_{phen}) as

$$\varepsilon = \int_{t_0}^{t} \dot{\varepsilon}dt = -B_{\mu T}\int_{t_0}^{t}\left(C\ln T + \frac{\varepsilon\alpha}{\kappa_T}\right)\frac{\dot{T}}{T}dt + B_W\int_{t_0}^{t} N_{dt}\frac{\sigma_N}{T} : \left[\dot{\varepsilon}_{eN} + \left(\frac{1-n'}{1+n'}\right)\dot{\varepsilon}_{pN}\right]dt. \tag{50}$$

For truly infinite life and assuming elastic work

$$\varepsilon = B_{W_e}\frac{\sigma_N}{T}\varepsilon_e\,. \tag{51}$$

If loading is strain-controlled, the measured stress response may become a cumulative degradation measure and similar relations developed.

5. Fatigue Experiments and Data Analysis—Instantaneous Characterization

Low-cycle fatigue data by Naderi, Amiri and Khonsari [15,18] will verify formulations. Details about equipment, procedures and data are in references [15,18]. Briefly, at sampling frequency 7.5 Hz, a high-resolution infra-red camera monitored temperature profiles of the SS 304 stainless steel fatigue specimen depicted in Figure 1, with material properties in Table 1.

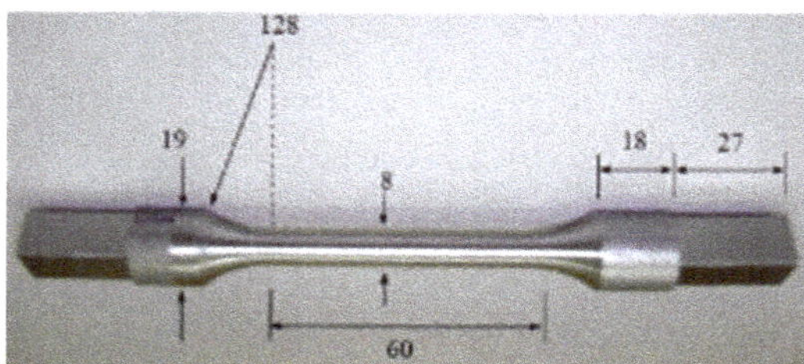

Figure 1. Torsion fatigue-tested steel sample SS 304 showing dimensions in mm, reproduced from [14].

Table 1. Material properties for SS 304 steel used in evaluating loading parameters [2,15,55,56].

Property	Bending	Torsion
Modulus, GPa	$E = 195$	$G = 82.8$
Fatigue strength coefficient, MPa	$\sigma\prime_f = 1000$	$\tau\prime_f = 709$
Fatigue strength exponent b	−0.114	−0.121
Fatigue ductility coefficient	$\varepsilon\prime_f = 0.171$	$\gamma\prime_f = 0.413$
Fatigue ductility exponent c	−0.402	−0.353
Cyclic strain hardening exponent n'	0.287	0.296
Specific heat capacity C, J/kg K	500	
Density ρ, kg/m^3	7900	
Coefficient of linear thermal expansion α	17.3×10^{-6}	

Displacement-controlled bending and torsional loads oscillated at 10 Hz. Plots in the upcoming figures, generated from Naderi et al.'s data, have "*a*" subfigures on the left pertaining to bending fatigue, and "*b*" subfigures on the right pertaining to torsional fatigue. Signs follow the thermodynamic convention of the formulations, e.g., boundary loading and MST energies and entropies are negative.

Figure 2a plots the constant cyclic stress amplitude obtained from $\sigma_a = \sigma\prime_f\left(2N_f\right)^b$, constant elastic strain amplitude from Hooke's law $\varepsilon_{ea} = \frac{\sigma_a}{E}$, constant plastic strain amplitude from Morrow's relation [53] $\varepsilon_{pa} = \varepsilon\prime_f\left(\frac{\sigma_a}{\sigma\prime_f}\right)^{1/n\prime}$ and measured temperature T versus number of cycles N. Torsional loading in part (b) of the figures employs shear stress τ and shear strain γ. In the rest of this article σ and ε will denote generalized stress and strain. Number of cycles accumulated at failure was N_f = 14,160 for bending, N_f = 16,010 for torsion [15].

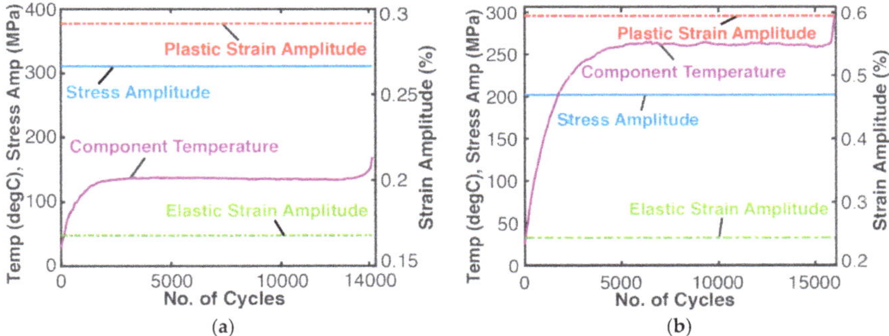

Figure 2. Parameters during cyclic (**a**) bending and (**b**) torsional fatigue of the SS 304 steel at a constant frequency 10 Hz and displacement loading δ = 45.72 mm and δ = 33.02 mm [15]. Temperatures and cyclic stress amplitude are on the left axis, and cyclic strain amplitude is on the right.

For bending, Figure 2a shows a constant normal stress amplitude σ_a = 311 MPa, a steady normal elastic strain amplitude ε_{ea} = 0.17% and steady normal plastic strain amplitude ε_{pa} = 0.29%. For torsion, Figure 2b shows a constant shear stress amplitude τ_a = 202 MPa, a steady elastic shear strain amplitude γ_{ea} = 0.24% and steady shear plastic strain amplitude γ_{pa} = 0.59% (this last value is high due to the high torsional fatigue ductility coefficient γ'_f found in literature, see Table 1). In both cases, a steep rise in temperature (purple curves in Figure 2) arose from high hysteresis dissipation from an initial rest state. After this initially transient response region (about 2000 cycles for bending and 5000 for torsion), pseudo-steady state temperature persists until a sudden rise occurs, followed by fatigue failure [14,15]. Substituting Naderi et al.'s data into Equations (42), (43) and (50), Table 2 was constructed. Units of %N, GJ/m^3 and MPa/K are used for cumulative strain, energy density and entropy density respectively (1 GPa = 1 GJ/m^3; 1 MPa/K = 1 MJ/m^{3}K) giving strain-based B coefficient units of %NK/MPa.

Table 2. Helmholtz energy-based DEG fatigue analysis results for bending and torsional loading to failure of the SS 304 steel specimen in Figure 1.

Load	ε_f, γ_f %N	A_W GJ/m^3	$A_{\mu T}$ GJ/m^3	S'_W MPa/K	$S'_{\mu T}$ MPa/K	B_W %NK/MPa	$B_{\mu T}$ %NK/MPa
Bending	130.1	−58.0	−7.8	−143.5	−18.8	−0.92	0.22
Torsion	268.5	−73.4	−12.3	−143.5	−24.1	−1.96	0.42

Table 2 column 1 lists fatigue loading types, bending and torsion. Section 4 formulations involved integrals over time. Trapezoidal quadratures with widths inverse to the data sampling frequency (7.5 Hz [15]) estimated time integrals. For a process occurring from t_0 to t, cumulative strain in Equation (41), Table 2 column 2, was estimated as

$$\varepsilon = \int_0^t \dot{\varepsilon} dt = \int_{t_0}^t (d\varepsilon_N / dt_N) dt \approx \left(\frac{1}{\Delta t_N}\right) \sum_1^m (\varepsilon_m) \Delta t = N_{\Delta t} \sum_1^m (\varepsilon_m) \tag{52}$$

where indices 1, 2, 3, ..., m correspond to times t_1, t_2, t_3, ..., t_m and $\Delta t = t_m - t_{m-1}$, period $\Delta t_N = 1/10$ [15], data sampling time increment $\Delta t = 1/7.5$, and total number of cycles within sampling time increment $N_{\Delta t} = 10/7.5$, see Equation (38). Finally, ε_m is strain range at t_m. Shear strain γ was similarly obtained for torsion. Via constant cyclic strain ranges [53] ε_N and γ_N, cumulative strains varied linearly with number of cycles N until sudden failure, with no indication of failure onset (Figure 3).

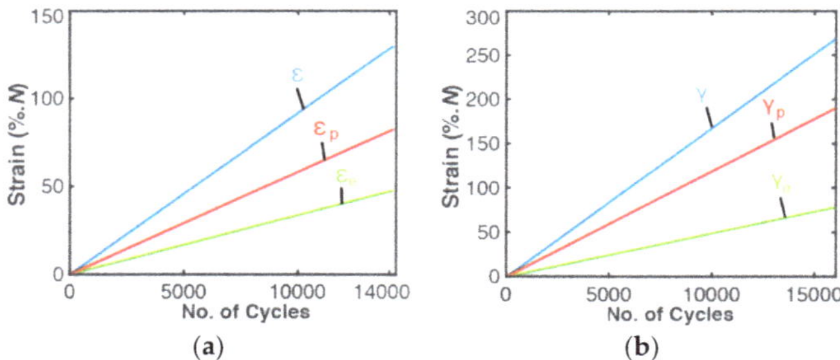

Figure 3. Cumulative strains—elastic (green), plastic (red) and total (blue) vs number of load cycles N for (**a**) bending—normal strain ε; (**b**) torsion—shear strain.

5.1. Instantaneous Evolution of Helmholtz Energy Density (Toughness) and Entropy Density

Table 2 lists components of Helmholtz toughness, Equation (40), $A_W = -N_{\Delta t}\sum_1^m \left(\sigma_m\left[\varepsilon_{em} + \varepsilon_{pm}\left(\frac{1-n\prime}{1+n\prime}\right)\right]\right)$ (column 3) and $A_{\mu T} = -\sum_1^m\left(\rho c \ln T_m + \frac{\alpha}{\kappa_T}\varepsilon_m\right)\Delta T_m$ (column 4) during bending and torsional fatigue of the steel member. Figure 4 plots the accumulated boundary/load (blue curves) and MST (red curves) entropy densities. In Figure 4, a near linear relationship is observed between load entropy, column 5 of Table 2,

$$S\prime_W = \int_0^t \frac{\sigma\dot{\varepsilon}}{T}dt = N_{\Delta t}\sum_1^m\left\{\frac{\sigma_m}{T_m}\left[\varepsilon_{em} + \varepsilon_{pm}\left(\frac{1-n\prime}{1+n\prime}\right)\right]\right\} \tag{53}$$

and accumulated strain for the assumed constant stress amplitude loading and constant strain amplitude response, with a slight curvature from the initial temperature rise (Figure 4). Table 2 shows the same failure value of 143.5 MPa/K for both bending and torsion, as previously observed by Naderi, Amiri and Khonsari [15,18–20], unlike load (strain) energy density A_W. MST entropy density (red curves), column 6,

$$S\prime_{\mu T} = \int_0^t -\left(\rho c \ln T + \frac{\alpha}{\kappa_T}\varepsilon\right)\frac{\dot{T}}{T}dt = -\sum_1^m\left(\rho c \ln T_m + \frac{\alpha}{\kappa_T}\varepsilon_m\right)\frac{\Delta T_m}{T_m} \tag{54}$$

shows a profile significantly influenced by the measured temperature profile but less steep than the latter due to the microstructural effect (second right side term in Equation (54), see Figure 4). Accurate determination of MST entropy includes effects of instantaneous temperature, especially for anisothermal conditions. Amiri and Khonsari [14] related fatigue life to the gradient of the initial temperature rise. Both MST energy and entropy densities are higher for torsion than bending. At every instant, load entropy $S\prime_W$ and an accompanying MST entropy $S\prime_{\mu T}$ are produced, both at the instantaneous boundary temperature. Figure 4 shows that with $S\prime_{\mu T}$ stabilizing with steady temperature, $S\prime_W$ quickly becomes more significant to total irreversible entropy, a desired feature (the boundary loading is the component's output work, hence the higher its contribution to total phenomenological entropy, the more optimal the component's response to loading). However, the sudden rise in magnitude of $S\prime_{\mu T}$ just before failure is not evident in load (boundary work) entropy.

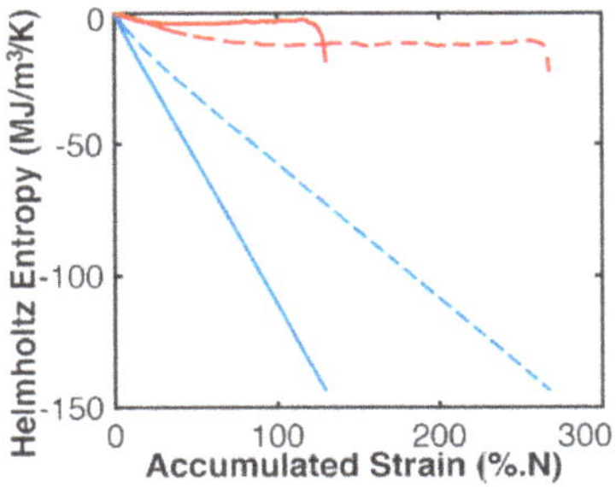

Figure 4. Phenomenological Helmholtz entropy density components—load entropy (blue plots) and MST entropy (red plots)—versus accumulated strain during bending (continuous curves) and torsion loading (dashed curves). Note 1 MJ/m^3/K = 1 MPa/K.

Figure 5 plots rates of phenomenological Helmholtz entropy generation components—load and MST entropies—versus number of cycles. Cyclic load entropy (blue curves) starts at a slightly higher rate and quickly steadies as quasi-steady temperature is reached. MST entropy rate (red curves in Figure 5, right axes label) shows more significant fluctuations with sudden discontinuity (large spike) just before failure. With measured non-constant strain response using appropriate equipment (particularly for variable and complex load types), the boundary work/load entropy characteristics could differ from those presented here in which constant stress and strain amplitudes were used, as often done in fatigue analysis [15,53–55].

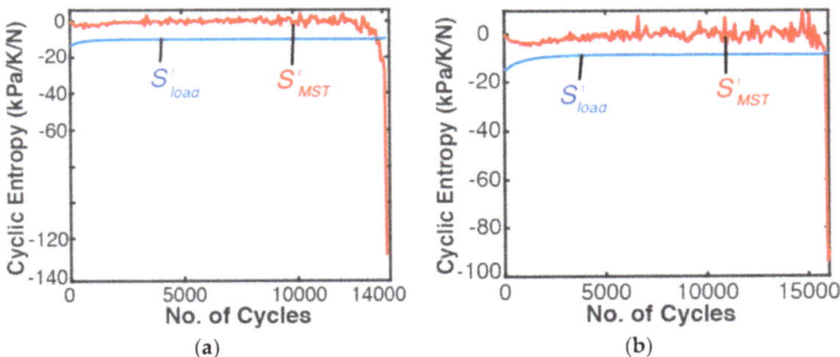

Figure 5. Cyclic phenomenological entropy generation components—load (blue) and MST (red) entropies—versus number of cycles N for (**a**) bending, (**b**) torsion of the SS 304 steel specimen.

5.2. DEG Analysis—Strain Versus Entropy (Linear Transformation)

By associating data from various time instants, accumulated strain ε from Equation (41) was plotted versus accumulated entropies $S\prime_W$ and $S\prime_{\mu T}$ in 3-dimensional Figure 6. Time is a parameter along curves: successive points from bottom to top on each curve correspond to later times along the fatigue evolution. Coincidence of measured data points with planar surfaces in Figure 6 has goodness of fit $R^2 = 1$, asserting a statistically perfect fit for all cases prior to impending failure. The end views emphasize the coincidence of points with the planes. This suggests a linear dependence of degradation/fatigue on both the actual output work/boundary loading and MST entropies at every instant of loading. The measured data points in the curves of Figure 6 that define the component's paths during loading—its Degradation-Entropy Generation (DEG) trajectories—lie on planar DEG surfaces. The orthogonal 3D space occupied by the DEG surfaces, the component's material-dependent DEG domain, appears to characterize the allowable regime in which the component can be loaded.

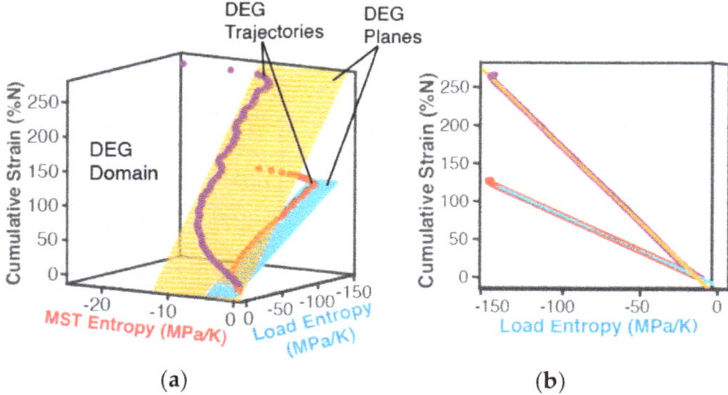

Figure 6. 3D plots and linear surface fits of cumulative strain vs load entropy and MST entropy during cyclic bending (red points, blue plane) and torsion (purple points, orange plane) of SS 304 steel sample, showing in (**b**) a goodness of fit of $R^2 = 1$, indicating a linear dependence on the 2 active processes. In (**a**), loading trajectories start from lowest corner. (Axes are not to scale and colors are for visual purposes only).

The dimensions of the DEG planes are determined by the accumulation of the entropy generation components before failure onset. As previously observed, bending and torsion have the same boundary work entropy dimension, indicating that this dimension is characteristic of the specimen material, not the process, further verifying Naderi, Amiri and Khonsari [15,18–20]. Overall, A_W and $S\prime_W$ are about 7 (6 for torsion) times $A_{\mu T}$ and $S\prime_{\mu T}$, respectively. MicroStructuroThermal (MST) dissipation accompanies boundary interaction/loading. Figure 6b also shows points of the trajectory not lying on the DEG plane. These points violate the linearity of Equation (50), suggesting another fundamentally different dissipative process at work. The pseudo-constant temperature region (see Figure 2) appears in the DEG domain as a pseudo-constant MST region, with fluctuations.

Degradation Coefficients B_i: Degradation coefficients B_W and $B_{\mu T}$, partial derivatives of fatigue measure—cumulative strain—with respect to loading and MST entropies respectively, Equation (49), were estimated from the orientations of the surfaces in Figure 6, see columns 7 and 8 of Table 2. For bending, $B_W = -0.92$ %K/MPa and $B_{\mu T} = 0.22$ %K/MPa, and for torsion, $B_W = -1.96$ %K/MPa and $B_{\mu T} = 0.42$ %K/MPa. A lower value for B implies lesser impact on fatigue degradation.

5.3. Phenomenological Transformation Versus Measured/Estimated Fatigue Parameter

Using constant B coefficients given in Table 2, instantaneous entropy transformations were projected onto the estimated fatigue or degradation parameter to determine phenomenological fatigue parameter, analogous to the previously defined phenomenological entropy generation. Figure 7a,c show reversible Helmholtz entropy $S\prime_{rev}$ (green curves), phenomenological entropy $S\prime_{phen}$ (purple curves) and boundary work/load entropy $S\prime_W$ (blue curves) during bending and torsion of the steel sample. In Figure 7b,d, DEG-evaluated phenomenological strains ε_{phen} and γ_{phen} (purple curves) and estimated strains $\varepsilon = \varepsilon_e + \varepsilon_p$ and $\gamma = \gamma_e + \gamma_p$ (blue curves) are plotted. The actual transient response of the component under load is unobservable in cyclic strains ε and γ estimated from currently available LCF analysis methods. The DEG methodology, via entropy which uses a component's instantaneous temperature, introduces more representative cyclic strains ε_{phen} and γ_{phen} which consistently show all instantaneous nonlinear transitions during loading including the initially high energy dissipation rate observable in Figure 7b,d.

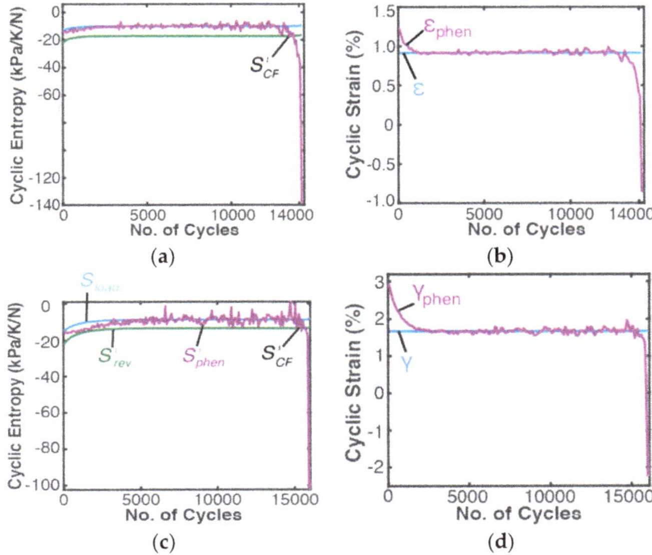

Figure 7. Cyclic entropy generation—load (blue), phenomenological (purple) and reversible (green)—as well as corresponding cyclic strain—estimated constant (blue) and phenomenological (purple)—during bending (**a**,**b**) and torsion (**c**,**d**) of the steel specimen. Region between S'_{phen} and S'_{rev} is entropy generation S' given by Equation (25). A similar critical failure entropy S'_{CF} is shown for both loading types.

Substituting coefficient values into Equation (48) gives the SS 304 steel sample's DEG cumulative strain-based fatigue life/degradation models for bending and torsion

$$\varepsilon_{phen} = \left(0.22S'_{\mu T} - 0.92S'_{W}\right) * 10^{-6} \tag{55}$$

$$\gamma_{phen} = \left(0.42S'_{\mu T} - 1.96S'_{W}\right) * 10^{-6}, \tag{56}$$

which linearly relate the phenomenological fatigue strains ε_{phen} and γ_{phen} to the phenomenological entropies $S'_{phen} = S'_{W} + S'_{\mu T}$ produced. Via the known relations between entropy production and the active variables of loads, materials and environment, Equations (55) and (56), in turn, relate the fatigue strains to the phenomenological variables.

Critical Failure Entropy S'_{CF}—MST Entropy and Fatigue Failure

A corollary of the DEG theorem: "if a critical value of degradation measure at which failure occurs exists, there must also exist critical values of accumulated irreversible entropies" [24]. Naderi, Amiri and Khonsari's extensive measurements [15,18–20] showed existence of a material-dependent fatigue fracture entropy *FFE* or S'_f evaluated as the load entropy (using constant plastic strain amplitude) accumulated at failure. The data of this article, obtained from references [15,18], verified similar magnitudes of cumulative S'_W for both bending and torsion of the SS 304 steel specimen. To anticipate onset of failure, Khonsari et al. empirically determined a normalized onset of failure entropy criterion $\frac{S'}{S'_f} \leq 0.9$ from several temperature profiles measured during loading [17]. Other common fatigue tools like σ—N and ε—N curves, with constant stress and strain amplitudes, do not exhibit the critical phenomenon. The DEG domain shows a distinct and consistent critical onset of failure. In Figure 7a,c, the abrupt drop in phenomenological Helmholtz entropy generation just before failure is attributed to the sudden rise in specimen temperature. Via the B coefficients, this abrupt drop is transferred to phenomenological strain, Figure 7b,d, introducing the critical feature to the hitherto steady fatigue measure, cumulative strain.

To understand the entropy generation critical value, reexamine Figure 7. The region between the reversible entropy $S\prime_{rev}$ and phenomenological entropy S_{phen} curves—the subtraction difference—is entropy generation. With the stable evolution criterion $S_{rev} \leq S_{phen} < 0$, the abrupt spike in S_{phen} resulted in the second law-prohibited negative entropy generation of Equations (25) and (43). The intersection of S_{rev} and S_{phen} marks the critical failure entropy $S\prime_{CF}$ (Figure 7a,c). With constant cyclic stress and strain, the cyclic load entropy (blue plots in Figure 7a,c) trends directly with measured temperature (Figure 2), accumulating linearly over time (Figure 4). Comparing Figures 5 and 7 shows that the downward spike in cyclic S_{phen}, also observed as the trajectory discontinuity in the DEG domains, is introduced by the microstructurothermal (MST) entropy composed of a thermal change- and microstructural change-induced internal entropy generation. If a pseudo-steady temperature was not attained, the MST entropies would have risen continuously and accelerated failures. Note that the initial temperature rise is less for bending fatigue than torsion [15], Figure 2, the effect of which is evident in the MST dimensions of the respective DEG planes. Hence, MST entropy measures a component's instantaneous instabilities and ultimate failure. In other forms of loading including thermal and chemical cycling of components, the significance of MST entropy is underscored by the limited safe operating temperature ranges specified by device manufacturers to prevent instabilities/runaway events.

5.4. Nonlinear Response

Via Morrow [53] and Lemaitre and Chaboche [7], this article assumed a constant cyclic strain response to constant stress loading, similar to Khonsari et al. However, for variable and complex asynchronous loading, a nonlinear response is typically observed.

6. Discussion and Contributions

Other experimental verification of the DEG methodology include nonlinear shear stress response to shear rate-controlled shearing of lubricant grease [32], (Figure 8), and abusive cycling of Li-ion batteries [40] have been demonstrated by Osara and Bryant. In Figure 8b, the DEG trajectories—independent datasets measured at different times and durations—all lie on the same DEG plane, characteristic of the grease.

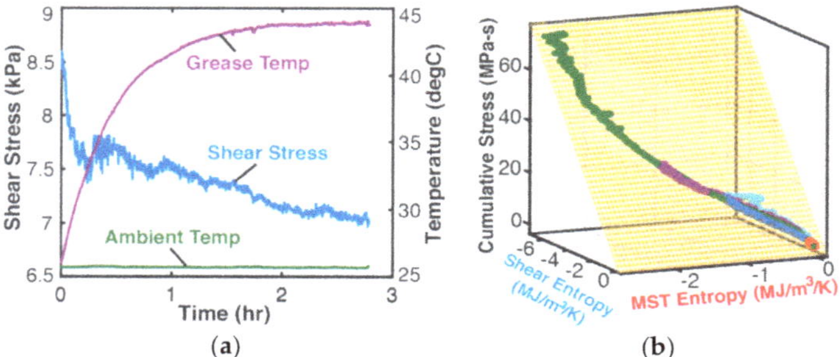

Figure 8. (**a**) Monitored parameters—shear stress and temperatures—and (**b**) DEG domain for mechanical shearing of high-consistency lubricant grease show multiple nonlinear shear stress trajectories coincident with the same DEG plane. Reproduced from [32].

Similar to Prigogine's successful extension of hitherto reversible thermodynamic formulations to irreversible and non-equilibrium processes and states [35,36,42,43], this study derived and verified a consistent utility-based, time-dependent system entropy generation. Based on Gibbs theory of thermodynamic stability of equilibrium states and the second law entropy balance, this article demonstrated that

- phenomenological entropy generation $S\prime_{phen}$ is the sum of boundary work/load entropy $S\prime_{W}$ and microstructurothermal MST entropy $S\prime_{MST}$;
- entropy generation is the difference between phenomenological $S\prime_{phen}$ and reversible $S\prime_{rev}$ Helmholtz entropies at every instant;
- entropy generation is always non-negative in accordance with the second law, whereas components $S\prime_{phen}$ and $S\prime_{rev}$ are directional, negative for a loaded system. This implies $|S'_{phen}| < |S'_{rev}|$ during load application in accordance with experience and thermodynamic laws. The actual work obtained from the system is always less than the maximum/reversible work.

Stress and strain (bending and torsional) were used as system conjugate variables to characterize energy dissipation and entropy generation in a loaded metal bar.

6.1. *Features of the DEG Methodology*

Basaran et al. [9–13] and Khonsari et al. [14–23] in several fatigue-entropy works demonstrated the robustness, consistency and ease of use of entropy generation-based damage/fatigue analysis. This article showed that the DEG methodology relates accumulated irreversibilities to the resulting damage in systems using entropy generation components. DEG theorem methods can accurately describe a system's fatigue level during operation in a fatigue measure versus entropy generation components space. Since the entropy generation depends on the load, materials and environment, the DEG methods in turn relate a material's fatigue measure to the working phenomenological variables of interest.

6.1.1. DEG Trajectories, Surfaces and Domains

Thermodynamics authors have consistently used multi-dimensional orthogonal spaces to describe thermodynamic states of reversible processes: Callen's thermodynamic configuration space [36], Messerle's energy surface [47] and Burghardt's equilibrium surface [41]. This study introduced the DEG domain, a multi-dimensional space that linearly characterizes a real system's nonlinear phenomenological transformation paths. Proper formulation of the governing entropies from the active dissipative processes is required to accurately determine fatigue degradation during loading.

DEG trajectories characterize loading conditions (torsion, bending, stress/strain amplitudes, etc.); DEG surfaces appear to characterize component material and process rates; and the DEG domain seems to define the normal operating/aging region and the failure region, fully characterizing the component's life for all loads and process rates. A component having a DEG domain with large accumulated fatigue measure span and small MST entropy span (relative to load entropy dimension) will accumulate more load strain (or do more work) before failure. Hence, the DEG fatigue methodology can directly compare designs and materials for manufacture and applications.

The out-of-plane points at the termini of the DEG trajectories of Figure 6 occurred at the onset of failure. Here, a crack in the fatigued specimen attains a critical length, which causes a catastrophic fracture crack growth that ruptures the specimen [56,59]. Fracture cracking as opposed to fatigue cracking involves fundamentally different dissipative processes and entropy generation [60]. The DEG model could add this effect via an additional term in Equation (50) for fracture entropy generation, similar to the fracture entropy formulated by Rice [61]. This third orthogonal entropy generation axis in Figure 6 would extend the plots to 4D: cumulative strain vs. load entropy, MST entropy and fracture entropy. Via the thermodynamic state principle [45] and the DEG theorem, other concurrent independent processes would append additional dimensions to the DEG domain.

6.1.2. DEG Coefficients

Unlike existing fatigue methods wherein stress-life and strain-life diagrams predict suitability of a component using extensive data from several failed samples, DEG coefficients can be obtained from one or two representative samples and applied to other components of the same material(s) undergoing similar processes. These coefficients show a component's response to prevalent interactions and conditions by quantifying the processes' contributions to fatigue failure.

Boundary work/load coefficient B_W is negative for positive evolution of fatigue measure—load entropy is negative during loading. MST coefficient $B_{\mu T}$ has varying sign characteristic. To understand $B_{\mu T}$ sign changes, rewrite Equation (48) as $w = w_{phen} = B_{\mu T} S'_{\mu T} + B_W S'_W$ and rearrange to get

$$B_{\mu T} = \frac{1}{S'_{\mu T}} \left(w_{phen} - B_W S'_W \right) \tag{57}$$

where phenomenological fatigue measure w_{phen} (e.g., ε_{phen}) fluctuates about the load-based measure $B_W S'_W$ (Figure 7b,d), making the parenthesis expression in Equation (57) fluctuate about zero during operation. It is also observed from Figure 5 that instantaneous MST entropy $S'_{\mu T}$ fluctuates about zero (more significantly for torsion).

6.2. *Entropy Generation vs Number of Cycles—A Linear Arrow of Time*

Describing entropy S as "time's arrow", Eddington [36] stated

$$\text{at } t \geq t_0,\ S \geq S_0 \tag{58}$$

for an isolated system where is entropy at initial/reference time t_0. Amiri et al. [15,18,23] via several experiments, observed an approximately linear relationship between normalized entropy and number of cycles. In Figure 9, normalized load entropy S'_W / S'_{Wf} (blue curves), microstructurothermal (MST) entropy $S'_{\mu T} / S'_{\mu T f}$ (red curves) and total phenomenological Helmholtz entropy generation $S'_{phen} / S'_{phen,f}$ (purple curves) vs. normalized number of cycles N/N_f are presented for bending fatigue (9a) and torsional fatigue (Figure 9b). An approximate linearity was observed in S'_W / S'_{Wf}. Figure 9 also shows that $S'_{\mu T} / S'_{\mu T f}$ and, consequently, $S'_{phen} / S'_{phen,f}$ do not evolve linearly with N/N_f; entropy generation as prescribed by the Helmholtz formulation for stress-strain loading, Equation (43), for an anisothermal process, includes a significant nonlinear microstructurothermal (MST) component.

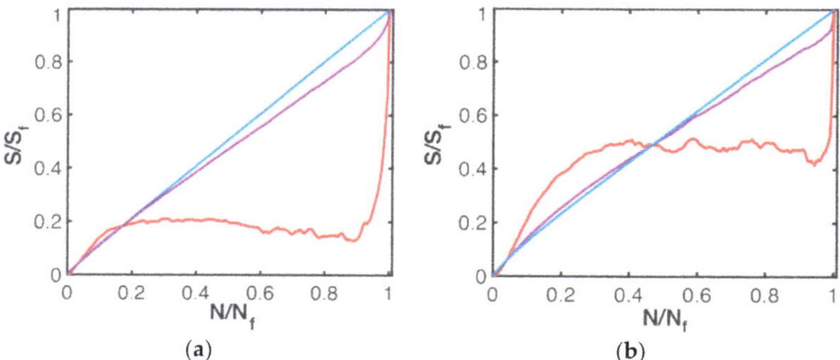

Figure 9. Normalized phenomenological entropy and components—load (blue), MST (red), phenomenological (purple) versus normalized cycles for (**a**) bending, (**b**) torsion of the SS 304 steel specimen.

Similar to Figure 6 which uses accumulated strain for component characterization via the DEG methodology, Figure 10 plots the components of phenomenological entropy generation S'_W and $S'_{\mu T}$ versus number of cycles N. Via Equation (38), N can be replaced by time t via $\int_{t_0}^{t_f} dt = \int_{N_0}^{N_f} \frac{dN}{h}$ where h is the load cycle frequency. For constant h and $N_0(t_0 = 0) = 0$, $t = \frac{N}{h}$. Considering the SS 304 steel torsional fatigue (last row of Table 2 and (b) figures in article), $N_f = 16010$ and $h = 10$ Hz give total time to failure Δt_f = 26.68 min. As depicted by Figure 10 using number of cycles, the DEG methodology linearizes the natural evolution of entropy generation over time: there exists a *linear* arrow of time.

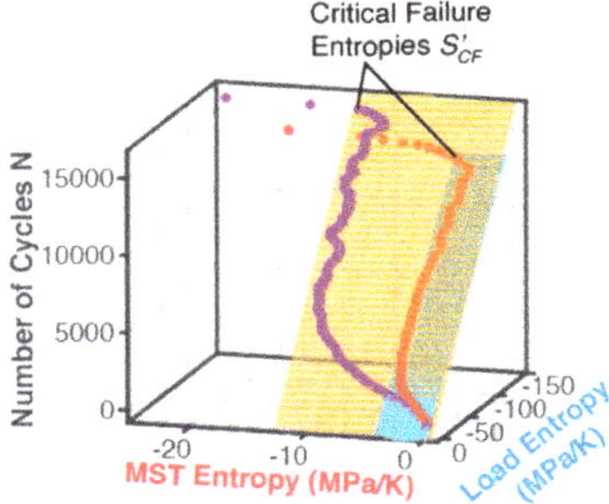

Figure 10. 3D plots and linear surface fits of numbers of cycles N vs load entropies and MST entropies during cyclic bending (red points, blue plane) and torsion (purple points, orange plane) of the SS 304 sample. Life trajectories start from lowest corner. (Axes are not to scale and colors are for visual purposes only).

Recall Equation (48) with $t_f = w$:

$$t_f = B_{\mu Tt} S\prime_{\mu T} + B_{Wt} S\prime_W. \tag{59}$$

From orientations of the Figure 10 DEG planes with $t_f = N_f/h$, $B_{\mu Tt}$ = 12.31 NK/MPa and $B_{Wt} = -99.78$ NK/MPa for bending. Equation (59) linearly relates entropy generation to degradation time, cycle life or time to failure for components under all types of load. Therefore, with a consistent evolution criterion, *entropy generation* via *the DEG theorem is a linear arrow of time*, Equation (59) and Figure 10. With DEG domains such as Figure 10, all systems undergoing cyclic or time-dependent loading can be fully and instantaneously characterized based on degradation or failure time t_f. The horizontal axes dimensions of the DEG domain (values of $S\prime_W$ and $S\prime_{\mu T}$ at N_f) can be directly correlated with other existing fatigue analysis methods that use N_f like the common σ—N and ε—N curves. The DEG approach appears universal and can be directly adapted to state of health and performance monitoring. The results in this article show that the DEG method can anticipate and potentially monitor and prevent fatigue failures accurately.

7. Summary and Conclusions

Fundamental irreversible thermodynamics and the degradation-entropy generation DEG theorem were applied to fatigue. The DEG theorem's fatigue/degradation model, which related a strain measure of fatigue to the load (boundary work) and MicroStructuroThermal entropies produced, was formulated and verified. A thermodynamic potential, the Helmholtz free energy, replaced steady state assumptions of previous DEG applications and employed the instantaneously applicable first and second laws of thermodynamics. The significance of the MicroStructuroThermal MST entropy and reversible Helmholtz entropy to total entropy generation and fatigue failure was demonstrated. Plots—DEG domains, Figures 6, 8 and 10—derived from published experimental data [15,18] showed the DEG-predicted linearity between fatigue/life measures and entropy generation components with goodness of fit R^2 = 1. Flexibility of fatigue parameter selection was also demonstrated. The DEG theorem provides a structured approach to component/system fatigue/degradation modeling, removing the need for many measurements, numerous curve fits and multiple analysis tools.

Author Contributions: Conceptualization, J.A.O. and M.D.B.; Data curation, J.A.O.; Formal analysis, J.A.O.; Investigation, J.A.O.; Methodology, J.A.O. and M.D.B.; Resources, J.A.O. and M.D.B.; Software, J.A.O.; Validation, J.A.O.; Visualization, J.A.O.; Writing—original draft, J.A.O.; Writing—review & editing, J.A.O. and M.D.B.

Funding: This research received no external funding.

Conflicts of Interest: The authors declare no conflict of interest.

Abbreviations

Nomenclature	Name	Unit
A	Helmholtz free energy density	J/m^3
B	DEG coefficient	%NK/MPa
C	heat capacity	J/K
N	number of cycles	
N, N_k	number of moles of substance	mol
P	dissipative process energy	J
P	pressure	Pa
Q	heat	J
S	entropy density or entropy content	J/m^{3}K or MPa/K
S'	entropy generation or production	J/m^{3}K or MPa/K
t	time	s
T	temperature	degC or K
U	internal energy	J
$\mathcal{V}$	volume	m^3
w	degradation measure	
W	work, strain energy density	J, J/m^3
Symbols		
α	thermal expansion coefficient	/K
κ_T	isothermal loadability	
μ	chemical potential	
ρ	density	Kg/m^3
σ	stress	MPa
ε	strain	%
ζ	phenomenological variable	
Subscripts & acronyms		
0	initial	
e	elastic	
MST, μT	Micro-Structuro-Thermal	
p	plastic	
rev	reversible	
irr	irreversible	
phen	phenomenological	
DEG	Degradation-Entropy Generation	

References

1. Vasudevan, A.K.; Sadananda, K.; Glinka, G. Critical parameters for fatigue damage. *Int. J. Fatigue* **2001**, *23*, 39–53. [CrossRef]
2. Callister, W.D., Jr. *Fundamentals of Materials Science and Engineering*; John Wiley & Sons: Hoboken, NJ, USA, 2001.
3. Timoshenko, S. *Strength of Materials (Part I)*, 2nd ed.; D. Van Nostrand Company Inc.: New York, NY, USA, 1940.
4. Goodno, B.J.; Gere, J.M. *Mechanics of Materials*, 9th ed.; Cengage Learning: Boston, MA, USA, 2016.
5. Hibbeler, R.C. *Mechanics of Materials*, 10th ed.; Pearson: London, UK, 2017.
6. Shigley, J.E.; Mischke, C.R. (Eds.) *Standard Handbook of Machine Design*, 2nd ed.; McGraw-Hill: New York, NY, USA, 1996.
7. Lemaitre, J.; Chaboche, J.-L. *Mechanics of Solid Materials*; Cambridge University Press: Cambridge, UK, 1990.
8. Vasudevan, A.K.; Sadananda, K.; Iyyer, N. Fatigue damage analysis: Issues and challenges. *Int. J. Fatigue* **2015**, *82*, 120–133. [CrossRef]

9. Gomez, J.; Basaran, C. A thermodynamics based damage mechanics constitutive model for low cycle fatigue analysis of microelectronics solder joints incorporating size effects. *Int. J. Solids Struct.* **2005**, *42*, 3744–3772. [CrossRef]
10. Gomez, J.; Basaran, C. Damage mechanics constitutive model for Pb/Sn solder joints incorporating nonlinear kinematic hardening and rate dependent effects using a return mapping integration algorithm. *Mech. Mater.* **2006**, *38*, 585–598. [CrossRef]
11. Basaran, C.; Lin, M.; Ye, H. A thermodynamic model for electrical current induced damage. *Int. J. Solids Struct.* **2003**, *40*, 7315–7327. [CrossRef]
12. Basaran, C.; Nie, S. An Irreversible Thermodynamics Theory for Damage Mechanics of Solids. *Int. J. Damage Mech.* **2004**, *13*, 205–223. [CrossRef]
13. Basaran, C.; Gomez, J.; Gunel, E.; Li, S. Thermodynamic Theory for Damage Evolution in Solids. In *Handbook of Damage Mechanics*; Voyiadjis, G., Ed.; Springer: New York, NY, USA, 2014; pp. 721–762.
14. Amiri, M.; Khonsari, M.M. Life prediction of metals undergoing fatigue load based on temperature evolution. *Mater. Sci. Eng. A* **2010**, *527*, 1555–1559. [CrossRef]
15. Naderi, M.; Khonsari, M.M. An experimental approach to low-cycle fatigue damage based on thermodynamic entropy. *Int. J. Solids Struct.* **2010**, *47*, 875–880. [CrossRef]
16. Naderi, M.; Khonsari, M.M. A thermodynamic approach to fatigue damage accumulation under variable loading. *Mater. Sci. Eng. A* **2010**, *527*, 6133–6139. [CrossRef]
17. Naderi, M.; Khonsari, M. Real-time fatigue life monitoring based on thermodynamic entropy. *Struct. Health Monit.* **2011**, *10*, 189–197. [CrossRef]
18. Amiri, M.; Naderi, M.; Khonsari, M.M. An Experimental Approach to Evaluate the Critical Damage. *Int. J. Damage Mech.* **2011**, *20*, 89–112. [CrossRef]
19. Naderi, M.; Khonsari, M.M. A comprehensive fatigue failure criterion based on thermodynamic approach. *J. Compos. Mater.* **2012**, *46*, 437–447. [CrossRef]
20. Naderi, M.; Khonsari, M.M. Thermodynamic analysis of fatigue failure in a composite laminate. *Mech. Mater.* **2012**, *46*, 113–122. [CrossRef]
21. Naderi, M.; Khonsari, M.M. On the role of damage energy in the fatigue degradation characterization of a composite laminate. *Compos. Part B Eng.* **2013**, *45*, 528–537. [CrossRef]
22. Amiri, M.; Modarres, M. An entropy-based damage characterization. *Entropy* **2014**, *16*, 6434–6463. [CrossRef]
23. Naderi, M.; Amiri, M.; Khonsari, M.M. On the thermodynamic entropy of fatigue fracture. *Proc. R. Soc. A Math. Phys. Eng. Sci.* **2010**, *466*, 423–438. [CrossRef]
24. Bryant, M.D.; Khonsari, M.M.; Ling, F.F. On the thermodynamics of degradation. *Proc. R. Soc. A Math. Phys. Eng. Sci.* **2008**, *464*, 2001–2014. [CrossRef]
25. Chaboche, J.L. Constitutive Equations for Cyclic Plasticity and Cyclic Viscoplasticity. *Int. J. Plast.* **1989**, *5*, 247–302. [CrossRef]
26. Chaboche, J.L. On some modifications of kinematic hardening to improve the description of ratchetting effects. *Int. J. Plast.* **1991**, *7*, 661–678. [CrossRef]
27. Callen, H.B. *Thermodynamics and an Introduction to Thermostatistics*; John Wiley & Sons, Ltd: Hoboken, NJ, USA, 1985.
28. Doelling, B.P.; Ling, K.L.; Bryant, F.F.; Heilman, M.D. An experimental study of the correlation between wear and entropy flow in machinery components. *J. Appl. Phys.* **2000**, *88*, 2999–3003. [CrossRef]
29. Duyi, Y.; Zhenlin, W. A new approach to low-cycle fatigue damage based on exhaustion of static toughness and dissipation of cyclic plastic strain energy during fatigue. *Int. J. Fatigue* **2001**, *23*, 679–687. [CrossRef]
30. Sosnovskiy, L.; Sherbakov, S. *Surprises of Tribo-Fatigue*; Magic Book: Minsk, Belarus, 2009.
31. Sosnovskiy, L.; Sherbakov, S. Mechanothermodynamic Entropy and Analysis of Damage State of Complex Systems. *Entropy* **2016**, *18*, 268. [CrossRef]
32. Osara, J.A.; Bryant, M.D. Thermodynamics of Grease Degradation. *Tribol. Int.* **2019**, *137*, 433–445. [CrossRef]
33. Strutt, J.W.; Rayleigh, B. *The Theory of Sound*; Macmillan & Co.: London, UK, 1877; Volume 2.
34. Onsager, L. Reciprocal Relations in Irreversible processes 1. *Am. Phys. Soc.* **1931**, *37*, 405. [CrossRef]
35. Nicolis, G.; Prigogine, I. *Self-Organization in Nonequilibrium Systems*; John Wiley & Sons: Hoboken, NJ, USA, 1977.
36. Kondepudi, D.; Prigogine, I. *Modern Thermodynamics: From Heat Engines to Dissipative Structures*; John Wiley & Sons: Hoboken, NJ, USA, 1998.

37. Bryant, M.D. On Constitutive Relations for Friction From Thermodynamics and Dynamics. *J. Tribol.* **2016**, *138*, 041603. [CrossRef]
38. Bryant, M.D. Entropy and Dissipative Processes of Friction and Wear. *FME Trans.* **2009**, *37*, 55–60.
39. Osara, J.A. *Thermodynamics of Degradation*; The University of Texas at Austin: Austin, TX, USA, 2017.
40. Osara, J.A.; Bryant, M.D. A Thermodynamic Model for Lithium-Ion Battery Degradation: Application of the Degradation-Entropy Generation Theorem. *Inventions* **2019**, *4*, 23. [CrossRef]
41. DeHoff, R.T. *Thermodynamics in Material Science*, 2nd ed.; CRC Press: Boca Raton, FL, USA, 2006.
42. de Groot, S.R. *Thermodynamics of Irreversible Processes*; North-Holland Publishing Company: Amsterdam, The Netherlands, 1951.
43. Prigogine, I. *Introduction to Thermodynamics of Irreversible Processes*; Charles C Thomas: Springfield, IL, USA, 1955.
44. Bejan, A. *Advanced Engineering Thermodynamics*, 3rd ed.; John Wiley & Sons: Hoboken, NJ, USA, 1997; Volume 70.
45. Moran, M.J.; Shapiro, H.N. *Fundamentals of Engineering Thermodynamics*, 5th ed.; Wiley: Hoboken, NJ, USA, 2004.
46. Burghardt, M.D.; Harbach, J.A. *Engineering Thermodynamics*, 4th ed.; HarperCollins College Publishers: New York, NY, USA, 1993.
47. Karnopp, D. Bond Graph Models for Electrochemical Energy Storage: Electrical, Chemical and Thermal Effects. *J. Frankl. Inst.* **1990**, *327*, 983–992. [CrossRef]
48. Bejan, A. The Method of Entropy Generation Minimization. In *Energy and the Environment*; Springer: Dordrecht, The Netherlands, 1990; pp. 11–22.
49. Pal, R. Demystification of the Gouy-Stodola theorem of thermodynamics for closed systems. *Int. J. Mech. Eng. Educ.* **2017**, *45*, 142–153. [CrossRef]
50. Glansdorff, P.; Prigogine, I. *Thermodynamic Theory of Structure, Stability and Fluctuations*; Wiley-Interscience: London, UK; New York, NY, USA, 1971.
51. Morris, J.W. Notes on the Thermodynamics of Solids, Chapter 16: Elastic Solids. 2007, pp. 366–411. Available online: http://www.mse.berkeley.edu/groups/morris/MSE205/Extras/Elastic.pdf (accessed on 10 July 2019).
52. Ramberg, W.; Osgood, W. *Description of Stress-Strain Curves by Three Parameters*; National Advisory Committee for Aeronautics: Washington, DC, USA, 1943.
53. Morrow, J. Cyclic Plastic Strain Energy and Fatigue of Metals. In *Internal Friction, Damping, and Cyclic Plasticit*; ASTM International: West Conshohocken, PA, USA, 1965.
54. Kim, K.; Chen, X.; Kim, K.S. Estimation methods for fatigue properties of steels under axial and torsional loading Estimation methods for fatigue properties of steels under axial and torsional loading. *Int. J. Fatigue* **2002**, *24*, 783–793. [CrossRef]
55. Socie, D. Multiaxial Fatigue Damage Models. *ASME Trans.* **1987**, *324–325*, 747–750. [CrossRef]
56. Budynas, R.G.; Nisbett, J.K. *Shigley's Mechanical Engineering Design*; McGraw-Hill: New York, NY, USA, 2015.
57. Meneghetti, G. Analysis of the fatigue strength of a stainless steel based on the energy dissipation. *Int. J. Fatigue* **2007**, *29*, 81–94. [CrossRef]
58. Osara, J.A. Thermodynamics of Manufacturing Processes—The Workpiece and the Machinery. *Inventions* **2019**, *4*, 28. [CrossRef]
59. Barsom, J.; Rolfe, S. Fracture and Fatigue Control. In *Structures: Applications of Fracture Mechanics*; ASTM International: West Conshohocken, PA, USA, 1999.
60. Bryant, M. Unification of friction and wear. *Recent Dev. Wear Prev. Frict. Lubr.* **2010**, *248*, 159–196.
61. Rice, J.R. Thermodynamics of the quasi-static growth of Griffith cracks. *J. Mech. Phys. Solids* **1978**, *26*, 61–78. [CrossRef]

Article

Effective Surface Nano-Crystallization of $Ni_2FeCoMo_{0.5}V_{0.2}$ Medium Entropy Alloy by Rotationally Accelerated Shot Peening (RASP)

Ningning Liang [1,†], Xiang Wang [1,†], Yang Cao [1,*], Yusheng Li [1], Yuntian Zhu [1,2] and Yonghao Zhao [1,*]

1 School of Materials Science and Engineering, Nanjing University of Science and Technology, Nanjing 210094, China; ningning623@126.com (N.L.); mumuchuntian@gmail.com (X.W.); liyusheng@njust.edu.cn (Y.L.); y.zhu@cityu.edu.hk (Y.Z.)
2 Department of Materials Science and Engineering, City University of Hong Kong, Hong Kong 999077, China
* Correspondence: y.cao@njust.edu.cn (Y.C.); yhzhao@njust.edu.cn (Y.Z.)
† Equal contribution.

Received: 31 August 2020; Accepted: 23 September 2020; Published: 24 September 2020

Abstract: The surface nano-crystallization of $Ni_2FeCoMo_{0.5}V_{0.2}$ medium-entropy alloy was realized by rotationally accelerated shot peening (RASP). The average grain size at the surface layer is ~37 nm, and the nano-grained layer is as thin as ~20 μm. Transmission electron microscopy analysis revealed that deformation twinning and dislocation activities are responsible for the effective grain refinement of the high-entropy alloy. In order to reveal the effectiveness of surface nano-crystallization on the $Ni_2FeCoMo_{0.5}V_{0.2}$ medium-entropy alloy, a common model material, Ni, is used as a reference. Under the same shot peening condition, the surface layer of Ni could only be refined to an average grain size of ~234 nm. An ultrafine grained surface layer is less effective in absorbing strain energy than a nano-grain layer. Thus, grain refinement could be realized at a depth up to 70 μm in the Ni sample.

Keywords: medium entropy alloy; deformation twinning; dislocation slip; surface nano-crystallization; shot peening

1. Introduction

After decades of fast development in physical metallurgy, dilute alloys and single-principal-element alloys have approached the limit of performance enhancement [1]. However, the trade-off between strength and ductility is still a thorny issue [2]. Different from conventional alloy design, high entropy alloys (HEAs) and medium entropy alloys (MEAs) have attracted immense attention [3–7]. Conventional alloys have configurational entropies, derived from mixing of the alloying components, less than 1R (R = 8.314 $J \cdot mol^{-1} \cdot K^{-1}$ is the gas constant); MEAs have configurational entropies in the range between 1R and 1.5R; HEAs have configurational entropies larger than 1.5R [8]. HEAs and MEAs may crystallize into single phase materials due to the configurational entropy maximization effect on solid-solution stabilization. Due to the unique atomic architecture and core effect, HEAs and MEAs exhibit exceptional mechanical properties, including high tensile strength [7,9], high ductility [2,10], excellent fatigue properties [11] and good fracture toughness at cryogenic temperatures [12]. Additionally, some noteworthy physical performances are also obtained for HEAs and MEAs, such as high thermal stability [13], irradiation resistance [14,15], corrosion resistance [8,16] and excellent mechanical behavior [17,18], as well as magnetic properties [19]. Thus, it is believed that both HEAs and MEAs have a huge potential in structural applications, especially for structures servicing in harsh environments.

The deformation mechanisms commonly found in conventional metallic materials, such as dislocation slip and deformation twinning, also play important roles in HEAs and MEAs. However, attributed to the low stacking-fault energy (SFE), short-range ordering effect and local elemental fluctuations, dislocation slip and deformation twinning can be very chaotic in HEAs and MEAs during plastic deformation [6,20–25]. It is well known that the propensity for deformation twinning is inversely proportional to the SFE. The SFE of the Cantor HEA is at the lower bond ~20–25 mJ/m^2 [25]. Thus, the presence of high densities of deformation twinning is found to be a major mechanism of the plastic strain in the Cantor HEA. Deng et al. [24] designed a face-centered cubic (FCC) single-phase $Fe_{40}Mn_{40}Co_{10}Cr_{10}$ HEA that has a large strain-hardening capacity attributed to the high densities of deformation twins and dislocations.

Similar to conventional alloys, single-phase coarse-grained FCC HEAs generally possess high tensile ductility but low yield strength [26,27]. For example, FeMnNiCoCr alloys having average grain sizes of 50 μm and 12 μm show low yield strengths of 95 MPa and 245 MPa, and elongations of 58% and 50%, respectively [27]. Considering the large strain hardening capacities of many high entropy alloys, severe plastic deformation (SPD) seems to be an ideal strategy for grain refinement and strength improvement. In the last 40 years, SPD techniques have been widely used to successfully prepare ultrafine grained (UFG) metals and alloys by means of grain refinement mechanisms [28,29]. According to the Hall–Petch relationship and experimental results, the UFG metals and alloys truly possess high yield strength but unfortunately low tensile ductility. Therefore, breaking the strength–ductility paradox and optimizing the strength–ductility combination are still hot research topics in the SPD field [30,31]. Up to now, many different SPD techniques have been developed, including equal-channel angular pressing (ECAP) [32], high-pressure torsion (HPT) [33], surface mechanical attrition treatment (SMAT) [34] and rotary swaging (RS) [35], etc. SMAT is an effective SPD method for generating a nano-structured surface layer [34,36–39]. Except for the global strength of SMAT-treated materials being effectively enhanced, wear resistance withstanding common failures on the surface and fatigue properties are also increased significantly [38,39].

Recently, Wu et al. used HPT to process an FeCoCrNi HEA, and significant grain refinement was realized via complicated concurrent nano-band subdivision and high-order hierarchical twinning mechanisms [25]. In addition, a high strain rate deformation may also facilitate grain refinement in medium-entropy alloy [40,41], but the relevant research is limited. Thus, the idea of using the SMAT method to improve the mechanical properties of MEAs has come through our mind. It is worth mentioning that some SMAT methods can break the size constraints of the traditional SPD method with an open specimen chamber, and thus process materials with possibly unlimited sizes. Therefore, an in-depth understanding of the SMAT of HEAs and MEAs is of significant importance to both engineering applications and scientific research. In this work, the same SMAT treatment was conducted on both $Ni_2FeCoMo_{0.5}V_{0.2}$ MEA (configurational entropy of 1.395R) and commercial purity Ni (CP-Ni). The gradient structures formed in the MEA are compared to that of the CP-Ni to reveal the uniqueness of the SMAT process in MEA from a microscopic point view.

2. Materials and Methods

Elemental Ni, Co, Fe, V and Mo were used as raw materials, each having a purity greater than 99.5%. The raw materials with the nominal composition of $Ni_2CoFeV_{0.5}Mo_{0.2}$ were alloyed via the arc-melting method under a high purity argon atmosphere. The compositional homogeneity of the alloy was analyzed by atom probe tomography (APT) [42], which revealed that all the alloying elements (Ni, Co, Fe, V, Mo) are homogenously distributed in a cylinder of ∅ 30 nm × 200 nm, indicating a random solid solution MEA without apparent elemental segregation or second phases. The X-ray diffraction (XRD) pattern revealed the simple FCC structure of the $Ni_2CoFeV_{0.5}Mo_{0.2}$ MEA [42]. A CP-Ni was purchased in the market. All sample materials were annealed at 950 °C for 10 h prior to rotationally accelerated shot peening (RASP) [43]. RASP was conducted at room temperature, for 10 min, using GCr15 bearing steel balls with a diameter of 3 mm and a velocity of 60 m/s.

The microstructures of samples before and after RASP were examined by a scanning electron microscope and a transmission electron microscope. Electron backscattering diffraction (EBSD) analysis was conducted using a field emission Carl Zeiss-Auriga SEM equipped with an Oxford Instruments EBSD system. TEM analysis was performed using a FEI T20 TEM operating at 200 kV. Prior to EBSD analysis, the specimens were carefully ground with SiC papers, and then mechanically polished with colloidal silica suspensions. Finally, all specimens were electro-polished (for polishing MEA:CH_3COOH:$HClO_4$ = 9:1, voltage: 50V; for polishing Ni:H_3PO_4:H_2O = 7:1, voltage: 7V). TEM specimens were mechanically polished to 40 μm thick foils and then prepared by a twin-jet polisher (for polishing MEA:C_2H_5OH:$HClO_4$ = 9:1; for polishing Ni:HNO_3:CH_3OH = 1:2) at −25 °C. The thin foil for EBSD and TEM was sectioned from the plane perpendicular to the treatment surface.

3. Results

The recrystallized equiaxed microstructures of the annealed FeCoNiMoV MEA are shown in Figure 1a,b. The average grain size of the annealed MEA is ~32 μm. Annealing twins are homogeneously distributed in the grains (Figure 1b). In contrast, the average grain size of the annealed CP-Ni is ~220 μm (Figure 1c), which is much larger than that of the MEA. This is because the diffusion kinetics of the MEA are comparatively low at 950 °C [13]. Annealing twins are also frequently found in the annealed CP-Ni (Figure 1d), but the twin density is clearly lower than that of the MEA (Figure 1b). This is because the CP-Ni possesses a much higher SFE and larger grain sizes than the FeCoNiMoV MEA.

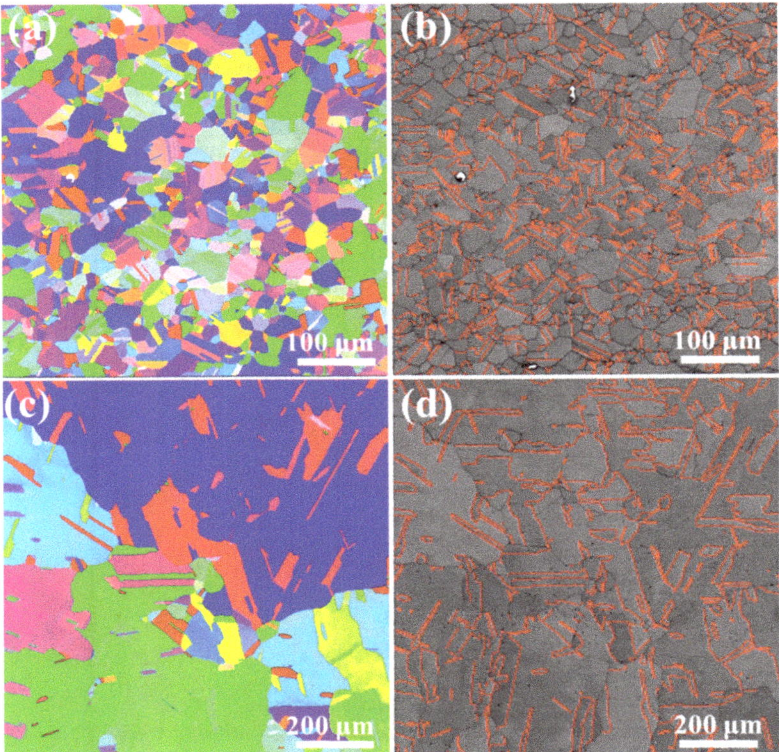

Figure 1. Electron backscattering diffraction (EBSD) maps of annealed sample materials: (**a**) inverse pole figure (IPF) map of $Ni_2FeCoMo_{0.5}V_{0.2}$ MEA, (**b**) twin boundaries in $Ni_2FeCoMo_{0.5}V_{0.2}$ MEA, (**c**) IPF map of CP-Ni, and (**d**) twin boundaries in CP-Ni.

RASP imposed both high strain and high strain rate on the surfaces of the sample materials. As a result, gradient nano-structures formed on the sample surfaces, as shown in Figure 2. The topmost surface of the sample experienced the highest strain and strain rate [43,44], and thus extreme grain refinement is expected at a depth of less than ~20 μm. However, the resolution of EBSD is insufficient for acquiring the actual nano-structure at the topmost surface. Hence, blurry EBSD images were obtained at the depth of ~20 μm, as shown in Figure 2(a-1,b-1,c-1,d-1). In spite of the limited quality of the images, defects in nano-scale are noticeable by the channeling contrast in Figure 2(b-1,d-1). Interestingly, the defects at the surface of the RASP-MEA appeared as straight dark lines under the channeling contrast, indicating that the nano-structures are related to confined dislocation slip and/or deformation twinning [44]. In contrast, the defects at the surface of the RASP-Ni appeared as cell-like structures, indicating that the nano-structures are related to dislocation sub-structures [45].

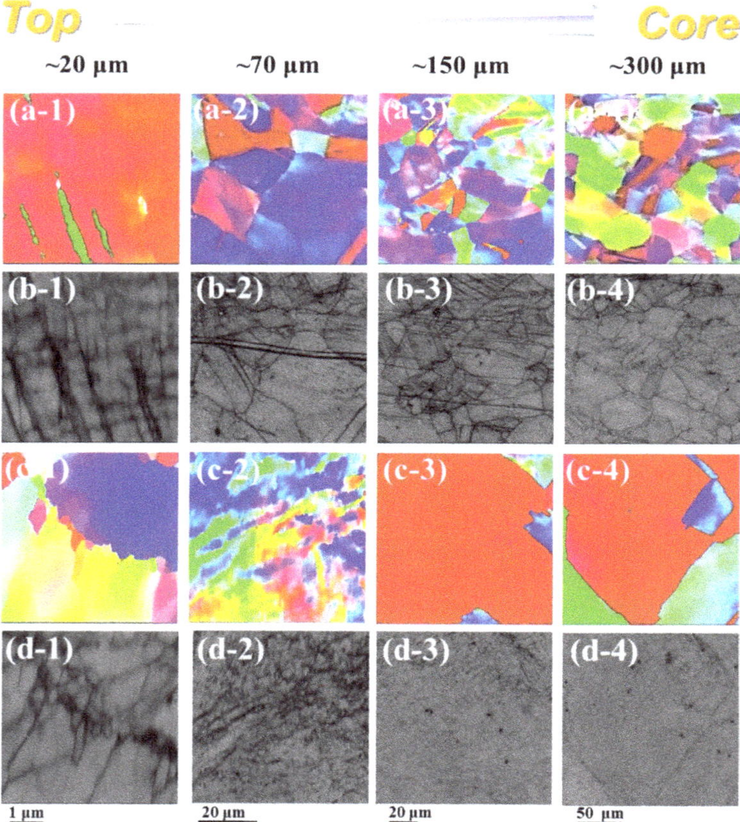

Figure 2. EBSD maps showing gradient microstructures at the depth range between ~20 μm and ~300 μm from the surfaces: (**a-1–a-4**) IPF map of RASP-MEA, (**b-1–b-4**) channeling contrast map of RASP-MEA, (**c-1–c-4**) IPF map of RASP-Ni, and (**d-1–d-4**) channeling contrast map of RASP-Ni.

At the depths of ~70 μm and 150 μm from the surface of the RASP-MEA, coarse grains and planar defects feature in the microstructure, as shown in Figure 2(a-2,a-3,b-2,b-3). This also indicates that grain refinement was only achieved at the depth of a few tens of micrometers, possibly less than 50 μm. In contrast, at the depth of ~70 μm from the surface of RASP-Ni, equiaxed sub-grains feature in the microstructure. Therefore, it can be concluded that under the same shot peening impact energy, the depth of grain refinement is shallower in MEA than in CP-Ni. In RASP-Ni, the coarse grains could be sustained at the depth of 150 μm. However, the moderate color variation within the grain shown in Figure 2(c-3) suggests that dislocation entanglement is pronounced at the grain interior.

At approximately 300 μm below the surface of the RASP-MEA, the defect densities are lower than those at the 150 μm depth, evidenced by the moderate variation of local misorientation (Figure 2(a-4)) and the low density of defect lines (Figure 2(b-4)). Similarly, the defect density at ~300 μm below the surface of the RASP-Ni is also comparatively low, in spite of the large grain size, as shown in Figure 2(c-4,d-4).

TEM analysis was conducted to reveal the microstructural details that are hardly resolved by EBSD. Figure 3a is a TEM image, and shows the corresponding selected area electron diffraction (SAED) pattern obtained at the surface layer (<20 μm from the surface) of the RASP-MEA. Uniformly distributed nano-grains, shown by the bright-field TEM image, and the diffraction rings shown by the SAED pattern together reveal that extreme grain refinement has been achieved at the surface of the RASP-MEA sample. Statistical analysis based on a series of TEM images produced the grain size distribution chart shown in Figure 3c. The average grain size at the surface is estimated to be 37 nm, which is towards the lower bond of the nano-crystalline regime [46–48]. In contrast, the grain refinement at the surface layer of RASP-Ni is not as significant as in RASP-MEA. Figure 3b shows ultrafine grains at the surface layer of RASP-Ni. Many of the ultrafine grains have diffused grain boundaries due to severe lattice distortion [49]. Figure 3d shows that the average grain size is ~230 nm at the surface layer of RASP-Ni. Thus, it can be concluded here that the effectiveness of surface nano-crystallization of FeCoNiMoV MEA under RASP processing is much higher than for single-phase materials with high SFE, such as Ni.

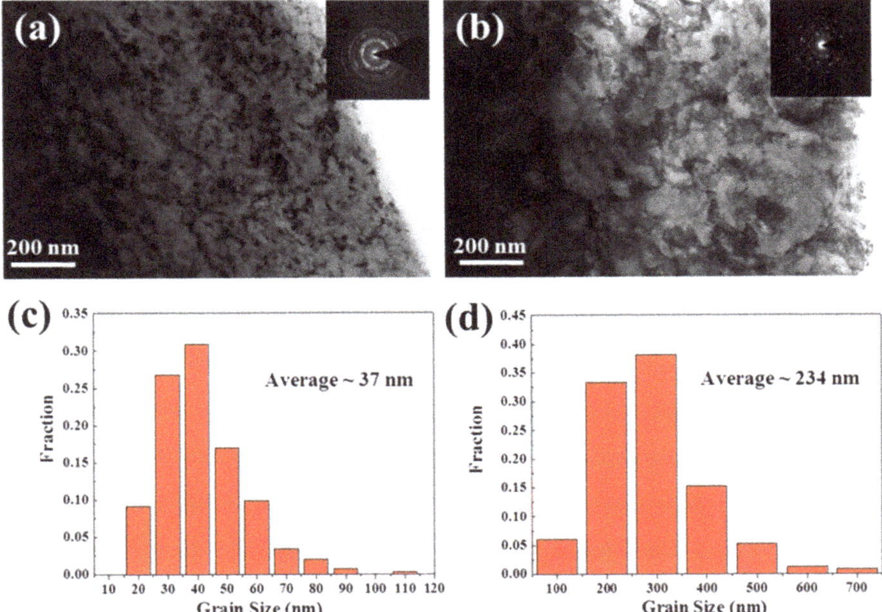

Figure 3. TEM images showing the microstructures at the surfaces of (**a**) the RASP-MEA sample and (**b**) the RASP-Ni sample (SAED patterns are provided as inserts). Charts showing grain size distributions at the surface regions of (**c**) the RASP-MEA sample and (**d**) the RASP-Ni sample.

At the depth of ~300 μm from the surface of the RASP-MEA sample, the planar dislocation slip along {111} is the major deformation structure, as shown in Figure 4a. Nano-twins are seldom found at this depth, indicating that the shear stress at the depth of 300 μm is insufficient to activate deformation twinning. Twin structures are found, as evidenced by Figure 4b, but they are just annealing twins preserved from the annealed state. For the purpose of comparison, typical microstructures at the depth of ~300 μm from the surface of RASP-Ni are provided in Figure 4c,d. Dislocation wall (Figure 4c) and cell structures (Figure 4d) are the major deformation structures at this depth. It is well known that planar dislocations, dislocation walls and cells form at the early stages of plastic deformation when the strain is very low [29]. Thus, it is believed that impact energy has been mostly absorbed at the depth of 300 μm for both MEA and Ni.

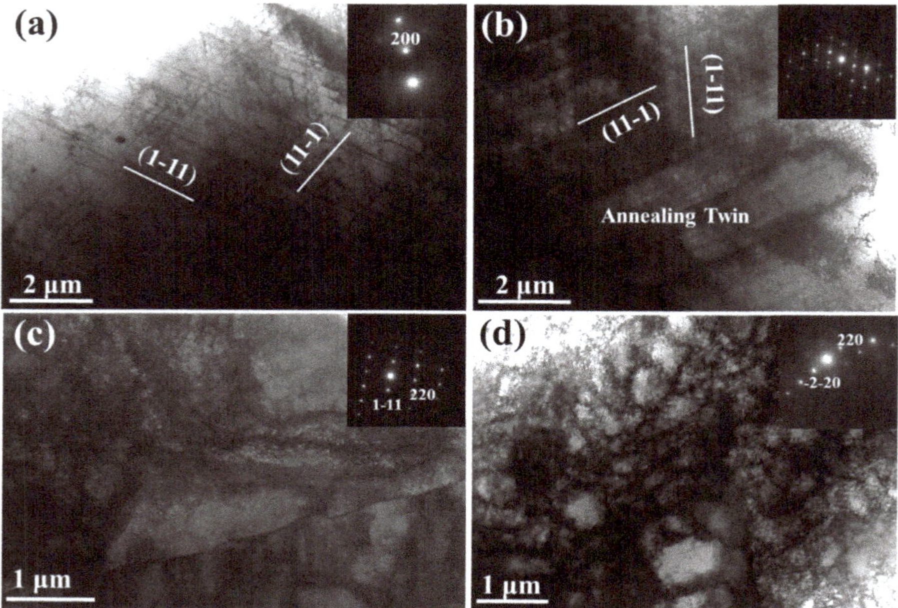

Figure 4. TEM images showing typical microstructures at the depth of ~300 μm from the surfaces of (**a**,**b**) RASP-MEA and (**c**,**d**) RASP-Ni (SAED patterns are provided as inserts).

4. Discussion

Microstructural characterization by both TEM and SEM reveals that planar dislocation slip is the major deformation mechanism when the FeCoNiMoV MEA was deformed by RASP. It is known that the SFE of $FeCoNi_2Mo_{0.2}V_{0.5}$ MEA is ~50 mJ/m^2 [50], which is comparable to copper. Thus, deformation twinning is also expected when strain and/or strain rate is sufficiently high [51]. Interestingly, deformation twins have only been observed at the depth range of 20 to 40 μm below the surface of RASP-MEA, as shown in Figure 5. This is because the SFE of the FeCoNiMoV MEA is still comparatively high. Hence, very high stress is required to activate deformation twinning. Both the shear stress and shear strain imposed by RASP are very high at the surface of impact, resulting in the quick formation of thin nano-crystalline layers, as shown in Figure 3. Once the nano-crystalline layer is formed, the material's surface is significantly hardened. The RASP-imposed shear stress decreased drastically when it transmitted through the hard nano-crystalline layer [52]. Thus, only a thin layer of a few tens of nanometers below the nano-crystalline layers experienced a high shear stress, which was just sufficient to activate deformation twinning. As the shear stress decreased further with increasing depth, only dislocation slips and stacking faults could be activated, as shown in Figure 4. Although the deformation twins existed only at the very surface of the RASP-MEA, it is still an important mechanism that facilitates grain refinement at a much faster rate than a grain refinement mechanism via dislocation activities [29]. This concept is also supported by the first-hand experimental results provided here. Ni is a representative high SFE material [29]. Under the same RASP condition, the microstructure of the material's surface was only refined to the ultrafine-grained regime. This is because high stacking fault energy facilitates cross slip and recovery [29,53,54]. At the very surface of the Ni sample, grain refinement and grain growth are balanced under the shot peening condition. Without any change of the shot peening condition, further grain refinement is impossible, similar to the equilibrium state under high-pressure torsion processing [31,53,55]. Clearly, the ultrafine grained surface layer is less effective in absorbing impact energy than the nanocrystalline layer. As such, the high strain energy is transmitted to a deeper region of 70 μm below the surface of RASP-Ni to cause grain refinement, as shown in Figure 2 (c-2,d-2).

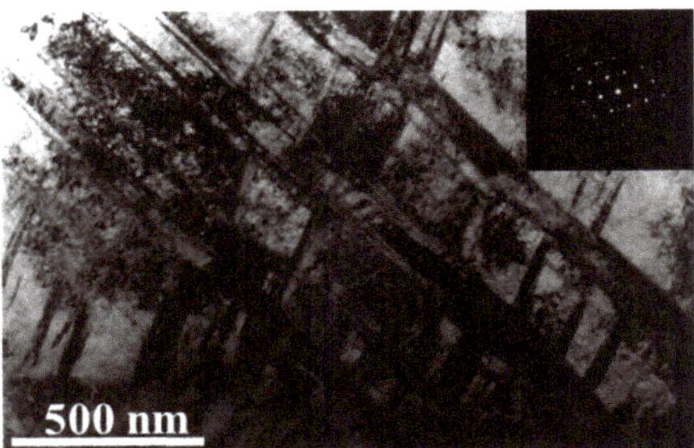

Figure 5. A TEM image showing deformation twins at the depth of 20–40 μm below the surface of RASP-MEA.

5. Conclusions

In summary, the $Ni_2FeCoMo_{0.5}V_{0.2}$ MEA and CP-Ni treated by RASP were characterized by EBSD and TEM. Nano-grains with an average size of 37 nm were obtained on the surface layer of the $Ni_2FeCoMo_{0.5}V_{0.2}$ MEA. Microstructural analysis shows that deformation twinning and dislocation activities are closely involved in the grain refinement mechanism. In contrast, only dislocation activities contributed to the grain refinement of CP-Ni, leading to the ultrafine grained surface layer. RASP exhibited a prominent structure refinement ability for MEA, and successfully produced gradient nano-structured MEA samples.

Author Contributions: Conceptualization, X.W. and Y.Z. (Yonghao Zhao); methodology and material preparation, X.W., Y.L. and N.L.; investigation and microstructural characterization, X.W. and N.L.; data curation, N.L.; writing—original draft preparation, N.L.; writing—review and editing, Y.C. and Y.Z. (Yonghao Zhao); project administration, Y.Z. (Yuntian Zhu); All authors have read and agreed to the published version of the manuscript.

Funding: This research was funded by the National Key R&D Program of China (Grant No. 2017YFA0204403), the National Natural Science Foundation of China (Grant No. 51301064, 51971112, 51225102), Natural Science Foundation of Jiangsu Province (BK20190478) and the Fundamental Research Funds for the Central Universities (Grant No. 30915012104, 30919011405).

Acknowledgments: The authors are thankful for the technical support from Jiangsu Key Laboratory of Advanced Micro&Nano Materials and Technology, and the Materials Characterization Facility of Nanjing University of Science and Technology.

Conflicts of Interest: The authors declare no conflict of interest.

References

1. Yang, T.; Zhao, Y.L.; Tong, Y.; Jiao, Z.B.; Wei, J.; Cai, J.X.; Han, X.D.; Chen, D.; Hu, A.; Kai, J.J.; et al. Multicomponent intermetallic nanoparticles and superb mechanical behaviors of complex alloys. *Science* **2018**, *362*, 933–937. [CrossRef]
2. Li, Z.; Pradeep, K.G.; Deng, Y.; Raabe, D.; Tasan, C.C. Metastable high-entropy dual-phase alloys overcome the strength-ductility trade-off. *Nature* **2016**, *534*, 227–230. [CrossRef] [PubMed]
3. Cantor, B.; Chang, I.T.H.; Knight, P.; Vincent, A.J.B. Microstructural development in equiatomic multicomponent alloys. *Mat. Sci. Eng. A Struct.* **2004**, *375*, 213–218. [CrossRef]
4. Yeh, J.W.; Chen, S.K.; Lin, S.J.; Gan, J.Y.; Chin, T.S.; Shun, T.T.; Tsau, C.H.; Chang, S.Y. Nanostructured high-entropy alloys with multiple principal elements: Novel alloy design concepts and outcomes. *Adv. Eng. Mater.* **2004**, *6*, 299–303. [CrossRef]
5. Otto, F.; Dlouhy, A.; Somsen, C.; Bei, H.; Eggeler, G.; George, E.P. The influences of temperature and microstructure on the tensile properties of a CoCrFeMnNi high-entropy alloy. *Acta Mater.* **2013**, *61*, 5743–5755. [CrossRef]
6. Zhang, Y.; Zuo, T.T.; Tang, Z.; Gao, M.C.; Dahmen, K.A.; Liaw, P.K.; Lu, Z.P. Microstructures and properties of high-entropy alloys. *Prog. Mater. Sci.* **2014**, *61*, 1–93. [CrossRef]
7. Gludovatz, B.; Hohenwarter, A.; Thurston, K.V.S.; Bei, H.; Wu, Z.; George, E.P.; Ritchie, R.O. Exceptional damage-tolerance of a medium-entropy alloy CrCoNi at cryogenic temperatures. *Nat. Commun.* **2016**, *7*, 10602. [CrossRef]
8. Miracle, D.B.; Miller, J.D.; Senkov, O.N.; Woodward, C.; Uchic, M.D.; Tiley, J. Exploration and Development of High Entropy Alloys for Structural Applications. *Entropy* **2014**, *16*, 494–525. [CrossRef]
9. Shi, P.; Ren, W.; Zheng, T.; Ren, Z.; Hou, X.; Peng, J.; Hu, P.; Gao, Y.; Zhong, Y.; Liaw, P.K. Enhanced strength-ductility synergy in ultrafine-grained eutectic high-entropy alloys by inheriting microstructural lamellae. *Nat. Commun.* **2019**, *10*, 489. [CrossRef]
10. Lei, Z.; Liu, X.; Wu, Y.; Wang, H.; Jiang, S.; Wang, S.; Hui, X.; Wu, Y.; Gault, B.; Kontis, P.; et al. Enhanced strength and ductility in a high-entropy alloy via ordered oxygen complexes. *Nature* **2018**, *563*, 546–550. [CrossRef]
11. Liu, K.M.; Komarasamy, M.; Gwalani, B.; Shukla, S.; Mishra, R.S. Fatigue behavior of ultrafine grained triplex Al0.3CoCrFeNi high entropy alloy. *Scr. Mater.* **2019**, *158*, 116–120. [CrossRef]
12. Gludovatz, B.; Hohenwarter, A.; Catoor, D.; Chang, E.H.; George, E.P.; Ritchie, R.O. A fracture-resistant high-entropy alloy for cryogenic applications. *Science* **2014**, *345*, 1153–1158. [CrossRef] [PubMed]
13. Schuh, B.; Mendez-Martin, F.; Volker, B.; George, E.P.; Clemens, H.; Pippan, R.; Hohenwarter, A. Mechanical properties, microstructure and thermal stability of a nanocrystalline CoCrFeMnNi high-entropy alloy after severe plastic deformation. *Acta Mater.* **2015**, *96*, 258–268. [CrossRef]
14. Barr, C.M.; Nathaniel, J.E.; Unocic, K.A.; Liu, J.P.; Zhang, Y.; Wang, Y.Q.; Taheri, M.L. Exploring radiation induced segregation mechanisms at grain boundaries in equiatomic CoCrFeNiMn high entropy alloy under heavy ion irradiation. *Scr. Mater.* **2018**, *156*, 80–84. [CrossRef]
15. Kumar, N.A.P.K.; Li, C.; Leonard, K.J.; Bei, H.; Zinkle, S.J. Microstructural stability and mechanical behavior of FeNiMnCr high entropy alloy under ion irradiation. *Acta Mater.* **2016**, *113*, 230–244. [CrossRef]
16. Qiu, X.W.; Zhang, Y.P.; He, L.; Liu, C.G. Microstructure and corrosion resistance of AlCrFeCuCo high entropy alloy. *J. Alloy. Compd.* **2013**, *549*, 195–199. [CrossRef]
17. Zhao, C.C.; Inoue, A.; Kong, F.L.; Zhang, J.Y.; Chen, C.J.; Shen, B.L.; Al-Marzouki, F.; Greer, A.L. Novel phase decomposition, good soft-magnetic and mechanical properties for high-entropy (fe0.25co0.25ni0.25cr0.125mn0.125)(100-x)b-x (x=9-13) amorphous alloys. *J. Alloys Compd.* **2020**, *843*, 155917. [CrossRef]
18. Brechtl, J.; Chen, S.; Lee, C.; Shi, Y.; Feng, R.; Xie, X.; Hamblin, D.; Coleman, A.M.; Straka, B.; Shortt, H.; et al. A review of the serrated-flow phenomenon and its role in the deformation behavior of high-entropy alloys. *Metals* **2020**, *10*, 1101. [CrossRef]
19. Chen, C.; Zhang, H.; Fan, Y.Z.; Zhang, W.W.; Wei, R.; Wang, T.; Zhang, T.; Li, F.S. A novel ultrafine-grained high entropy alloy with excellent combination of mechanical and soft magnetic properties. *J. Magn. Magn. Mater.* **2020**, *502*, 166513-1–166513-5. [CrossRef]

20. Ding, Q.; Zhang, Y.; Chen, X.; Fu, X.; Chen, D.; Chen, S.; Gu, L.; Wei, F.; Bei, H.; Gao, Y.; et al. Tuning element distribution, structure and properties by composition in high-entropy alloys. *Nature* **2019**, *574*, 223–227. [CrossRef]
21. Zhang, R.; Zhao, S.; Ding, J.; Chong, Y.; Jia, T.; Ophus, C.; Asta, M.; Ritchie, R.O.; Minor, A.M. Short-range order and its impact on the CrCoNi medium-entropy alloy. *Nature* **2020**, *581*, 283–287. [CrossRef]
22. Gao, X.Z.; Lu, Y.P.; Zhang, B.; Liang, N.N.; Wu, G.Z.; Sha, G.; Liu, J.Z.; Zhao, Y.H. Microstructural origins of high strength and high ductility in an AlCoCrFeNi2.1 eutectic high-entropy alloy. *Acta Mater.* **2017**, *141*, 59–66. [CrossRef]
23. Liu, J.B.; Chen, C.X.; Xu, Y.Q.; Wu, S.W.; Wang, G.; Wang, H.T.; Fang, Y.T.; Meng, L. Deformation twinning behaviors of the low stacking fault energy high-entropy alloy: An in-situ TEM study. *Scr. Mater.* **2017**, *137*, 9–12. [CrossRef]
24. Deng, Y.; Tasan, C.C.; Pradeep, K.G.; Springer, H.; Kostka, A.; Raabe, D. Design of a twinning-induced plasticity high entropy alloy. *Acta Mater.* **2015**, *94*, 124–133. [CrossRef]
25. Wu, W.; Song, M.; Ni, S.; Wang, J.; Liu, Y.; Liu, B.; Liao, X. Dual mechanisms of grain refinement in a FeCoCrNi high-entropy alloy processed by high-pressure torsion. *Sci. Rep.* **2017**, *7*, 46720. [CrossRef] [PubMed]
26. Gu, J.; Song, M. Annealing-induced abnormal hardening in a cold rolled CrMnFeCoNi high entropy alloy. *Scr. Mater.* **2019**, *162*, 345–349. [CrossRef]
27. Yao, M.J.; Pradeep, K.G.; Tasan, C.C.; Raabe, D. A novel, single phase, non-equiatomic FeMnNiCoCr high-entropy alloy with exceptional phase stability and tensile ductility. *Scr. Mater.* **2014**, *72–73*, 5–8. [CrossRef]
28. Valiev, R.Z.; Islamgaliev, R.K.; Alexandrov, I.V. Bulk nanostructured materials from severe plastic deformation. *Porgress Mater. Sci.* **2000**, *45*, 103–189. [CrossRef]
29. Cao, Y.; Ni, S.; Liao, X.Z.; Song, M.; Zhu, Y.T. Structural evolutions of metallic materials processed by severe plastic deformation. *Mater. Sci. Eng. R-Rep.* **2018**, *133*, 1–59. [CrossRef]
30. Valiev, R.Z.; Langdon, T.G. Principles of equal-channel angular pressing as a processing tool for grain refinement. *Porgress Mater. Sci.* **2006**, *51*, 881–981. [CrossRef]
31. Zhilyaev, A.P.; Langdon, T.G. Using high-pressure torsion for metal processing: Fundamentals and applications. *Porgress Mater. Sci.* **2008**, *53*, 893–979. [CrossRef]
32. Ovid'ko, I.A.; Valiev, R.Z.; Zhu, Y.T. Review on superior strength and enhanced ductility of metallic nanomaterials. *Porgress Mater. Sci.* **2018**, *94*, 462–540. [CrossRef]
33. Zhao, Y.H.; Zhu, Y.T.; Lavernia, E.J. Strategies for improving tensile ductility of bulk nanostructured materials. *Adv. Eng. Mater.* **2010**, *12*, 769–778. [CrossRef]
34. Tao, N.R.; Wang, Z.B.; Tong, W.P.; Sui, M.L.; Lu, J.; Lu, K. An investigation of surface nanocrystallization mechanism in Fe induced by surface mechanical attrition treatment. *Acta Mater.* **2002**, *50*, 4603–4616. [CrossRef]
35. Wan, Y.C.; Tang, B.; Gao, Y.H.; Tang, L.L.; Sha, G.; Zhang, B.; Liang, N.N.; Liu, C.M.; Jiang, S.N.; Chen, Z.Y.; et al. Bulk nanocrystalline high-strength magnesium alloys prepared via rotary swaging. *Acta Mater.* **2020**, *200*, 274–286. [CrossRef]
36. Wang, K.; Tao, N.R.; Liu, G.; Lu, J.; Lu, K. Plastic strain-induced grain refinement at the nanometer scale in copper. *Acta Mater.* **2006**, *54*, 5281–5291. [CrossRef]
37. Lu, K. The future of metals. *Science* **2010**, *328*, 319–320. [CrossRef] [PubMed]
38. Lu, K.; Lu, J. Nanostructured surface layer on metallic materials induced by surface mechanical attrition treatment. *Mater. Sci. Eng. A* **2004**, *375–377*, 38–45. [CrossRef]
39. Zhang, Y.S.; Han, Z.; Wang, K.; Lu, K. Friction and wear behaviors of nanocrystalline surface layer of pure copper. *Wear* **2006**, *260*, 942–948. [CrossRef]
40. Won, J.W.; Lee, S.; Park, S.H.; Kang, M.; Lim, K.R.; Park, C.H.; Na, Y.S. Ultrafine-grained CoCrFeMnNi high-entropy alloy produced by cryogenic multi-pass caliber rolling. *J. Alloys Compd.* **2018**, *742*, 290–295. [CrossRef]
41. Ma, Y.; Yuan, F.P.; Yang, M.X.; Jiang, P.; Ma, E.; Wu, X.L. Dynamic shear deformation of a CrCoNi medium-entropy alloy with heterogeneous grain structures. *Acta Mater.* **2018**, *148*, 407–418. [CrossRef]
42. Gao, X.Z.; Jiang, L.; Lu, Y.P.; Cao, Z.Q.; Wen, B.; Gang, G.; Liang, N.N.; Wang, T.M.; Li, T.J.; Zhao, Y.H. Extraordinary hardening and ductility in $Ni_2Co_1Fe_1V_{0.5}Mo_{0.2}$ high entropy alloy by unique nano-scale planar slips. *Nature* **2020**, in press.

43. Wang, X.; Li, Y.S.; Zhang, Q.; Zhao, Y.H.; Zhu, Y.T. Gradient Structured Copper by Rotationally Accelerated Shot Peening. *J. Mater. Sci. Technol.* **2017**, *33*, 758–761. [CrossRef]
44. Wang, Z.W.; Guo, L.; Xia, W.Z.; Yuan, Z.; Cao, Y.; Ni, S.; Song, M. An SEM-based approach to characterize the microstructural evolution in a gradient CoCrFeNiMo0.15 high-entropy alloy. *Mater. Charact.* **2020**, *161*, 110169. [CrossRef]
45. Cao, Y.; Wang, Y.B.; An, X.H.; Liao, X.Z.; Kawasaki, M.; Ringer, S.P.; Langdon, T.G.; Zhu, Y.T. Concurrent microstructural evolution of ferrite and austenite in a duplex stainless steel processed by high-pressure torsion. *Acta Mater.* **2014**, *63*, 16–29. [CrossRef]
46. Zhu, Y.T.; Liao, X.Z.; Wu, X.L. Deformation twinning in nanocrystalline materials. *Pro. Mater. Sci.* **2012**, *57*, 1–62. [CrossRef]
47. Zhou, X.; Li, X.Y.; Lu, K. Enhanced thermal stability of nanograined metals below a critical grain size. *Science* **2018**, *360*, 526–530. [CrossRef]
48. Lee, D.H.; Choi, I.C.; Yang, G.H.; Lu, Z.P.; Kawasaki, M.; Ramamurty, U.; Schwaiger, R.; Jang, J.I. Activation energy for plastic flow in nanocrystalline CoCrFeMnNi high-entropy alloy: A high temperature nanoindentation study. *Scr. Mater.* **2018**, *156*, 129–133. [CrossRef]
49. Qiang, J.; Tsuchiya, K.; Diao, H.Y.; Liaw, P.K. Vanishing of room-temperature slip avalanches in a face-centered-cubic high-entropy alloy by ultrafine grain formation. *Scr. Mater.* **2018**, *155*, 99–103. [CrossRef]
50. Jiang, L.; Lu, Y.P.; Song, M.; Lu, C.; Sun, K.; Cao, Z.Q.; Wang, T.M.; Gao, F.; Wang, L.M. A promising CoFeNi2V0.5Mo0.2 high entropy alloy with exceptional ductility. *Scr. Mater.* **2019**, *165*, 128–133. [CrossRef]
51. Li, Y.S.; Tao, N.R.; Lu, K. Microstructural evolution and nanostructure formation in copper during dynamic plastic deformation at cryogenic temperatures. *Acta Mater.* **2008**, *56*, 230–241. [CrossRef]
52. Liu, Y.F.; Cao, Y.; Zhou, H.; Chen, X.F.; Liu, Y.; Xiao, L.R.; Huan, X.W.; Zhao, Y.H.; Zhu, Y.T. Mechanical Properties and Microstructures of Commercial-Purity Aluminum Processed by Rotational Accelerated Shot Peening Plus Cold Rolling. *Adv. Eng. Mater.* **2020**, *22*, 1900478. [CrossRef]
53. Cao, Y.; Wang, Y.B.; Figueiredo, R.B.; Chang, L.; Liao, X.Z.; Kawasaki, M.; Zheng, W.L.; Ringer, S.P.; Langdon, T.G.; Zhu, Y.T. Three-dimensional shear-strain patterns induced by high-pressure torsion and their impact on hardness evolution. *Acta Mater.* **2011**, *59*, 3903–3914. [CrossRef]
54. Liu, Y.F.; Cao, Y.; Mao, Q.Z.; Zhou, H.; Zhao, Y.H.; Jiang, W.; Liu, Y.; Wang, J.T.; You, Z.S.; Zhu, Y.T. Critical microstructures and defects in heterostructured materials and their effects on mechanical properties. *Acta Mater.* **2020**, *189*, 129–144. [CrossRef]
55. Liu, Y.F.; Wang, F.; Cao, Y.; Nie, J.F.; Zhou, H.; Yang, H.B.; Liu, X.F.; An, X.H.; Liao, X.Z.; Zhao, Y.H.; et al. Unique defect evolution during the plastic deformation of a metal matrix composite. *Scr. Mater.* **2019**, *162*, 316–320. [CrossRef]

MDPI
St. Alban-Anlage 66
4052 Basel
Switzerland
Tel. +41 61 683 77 34
Fax +41 61 302 89 18
www.mdpi.com

Entropy Editorial Office
E-mail: entropy@mdpi.com
www.mdpi.com/journal/entropy

www.ingramcontent.com/pod-product-compliance
Lightning Source LLC
LaVergne TN
LVHW070056170726
843515LV00007B/1548

* 9 7 8 3 0 3 9 4 3 8 0 7 5 *